ововs # Electronic Meters

Electronic Meters
Techniques and Troubleshooting

MILES RITTER-SANDERS, JR.

RESTON PUBLISHING COMPANY, INC.
A Prentice-Hall Company
Reston, Virginia

Library of Congress Cataloging in Publication Data
Ritter-Sanders, Miles, Jr.
  Electronic meters.
  Includes index.
  1. Electric meters. 2. Electronic measurements.
3. Electric measurements. I. Title.
TK301.M47   621.37'4   76-30480
ISBN 0-87909-223-8

© 1977 by Reston Publishing Company, Inc.
A Prentice-Hall Company
Reston, Virginia 22090

All rights reserved. No part of this book may be reproduced in any way, or by any means, without permission in writing from the publisher.

10 9 8 7 6 5 4 3 2 1

PRINTED IN THE UNITED STATES OF AMERICA

Contents

Preface   ix

Introduction   1

1 Performance Verification and Calibration   9
    1-1   General Considerations   9
    1-2   Basic Meter Circuitry   12
    1-3   Accuracy Check of Lower DC-Voltage Ranges with Mercury Batteries   15
    1-4   Check or DC-Voltage Calibration with Weston Cell   18
    1-5   Analysis of Incorrect Readings   20
    1-6   Calibration Cross-checks Against Other Voltmeters   21
    1-7   AC-voltage Calibration Checks   22
    1-8   Ohmmeter Checkout   24
    1-9   DC Current Calibration   28
    1-10 Waveform Errors   30
    1-11 Calibration Procedure   32
    1-12 Notes on Digital Voltmeters   33
    1-13 Notes on Analog Meter Circuit Tolerances   35

2 DC-Voltage Measuring Techniques   37
    2-1   General Considerations   37
    2-2   Instrument Accuracy Ratings   41
    2-3   Polarity Considerations   41
    2-4   Measurement of High DC Voltages   43
    2-5   Measurement of Low DC Voltages   44
    2-6   Measurement of Difference Voltages   45
    2-7   Use of Isolation Transformer   46

vi   Contents

    2-8   Measurement of DC Voltage Accompanied by High-voltage AC Pulses   47
    2-9   DC Voltage Indication in Conona Field   48
    2-10   Line-operated Voltmeters Must Be Properly Grounded   48
    2-11   Quick Check for Circuit Loading With Multimeter   48
    2-12   DC Voltmeter Response to Pulse Voltage   49
    2-13   DC Voltage Measurements in Nonlinear Circuits   50
    2-14   Conduction in Reverse-biased Transistor   53
    2-15   Measurement of Battery Voltage   54
    2-16   Open-circuit Testing Precaution   54
    2-17   Voltmeter Response to DC Pulses   55
    2-18   Application of Zero-center Indication   55
    2-19   Measuring Dwell Time of Contactor   7
    2-20   Photovoltaic Cell Output   57
    2-21   Zener Diode Checkout   58
    2-22   Measurement of DC Voltage in HF Circuits   58

3   Resistance Measuring Techniques   63
    3-1   General Considerations   63
    3-2   In-circuit Resistance Measurements   68
    3-3   High-voltage Ohmmeter Application   71
    3-4   Low-ohms Adapter for Multimeter   73
    3-5   Checking Special Types of Resistors   74
    3-6   Resistance Charts   76
    3-7   Measurement of Resistance of Meter Movement   76
    3-8   Transistor Test With Ohmmeter   76
    3-9   Testing Integrated Circuits With an Ohmmeter   86
    3-10   Ohmmeter Test of Circuit Boards   91

4   DC Current Measurement Methods   93
    4-1   General Considerations   93
    4-2   Measurement of High DC Current Values   94
    4-3   Clip-around DC Current Probe   95
    4-4   Ultrasensitive DC Microammeter   95
    4-5   Direct Current Galvanometer   96
    4-6   Internal Resistance Considerations   97
    4-7   Current Measurement in Rectangular Waveform   98
    4-8   Measuring True RMS Value of Pulsating DC Waveform   99
    4-9   Application of Kirchhoff's Current Law   101
    4-10   Constant-current Principle   102
    4-11   Checking Current Balance   103
    4-12   Transistor Electrode Currents   104
    4-13   Transistor Current Measurements   105

Contents   vii

5   AC Voltage Measurement Procedures   116
- 5-1   General Considerations   116
- 5-2   Measurement of AC Voltages at High Frequencies   118
- 5-3   Decibel Measurements With AC Voltmeters   119
- 5-4   Measurement of Small AC Voltages   126
- 5-5   Volume Unit Measurements   128
- 5-6   Tuned AC Voltmeters   129
- 5-7   Response of Basic AC Voltmeters to Complex Waveforms   133
- 5-8   RMS Values of Basic Complex Waveforms   134
- 5-9   AC Power Measurements   134
- 5-10 Test for "Above-ground" Soldering Gun   136
- 5-11 AC Converters in Service-type Voltmeters   137

6   Hi-fi Stereo Troubleshooting   139
- 6-1   General Considerations   139
- 6-2   Amplifier Frequency Response   139
- 6-3   Harmonic and Intermodulation Distortion   141
- 6-4   Classes of Amplifier Operation   148
- 6-5   Amplifier Voltage Gain and Sensitivity   148
- 6-6   Intermodulation Distortion Measurement   151
- 6-7   Pinpointing Defective Components and Devices   152
- 6-8   Quasi-Complementary Amplifier   155
- 6-9   Stereo-multiplex Separation Test   159
- 6-10 Applications of a Stereo Analyzer   163
- 6-11 Applications of an Impedance Bridge   168
- 6-12 Varistor Response to AC Voltage   171
- 6-13 Notes on Tools and Accessories   171

7   Radio Receiver Troubleshooting   174
- 7-1   General Considerations   174
- 7-2   Current Drain and Operating Voltages   175
- 7-3   Systematic Troubleshooting Procedures   178
- 7-4   Signal-injection Tests   181
- 7-5   AM Alignment Procedures   184
- 7-6   Troubleshooting Distortion and/or Regeneration   186
- 7-7   Troubleshooting FM Receivers   188
- 7-8   Weak Output in AM and FM Receivers   191
- 7-9   Troubleshooting CB Transmitters and Receivers   192
- 7-10 Optimum Value of Oscillator Injection Voltage   197
- 7-11 Diode and Triode Transistor Detectors   198
- 7-12 Stage Gain Values for an FM Broadcast Receiver   201
- 7-13 Automatic Frequency Control   201

viii   Contents

8   Black-and-White Television Troubleshooting   206
    8-1   General Considerations   206
    8-2   Evaluation of Trouble Symptoms   210
    8-3   IF Amplifier Troubleshooting Techniques   216
    8-4   AGC Section Troubleshooting   220
    8-5   Multiple Faults   222
    8-6   Video-amplifier Troubleshooting   222
    8-7   DC Voltages Under Signal and No-signal Conditions   225
    8-8   Signal Tracing the AFC and Oscillator Sections   227
    8-9   Horizontal-output Section   229
    8-10   Case-history Approach   235

9   Color Television Troubleshooting   240
    9-1   General Considerations   240
    9-2   Automatic Tint Control Troubleshooting   243
    9-3   Color Demodulator Tests   246
    9-4   Silicon Controlled Rectifier Sweep Circuit   247
    9-5   Fail-safe Circuit Operation   255
    9-6   Deflection Linearity Correction   256
    9-7   Summary of Trouble Localization   257
    9-8   Notes on Automatic Frequency-phase Control Troubleshooting   260
    9-9   Automatic Fine Tuning   267

10   Miscellaneous Applications   271
    10-1   Oscilloscope Troubleshooting   271
    10-2   CCTV Troubleshooting   281
    10-3   Fuel-vapor Detector Troubleshooting   283

Appendix I
    Resistor Color Codes   289

Appendix II
    Capacitor Color Codes   290

Appendix III
    Diode Polarity Identification   292

Index   293

Preface

It is the purpose of *Electronic Meters: Techniques and Troubleshooting* to provide electronics technicians and students with a completely practical guide to trouble localization in a broad spectrum of modern electronic equipment. Preliminary analysis of trouble symptoms is followed by explanation of suitable techniques that narrow down the possible malfunctions, and lead to pinpointing of the defective component or device. This is essentially a "how to" book that addresses itself primarily to the working technician and the vocational student. It minimizes theoretical considerations; the author assumes that the reader has completed basic training in electrical and electronics theory.

The reader need not have any previous on-the-job troubleshooting experience. This text is intended to bridge the gap between classroom instruction and field or shop hands-on activities. Accordingly, *Electronic Meters* will find equal utility in the technical institute and in the servicing industry. The text has been prepared with the home-study student in mind, as well as the junior-college or trade-school student. The book starts with a brief overview of basic meter movements, followed by a practical explanation of performance verification and calibration. Direct-current-voltage measurements are then discussed, with detailing of good practices and cautions against common pitfalls.

Resistance measuring techniques are explained for both conventional and low-power ohmmeters. Careful distinction is made between linear and nonlinear resistance parameters. Direct-current measurement methods are detailed, including both direct and indirect measuring techniques. The clip-around DC-current probe is called to the reader's attention. Common errors owing to unexpected disturbance of circuit action are pointed out. Alternating-current-voltage measurement procedures are explained in appropriate detail, and the reader is introduced to harmonic-distortion and intermodulation-distortion measurements. Troubleshooting of hi-fi stereo units and systems, radio receivers, black-and-white television receivers, and color-television receivers is explained from the viewpoint of electronic meter application, with notes of various incidental requirements.

Mathematics has been held to a minimum, and graphical treatments of quantitative considerations have been utilized in this book. Nevertheless, it will be considerably easier for the student to understand the reasoning underlying troubleshooting procedures and data evaluation if he has completed courses in algebra, geometry, and elementary trigonometry. The author is indebted to numerous manufacturers, as credited throughout the text, for illustrative material and technical data. It is appropriate that this work be dedicated as a teaching tool to the instructors and students of our technical schools and junior colleges.

M. R–S.

Electronic Meters

Introduction

In the field of electricity and electronics, as in all of the other physical sciences, accurate quantitative measurements are essential. These involve two important items—numbers and units. Simple arithmetic is used in most cases, and the units are well-defined and easily understood. The standard units of current, voltage, and resistance, as well as other units, are defined by the National Bureau of Standards (NBS). In an instrument factory, various instruments are calibrated by comparing them with established standards.

A technician commonly works with ammeters, voltmeters, ohmmeters, and semiconductor analyzers. He may also have various occasions to use wattmeters, watt-hour meters, power-factor meters, synchroscopes, frequency meters, and resistance-capacitance-inductance (RCL) bridges. Electrical and electronic equipment is designed to operate in accordance with published specifications. To aid the technician in maintaining and troubleshooting electrical and electronic equipment, standard service data are published, such as the Howard W. Sams & Co.'s "Photofacts."

To the technician, a good understanding of the functional design and operation of electrical instruments is important. In electrical or electronic servicing procedures, one or more of the following methods are commonly used to determine if the circuits in a unit of equipment are functioning normally.

1. Use a voltmeter to determine the voltage existing between two points in a circuit.
2. Use a microammeter, milliammeter, or ammeter to measure the amount of current flowing in a circuit.
3. Use an ohmmeter or a megohmmeter to check for circuit continuity and to measure total or partial circuit resistance.

A technician may also find it necessary to utilize a wattmeter to determine the total power being consumed by a unit of equipment. Sometimes he needs to measure the energy consumed by the equipment over a period of time. In turn, a watt-hour or kilowatt-hour meter is employed. For measuring other quantities, such as power factor, frequency, and so on, a technician utilizes appropriate basic or specialized instruments. In each case, the instrument indicates the value of the quantity measured, or the value of one of its components. Thus, calculation is sometimes required to complete a measurement. In turn, the technician interprets the quantitative data in a manner that will help him to understand the malfunction that is causing the trouble symptom. Occasionally, the technician will need to determine the value of an inductor or of a capacitor. Inductance or capacitance bridges are generally employed for this purpose.

d'Arsonval Meter

The stationary permanent-magnet moving-coil meter is the basic movement used in most measuring instruments for troubleshooting electrical and electronic equipment. It consists basically of a permanent magnet and a movable coil, as depicted in Fig. i. When current flows through the coil, the resulting magnetic field reacts with the field of the permanent magnet and causes the coil to rotate. The greater the amount of current flow through the coil, the stronger is the magnetic field produced and the greater is the rotation of the coil. This arrangement is often called a *galvanometer*. The movable coil is suspended between the poles of the magnet by means of thin flat ribbons of phosphor bronze. These ribbons provide the conducting path for the current between the circuit under test and the movable coil.

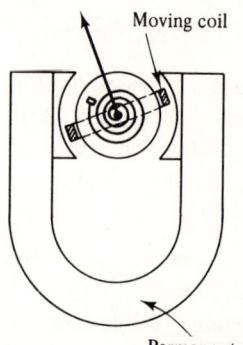

Figure i   Permanent-magnet moving-coil instrument.

The ribbon springs also provide the restoring force for the coil. This restoring force balances the force of the coil's magnetic field in order to obtain an indication of the current strength. Thus, the ribbon springs tend to oppose the motion of the coil, and will yield by an angle that is proportional to the force applied to the coil by the action of the coil's magnetic field in combination with the permanent field. When the driving force of the coil current is removed, the restoring force returns the coil to its zero position. In modified form, the basic d'Arsonval movement has the highest sensitivity of any of the various types of meters in present-day use. Most of the movements used in service shops are microammeters, and have a full-scale current value of 50 μA. However, movements with a full-scale current value of 10 μA are also in use. A few movements with a full-scale current value of 1 μA are utilized by electronic technicians.

Dynamometer Movement

The most common *high-accuracy* ac instrument types use the dynamometer movement. The magnet of the permanent-magnet moving-coil type of instrument is replaced by a pair of field coils, which provide the magnetic field in which the moving coil rotates, as depicted in Fig. ii.

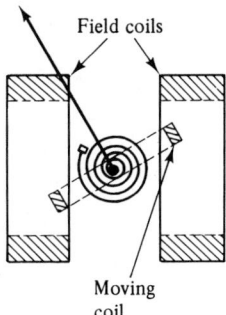

Figure ii Plan of the dynamometer movement.

The two fixed coils are connected in series and positioned coaxially, with a space between them. The two movable coils are also positioned coaxially, and are connected in series. These two pairs of coils (fixed and movable) are further connected in series with each other. Note that the movable-coil unit is pivot-mounted between the fixed coils. In turn, the central shaft on which the movable coils are mounted is restrained by spiral springs, which hold the pointer at zero when no current is flowing through the coil. These springs also serve as conductors for delivering current to the movable coils. Since these conducting springs are very small, the meter cannot carry a very heavy current. Although electrodynamometer-type meters are very accurate, they do not have the sensitivity of the d'Arsonval movement. For this reason, they are not widely used outside of the laboratory, except for measurement of line voltage, current drain, and power demand of appliances in servicing operations.

Moving Iron-vane Meter

The moving iron-vane meter is another basic type of meter. Unlike the d'Arsonval movement, which employs permanent magnets, the moving iron-vane meter depends on induced magnetism for its operation. It utilizes the principle of repulsion between two concentric iron vanes, one fixed and one movable, as shown in Fig. iii. A pointer is attached to the movable vane. When current flows through the coil, the two iron vanes become magnetized with north poles at their upper ends, and south poles at their lower ends, for one direction of current through the coil, as shown in the diagram. Because like poles repel, the unbalanced

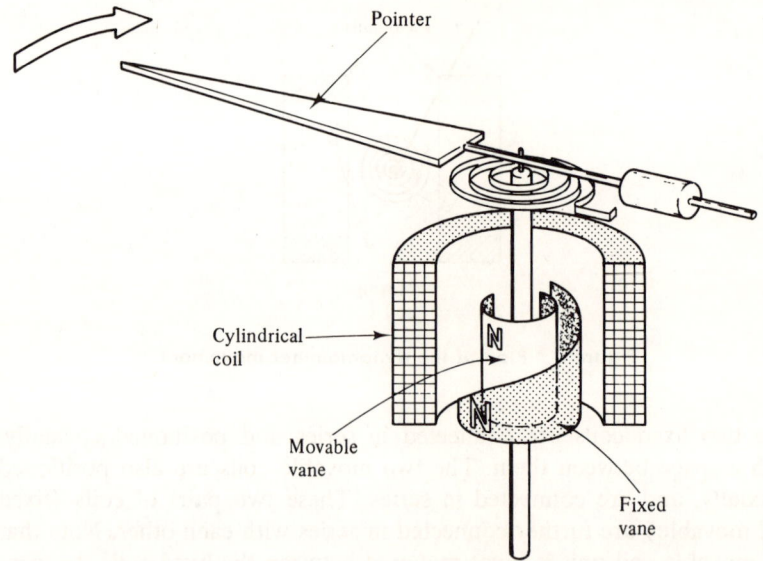

Figure iii Simplified diagram of a moving iron-vane meter.

component of force tangent to the movable element causes it to turn against the force exerted by the springs.

The movable vane is rectangular in shape, and the fixed vane is tapered. This design permits the use of a relatively uniform scale. When no current flows through the coil, the movable vane is positioned so that it is opposite the larger portion of the tapered fixed vane, and the scale indication is zero. The amount of magnetization of the vanes depends on the strength of the field, which, in turn, depends on the amount of current flowing through the coil. The force of repulsion is greater opposite the larger end of the fixed vane than it is nearer the smaller end. Therefore, the movable vane moves toward the smaller end through an angle that is proportional to the magnitude of the coil current. The movement ceases when the force of repulsion is balanced by the restoring force of the spring.

Because the repulsion is always in the same direction (toward the smaller end of the fixed vane) regardless of the direction of current flow through the coil, the moving iron-vane instrument operates on either dc or ac. However, when used to measure dc values, the moving iron-vane instrument has a residual error resulting from residual magnetism in the vanes. This error may be minimized by reversing the meter leads and averaging the readings. Because of its simplicity, its relatively low cost, and the fact that no current is conducted to the moving element, this type of movement is used extensively to measure current and voltage in ac power circuits. However, because the reluctance of the magnetic circuit is high, the moving iron-vane meter requires much more power to produce full-scale deflection than is required by a d'Arsonval meter of the same range. Therefore, the moving iron-vane meter is unsuitable for use in high-resistance low-power circuits.

Thermocouple-type Meter

If two of the ends of two dissimilar metals are welded together and this junction is heated, a dc voltage is developed across the two open ends. This voltage depends on the material of which the wires are made and on the difference in temperature between the heated junction and the open ends. In one type of instrument, the junction is heated electrically by the flow of current through a heater element. It does not matter whether the current is ac or dc, because the heating effect is independent of the current direction. The maximum current that may be measured depends on the current rating of the heater, on the heat that the thermocouple can withstand without being damaged, and on the current rating of the meter movement used with the thermocouple.

A simplified diagram of one type of thermocouple is shown in Fig. iv. The input current flows through the heater strip via the terminal blocks. In turn, the function of the heater is to raise the temperature of the thermocouple, which consists of a junction of two dissimilar wires welded to the heater strip. The open ends of these wires are connected to the center of two copper compensating strips. The function of these strips is to radiate heat so that the open ends of the wires will be much cooler than the junction end of the wires. This permits a higher voltage to be developed across the open ends of the thermocouple. The compensating strips are thermally and electrically insulated from the terminal blocks.

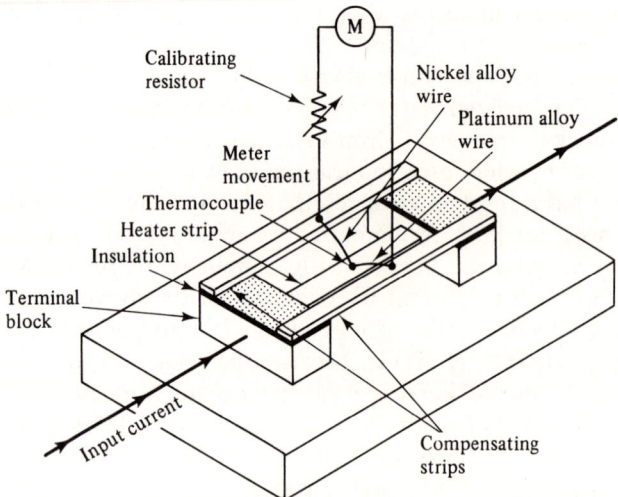

Figure iv Simplified diagram of one type of thermocouple.

Note that the heat produced by the flow of line current through the heater strip is proportional to I^2R. Because the voltage appearing across the two open terminals is proportional to the temperature, the movement of the meter element connected across these terminals is proportional to the square of the current flowing through the heater element. In turn, the scale of the meter is crowded near the zero end, and is progressively less crowded near the maximum end of the scale. Because the lower end of the scale is crowded, the reading is necessarily less accurate in this region. A more nearly uniform meter scale may be obtained if the permanent magnet in the movement is constructed so that as the coil rotates (needle moves up scale), it moves into a magnetic field of less and less density. The torque then increases approximately as the first power of the current instead of as the square of the current, and a more nearly linear scale is achieved.

Digital Voltmeters

Digital voltmeters (DVM's) provide several advantages over analog types of voltmeters. The lab-type digital voltmeter responds more rapidly, has comparatively high accuracy and resolution, reduces operator errors, and provides automatic measurements in systems applications. A DVM displays electrical values as discrete numerals instead of a pointer position on a scale. With reference to Fig. v, the chief features of a DVM are as follows:

1. FUNCTION switch
   Selects the type of measurement to be performed. Also used to turn the instrument on and off.
2. RANGE switch
   Selects the desired range for the measurement selected by the FUNCTION switch.
3. 1A jack
   This jack is used when measuring dc and ac currents on the 1A range.
4. COM jack
   This is the common connection to the instrument input.
5. (+) jack
   This is the "hot" input jack. In dc measurements the test lead connected to this jack is normally applied to the positive potential point in the circuit under test when FUNCTION switch is in the +DCV position.
6. NUMERICAL READOUT DIGIT NUMBER 3
   An indication between 0 and 9 is obtained, depending upon the value of the reading displayed. This digit is identified as the "third" digit.
7. NUMERICAL READOUT DIGIT NUMBER 2
   A reading of 0 to 9 is obtained, depending upon the value of the measurement performed. This digit is identified as the "second" digit.
8. NUMERICAL READOUT DIGIT NUMBER 1
   Either an off (no display) or a "1" is displayed, depending upon the value of the measurement performed. This digit is used when readings corresponding to full scale and over range are performed, and is identified as the "first" digit.

9. Adjustable Handle

This handle can be used for carrying the instrument and can also be used as a tilt stand to raise the front panel to a convenient viewing angle when desired. The handle can be locked into any desired position by rotating it until it snaps into the position desired.

10. Test Probe

This probe is used for all measurements performed. This probe is provided with a selectable 100-KΩ resistor which is used when performing dc measurements in high-impedance or high-frequency circuits. The resistor isolates the cable and input capacitance of the meter from the circuit in which the measurements are performed. This reduces capacitive loading that may result in erroneous reading.

The operation of a DVM is based on an electronic counter. An input voltage to be measured starts the counter action, and this counting action continues for a time that is required to charge up a capacitor to the value of the input voltage. At this instant the counter stops. (The oscillator ceases operation.) A digital-logic arrangement functions between the oscillator and the readout device to display numerals corresponding to the input voltage level. In other words, the input voltage starts a pulse generator, and pulses are generated for a period of time that corresponds to the voltage level to be measured. An electronic counter totals these pulses, and their sum is converted to a digital readout by logic circuitry.

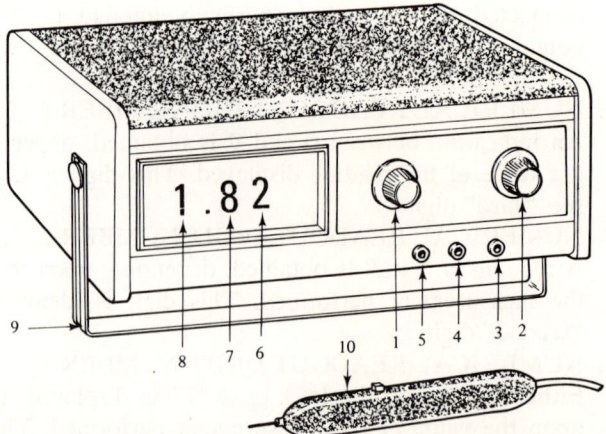

Figure v A digital voltmeter displays discrete numerals. (*Courtesy of B & K Precision, Division of Dynascan Corp.*)

1 Performance Verification and Calibration

1-1 General Considerations

Voltage, current, and resistance measurements can be made easily, rapidly, and accurately with analog electronic instruments using meter movements. Although digital readout is also utilized to a considerable extent, meter-movement readout is most extensively employed, since it is economical and suitable for many jobs. Analog indication also lends itself well to special, nonlinear scales such as decibel scales.

Before *voltmeter accuracy* considerations can be explained, the reader should have a familiarity with the various meter scales that are available. Many instruments have meter scales that are marked both in volts and decibel (dB) units. Decibel units and voltage units are complements of each other. In other words, if a linear voltage scale is provided, the dB scale on the same meter face will have a logarithmic (nonlinear) form. Similarly, if a meter is designed to utilize a linear dB scale, the voltage scale on the same meter face will have a nonlinear form. The Hewlett-Packard Company notes as follows:

The term "linear-log" scale denotes that an instrument has a linear dB scale and, in turn, a nonlinear voltage scale. Examples of various meter scales are shown in Fig. 1-1. With reference to Fig. 1-2, analog meters usually have nonlinearities and/or offsets inherent in the attenuators and amplifiers. In addition, the meter movement itself can have nonlinearities—in spite of the fact that the scales may have been individually calibrated. Nonlinearity factors cause *percent-of-reading* errors, and offset factors cause *percent-of-full-scale* errors. Percent-of-reading errors are constant, no matter where the pointer may fall on the scale. On the other hand, percent-of-full-scale error increases in percent-of-reading value as the pointer rests farther down on the scale.

10   Performance Verification and Calibration

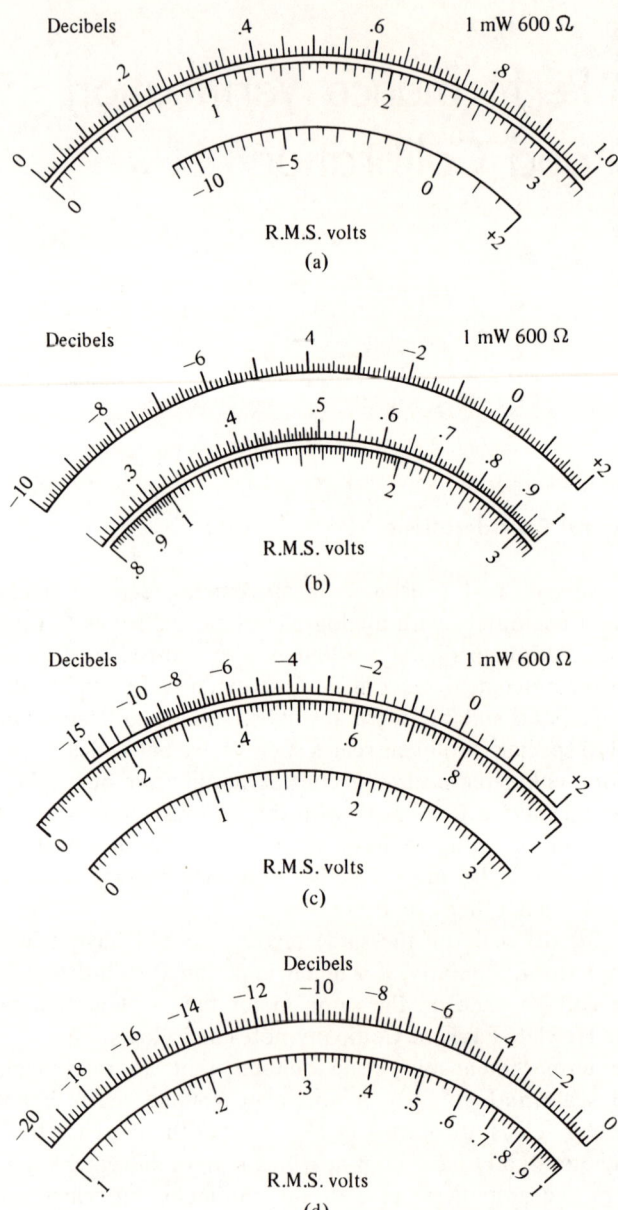

Figure 1-1 Typical meter scales: **(a)** linear 0-3 V scales, plus a dB scale; **(b)** linear dB scale, plus nonlinear (logarithmic) voltage scales; **(c)** dB scale placed on larger arc for greater resolution; **(d)** linear -20 to 0dB scale for acoustical and communications applications. (*Courtesy of Hewlett-Packard*)

1-1 General Considerations 11

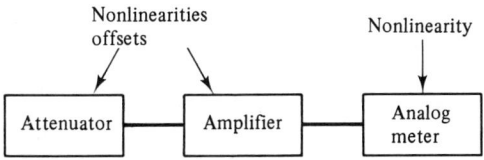

Figure 1-2 Nonlinear factors cause percent-of-reading errors; offset factors cause percent of full-scale errors. *(Courtesy of Hewlett-Packard)*

With reference to instrument specification sheets, accuracy specifications are usually expressed in one of three ways: (1) percent of the full-scale value, (2) percent of the reading, and (3) percent of reading plus percent of full-scale value. The first specification is most widely used. The second specification is more commonly applied to meters that have a logarithmic scale. The third specification has been used in recent practice to provide a tighter accuracy specification on instruments that have linear scales. For a thorough evaluation of instrument accuracy, the following questions should be considered: Does it apply at all input-voltage levels up to the maximum overrange point? (This consideration may be qualified by means of linearity specifications.) Does it apply to all frequencies throughout its specified bandwidth? Does it apply on all ranges of the instrument? Does it apply over a useful temperature range? If temperature variations are not included in the accuracy rating, is the temperature coefficient specified?

If dc measurements are of primary concern, it is advisable to select an instrument that has the broadest capability in meeting the requirements. If ac measurements involving sine waves with only moderate amounts of distortion (<10%) are of primary concern, a voltmeter with an average response can perform over a bandwidth extending to several megahertz. In the case of high-frequency measurements (>10 MHz), a peak-responding voltmeter with a diode-probe input is the most economical choice. Peak-responding circuits are satisfactory if inaccuracies caused by distortion in the input waveform can be tolerated. Again, when measurements are required to determine the effective power of nonsinusoidal waveforms, a true-rms-responding voltmeter is the best choice. In general, true-rms meters can indicate the rms value of ac voltages only, because they are usually ac-coupled. Most true-rms voltmeters have a frequency cutoff in the vicinity of 20 Hz. This restriction prevents a true-rms voltmeter from indicating the value of any subsonic dc voltage or of the dc component in a waveform.

Note in passing that ac voltmeters are often combined with other types of instruments, both in service-type and lab-type equipment. For example, the audio tester illustrated in Fig. 1-3 consists of an ac volt-

12 Performance Verification and Calibration

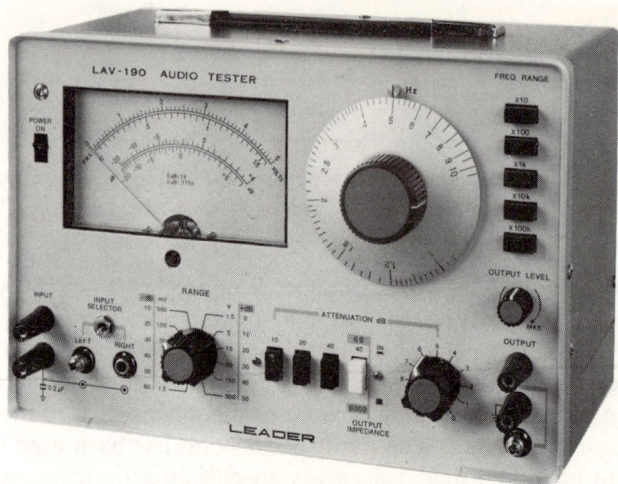

Figure 1-3 Signal-level meter provided in an audio oscillator with a wide-range attenuator. (*Courtesy of Leader Electronics*)

meter with rms and decibel scales combined with a sine-wave audio oscillator. The meter serves as a signal-level indicator, so that the technician knows precisely how much signal voltage is being applied to the audio unit under test. A frequency range from 1 Hz to 1 MHz is provided, at any level between -60 and $+60$ dB.

1-2 Basic Meter Circuitry

All analog meters employ instrument circuitry that energizes a meter movement, as exemplified in Fig. 1-4. The basic d'Arsonval meter movement is both a current meter and a voltmeter. In other words, the movement has a certain full-scale current value, and since it has a given internal resistance, the movement also has a specified full-scale voltage value. As an illustration, if the movement depicted in Fig. 1-5(a) has a full-scale current value of 50 μA and an internal resistance of 2000 ohms, it then has a full-scale voltage of 100 mV. In turn, the movement can be used directly as a 50-μA microammeter, or as a 100-mV millivoltmeter. These direct-indication functions are exploited in various multimeters.

Next, a higher voltage-measuring range is obtained by connecting a multiplier resistor in series with the movement, as depicted in Fig.

Figure 1-4 A 200-microampere dc meter movement (*Courtesy of Simpson Electric Co.*)

1-5(b). To continue the previous example, if a 2000-ohm multiplier resistor were utilized, the meter would provide a full-scale voltage indication of 200 mV. Or, if a 20,000-ohm multiplier resistor were used, the meter would provide a full-scale voltage indication of 2 volts. Although this instrument arrangement can be used to measure current, it would be useful only in exceptional applications wherein the internal resistance of the source happened to be very high. Otherwise, the circuit action would be disturbed excessively because of insertion of excessive instrument resistance.

Next, a higher current range than that provided in Fig. 1-5(a) is obtained by inclusion of a resistive shunt, as shown in Fig. 1-5(c). To continue the original example, if a 2000-ohm shunt were employed, the full-scale current indication would be 100 μA. If a 105.26-ohm shunt were used, the full-scale current indication would be 1 mA. This arrangement would be practically useless in voltmeter applications, of course, because of the excessive circuit loading that would result from the low input resistance of the instrument. Ideally, a voltmeter would have infinite input resistance, and an ammeter would have zero input resistance.

Since a current meter has an input resistance that is greater than

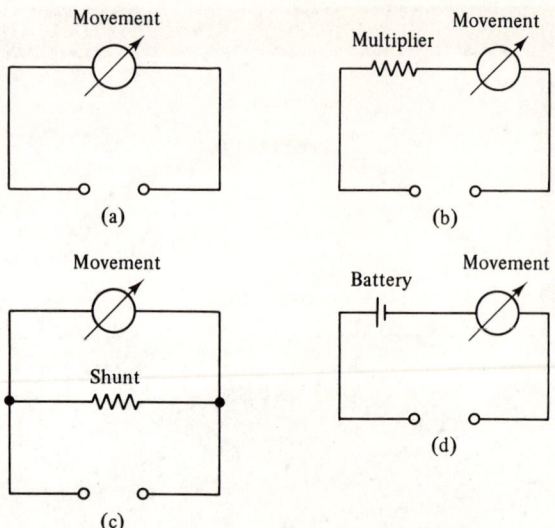

Figure 1-5 Skeleton meter circuits: **(a)** movement alone operates as current or voltage meter; **(b)** series multiplier resistor provides voltmeter function; **(c)** shunt resistor provides ammeter function; **(d)** series battery provides ohmmeter function.

zero, an IR drop occurs across the meter input terminals whenever a current value is indicated. This voltage drop is called the *voltage burden* of the current meter. A typical service-type multimeter has a voltage burden on its current-measuring ranges as follows: 200 mV on 200-μA, 2-, and 20-mA ranges; 300 mV on its 200-mA range; 1 V on its 2-amp range. The voltage burden of a milliammeter or ammeter can be easily measured by connecting a voltmeter across the input terminals of the current meter. In most cases, the voltage burden is stated for full-scale deflection of the current meter.

An ohmmeter function is basically provided by a battery connected in series with a meter movement, as depicted in Fig. 1-5(d). Suppose, for example, that the battery has a potential of 1.5 volts, and that the movement has a full-scale current value of 1 mA, with an internal resistance of 1500 ohms. In turn, the pointer will deflect full-scale when the ohmmeter test leads are short-circuited together. Next, if the test leads were connected across a 1500-ohm resistor, the pointer would deflect to the half-of-full-scale point. It is evident that the instrument can be provided with a direct-reading ohms scale, which will be accurate as long as the ohmmeter battery maintains correct terminal voltage.

1-3 Accuracy Check of Lower DC-Voltage Ranges with Mercury Batteries

Mercury batteries such as that illustrated in Fig. 1-6 are often used for checking the indication accuracy of service-type dc voltmeters. A mercury battery has a comparatively precise terminal voltage until its useful life is ended. This terminal voltage has a higher precision than the rated accuracy of typical service-type meters. The mercury battery shown in Fig. 1-5 has output voltages of 1.35, 2.70, 4.05, 5.40, 6.75, 8.10, 9.45, and 10.80 volts. The EMF (open-circuit voltage) of a mercury battery decreases slightly with shelf-life time as shown in Fig. 1-7(a). Its EMF and voltage under load vary with temperature, as depicted in Fig. 1-7(b). In general, a mercury battery has an accuracy of ±½ percent with respect to its nominal value.

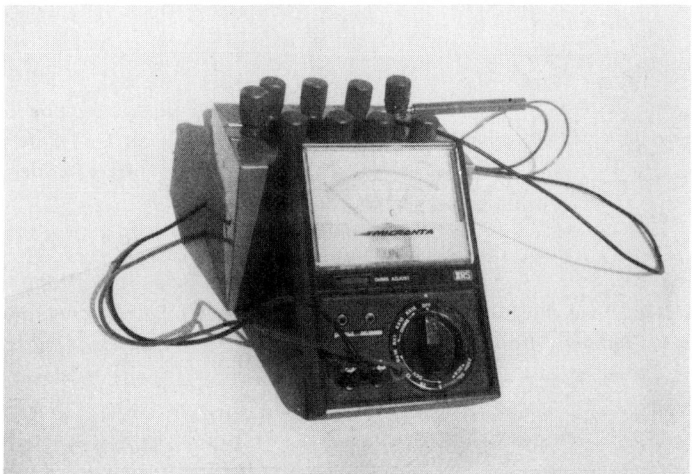

Figure 1-6 A shop-assembled mercury battery used to check the indication accuracy of a service-type multimeter.

To check the higher dc-voltage ranges of a multimeter, precision resistors may be employed as an electrical "stepladder." Precision resistors rated ±1 percent accuracy are commonly available with ½ watt power dissipation rating. Note that typical service-type multimeters are rated for an accuracy of ±1.5 percent of full scale. (This rating corresponds to an indication accuracy of ±3 percent at half scale.) Precision resistors may be connected in series with a power supply and a multimeter, as shown in Fig. 1-8. Ohm's law is applied, as in the following accuracy check procedure:

16   Performance Verification and Calibration

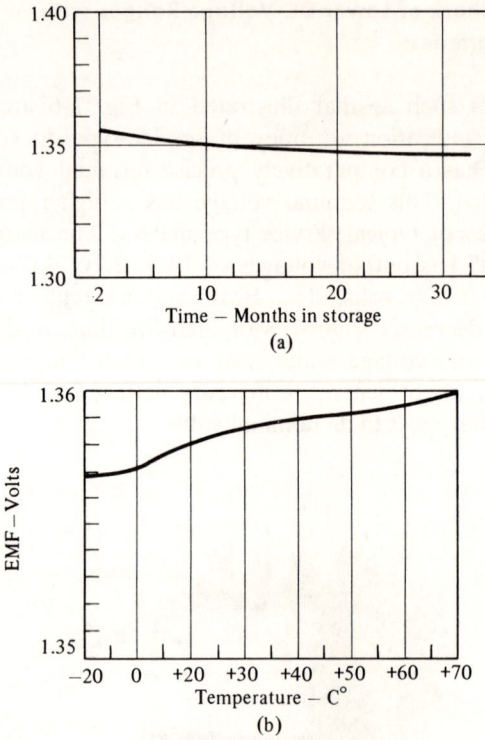

Figure 1-7 Basic mercury battery characteristics: **(a)** EMF versus age; **(b)** EMF versus temperature.

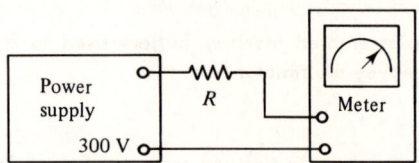

Figure 1-8 DC voltage measurement with a precision series resistor.

1. A 20,000 ohms-per-volt multimeter is under test in this example. When checked with a mercury battery, the first range of the multimeter indicates with satisfactory accuracy. The next question is whether the higher ranges of the multimeter have satisfactory indication accuracy.

1-3 Accuracy Check of Lower DC-Voltage Ranges with Mercury Batteries 17

2. The multimeter has a first range of 2.5 volts, in this example. In turn, its normal input resistance on its first range is 50,000 ohms. With reference to Fig. 1-8 the power supply has an output of approximately 300 volts. We proceed to measure the power-supply voltage precisely, using the first range of the multimeter with a suitable series resistor. Thus, the value of R in Fig. 1-8 will be selected to apply 1/150 of the power-supply voltage to the multimeter.

3. As shown in Fig. 1-9, the value of R is chosen so that a voltage division of 150-to-1 is provided. Since the input resistance of a 20,000 ohms-per-volt meter is 50,000 ohms on its 2.5-volt range, the value of R is calculated as follows:

$$1/150 = 50,000/(R + 50,000)$$

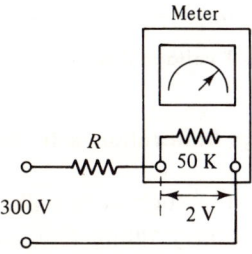

Figure 1-9 Resistor R has a value that provides a 150-to-1 voltage ratio.

4. Thus, R will have a value of 7,450,000 ohms. Several precision resistors can be connected in series to obtain this value. For example, three 1.96-megohm resistors may be connected in series with a 1.47-megohm resistor and a 100-kΩ resistor.

5. In turn, the power-supply voltage can be precisely measured on the first range of the multimeter. As an illustration, if the meter indicates two volts, the power-supply voltage is 300 volts. In any case, the power-supply voltage will be equal to 150 times the meter reading.

6. Next, the multimeter is switched to its 500-volt range, and the power-supply voltage is measured directly. In this example, the reading should be 300 volts, within the accuracy rating of the instrument.

7. To check the indication accuracy of the multimeter on its 50-volt range, the power-supply voltage must be dropped to a suitable value, such as 40 volts. If a 300-volt power supply is utilized, a 7.5-to-1 voltage divider is appropriate.

18   Performance Verification and Calibration

8. A 20,000 ohms-per-volt multimeter has an input resistance of 1 megohm on its 50-volt range. As depicted in Fig. 1-10, a series resistance of 6.5 megohms provides a voltage division of 7.5 to 1. If the power-supply voltage is precisely 300 volts, the multimeter should read 40 volts to an accuracy within its rating.
9. This same "stepladder" method can be used to check indication accuracy on other voltage ranges of the multimeter.

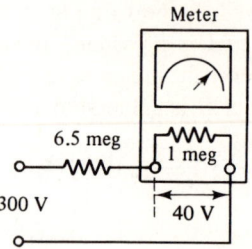

Figure 1-10   A 7.5-to-1 voltage division is provided by this circuit.

1-4 Check of DC-Voltage Calibration with Weston Cell

Digital multimeters (DVM's) such as that illustrated in Fig. 1-11 have a considerably higher accuracy rating than conventional multimeters. Note that service-type digital voltmeters do not provide decibel ranges, as is customary with analog voltmeters. For example, an accuracy of ±0.5 percent of reading, ±1 digit might be specified. In turn, the calibration accuracy cannot be checked properly with a mercury battery standard. Instead, high-accuracy dc voltmeters are often checked with a Weston cell, such as that shown in Fig. 1-12. Two types of Weston cells are available. A Weston normal cell provides a highly accurate potential of 1.0183 volts at 20°C. Normal cells are used chiefly in laboratories, with elaborate potentiometric calibrating arrangements. Another type of Weston cell, called a *student* cell or a *shop* cell, provides an output potential from 1.0185 to 1.0190 volts at 20°C. Its accuracy is not as great as that of a normal cell. However, a shop cell is comparatively rugged, and its accuracy is adequate for checking the calibration of any service-type voltmeter.

One advantage of a shop cell is that it does not have to be used with a potentiometric calibrating arrangement. In other words, a voltmeter can be connected directly to a shop cell, provided that the current demand is not greater than 100 μA. For example, the digital multimeter illustrated in Fig. 1-11 has an input resistance of 10 megohms on its dc-voltage ranges. Accordingly, this meter will draw approximately

1-4 Check of DC-Voltage Calibration with Weston Cell 19

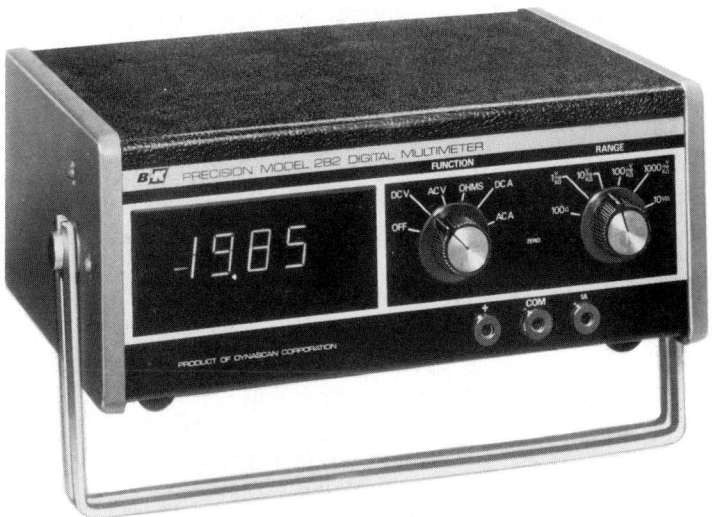

Figure 1-11 Digital multimeter. (*Courtesy of Dynascan Corp.*)

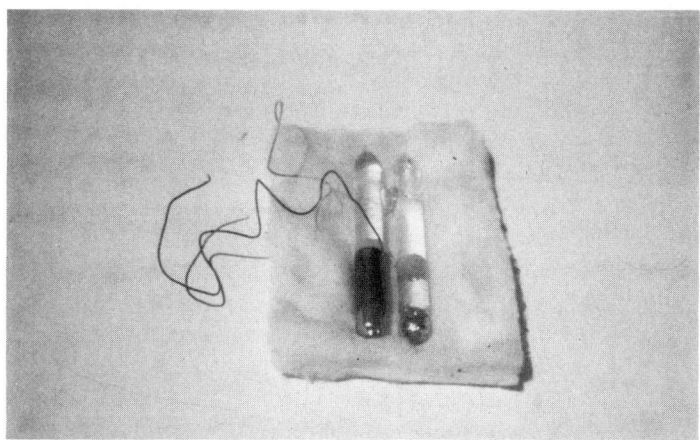

Figure 1-12 A Weston shop cell.

0.1 μA from a shop cell. By way of comparison, a 1000 ohms-per-volt multimeter operated on its 2.5-volt range would draw approximately 0.4 mA from a shop cell. Although the digital multimeter does not load the cell appreciably, the 1000 ohms-per-volt multimeter draws four times the maximum tolerable current from the cell.

1-5 Analysis of Incorrect Readings

If a conventional multimeter is off calibration on its dc-voltage ranges, the trouble will be found in the resistive multiplier or in the meter movement. Circuitry on the dc-voltage function is comparatively simple, as exemplified in Fig. 1-13. When this instrument is operated on its 2.5-volt dc range, the measuring circuit comprises a 48-kΩ multiplier resistor and the meter movement. Note the following possibilities of malfunction:

1. If the scale indication is too low on the first range, the 48-kΩ resistor may have increased in value, or the permanent magnet in the meter movement may have weakened.
2. A defective meter movement gives the same percentage error on all of the dc-voltage ranges. For example, if the indication is 10 percent low on the first range, the indication on any other range will also be 10 percent low, if the multiplier resistors are within tolerance.
3. If indication on the first range is out of rated accuracy, it is pointless to make a "stepladder" test of the higher ranges. Replace the 48-kΩ resistor to determine whether this will eliminate the error.
4. Unless rated accuracy can be restored by replacement of the multiplier resistor, additional troubleshooting is impractical for the usual service shop. In other words, the meter will be found faulty and it must be repaired by the manufacturer or in a repair depot.
5. Defective test leads rarely develop effective series resistance that causes subnormal voltage indication. However, if this possibility is suspected, replace the test leads, or connect the mercury cell directly to the multimeter input jacks.
6. As in the case of test leads, meter switches rarely develop effective series resistance that causes subnormal voltage indication. If this possibility is suspected, the switch path can be temporarily short-circuited.
7. Some multimeters employ printed circuit boards. It is possible for a PC conductor to become cracked and open-circuited. The conductors may be inspected under a magnifying glass. Also, the board may be tapped or flexed slightly to determine whether there is an intermittent open circuit.
8. Sometimes a conventional multimeter indicates correctly when lying flat, but develops objectionable indication error while standing upright. This malfunction is caused by an unbalanced

meter movement, and must be corrected at the meter factory or in a repair depot.

9. Occasionally the pointer in a meter movement does not swing freely, and correct indication is obtained only after the meter is tapped lightly. This condition is caused by bearing friction. It must be corrected at the meter factory or in a repair depot.

10. If the pointer cannot be brought to zero on the scale by adjustment of the zero-set screw, the pointer arm has probably been bent owing to excessive overload. This, too, is a specialized repair project.

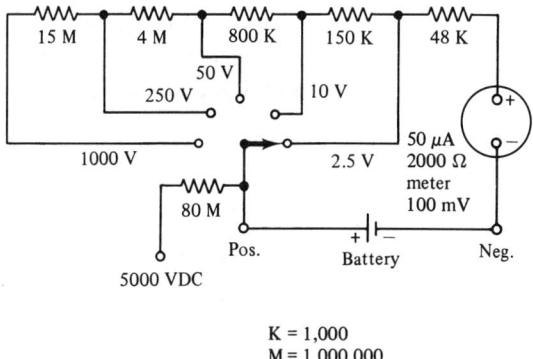

Figure 1-13 Circuitry for a typical multimeter operated on its dc-voltage function.

1-6 Calibration Cross-checks Against Other Voltmeters

Tests of dc-voltage indication accuracy with standard-cell and "step-ladder" techniques do not require the use of auxiliary voltmeters. When other voltmeters are available, cross-checks of indication accuracy can be informative. For example, if several dc voltmeters agree within rated indication accuracy, the probability is quite high that all of the meters are in calibration. Again, if one dc voltmeter out of a group of several indicates inaccurately on one or more ranges, whereas the others agree within rated accuracy on all ranges, the probability is very high that all of the meters are in calibration with the exception of the one that indicates inaccurately. Similar cross-checks may be utilized to check multimeters or digital voltmeters on their ac-voltage, resistance, and dc-current ranges.

1-7 AC-Voltage Calibration Checks

Multimeters use semiconductor rectifiers in their ac-voltage circuitry to change the ac to dc for energizing the meter movement. Either half-wave or full-wave rectification may be employed. The first ac-voltage range of a multimeter can be checked with a mercury cell. Normal scale indication depends upon whether half-wave or full-wave rectification is utilized. If a half-wave rectifier is used, the normal scale indication will be equal to 2.22 times the applied dc-voltage value. On the other hand, if a full-wave rectifier is used, the normal scale indication will be equal to 1.11 times the applied dc-voltage value. Circuitry for the ac-voltage section of one type of multimeter is shown in Fig. 1-14. Although two instrument rectifiers are connected into the circuit, only one of the rectifiers conducts meter current. The shunt rectifier merely bypasses the unused half-cycles around the meter movement, thereby reducing the back-voltage across the series rectifier and improving the indication accuracy.

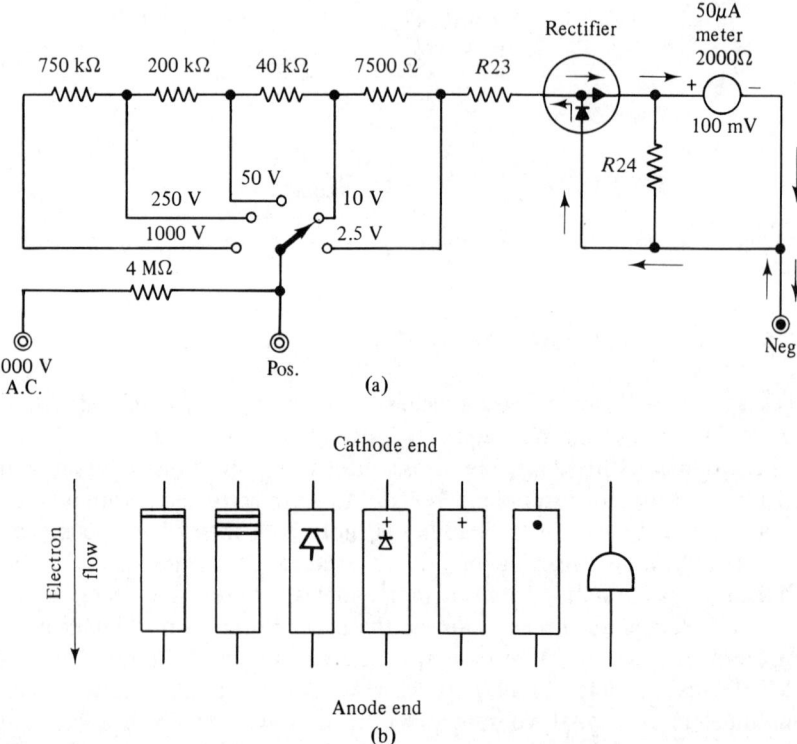

Figure 1-14 Typical ac-voltage circuitry for a multimeter: (a) configuration; (b) diode symbols and electron flow.

If 2.7 volts are applied to the ac-voltage measuring circuit depicted in Fig. 1-14, the scale reading will normally be 2.22 as great, or 6 volts. Observe that the dc test voltage must be applied to forward-bias the series diode. If the test voltage is applied in the opposite polarity, the scale indication will be zero. The rated accuracy of a multimeter is generally less on its ac-voltage function than on its dc-voltage function. For example, if the instrument is rated for an accuracy of ±1.5 percent of full scale on its dc-voltage function, it will be typically rated for an accuracy of ±3 percent of full scale on its ac-voltage function. Also, the input resistance of a multimeter is generally less on its ac-voltage function than on its dc-voltage function. As an illustration, if a meter is rated for 20,000 ohms per volt on direct current, it may be rated for 5000 ohms-per-volt on ac. This point must be kept in mind when making "stepladder" calibration checks on the ac-voltage function.

Some multimeters utilize a full-wave bridge instrument-rectifier configuration, as shown in Fig. 1-15. Both half-cycles of an applied ac voltage are used to energize the meter movement. In turn, the indication for an applied dc voltage is only half as great as if half-rectification were used. In other words, the scale indication will normally be equal to 1.11 times the applied dc-voltage value when full-wave instrument rectification is used. Still other multimeters utilize a half-bridge instrument rectifier configuration, as shown in Figure 1-16. This arrangement responds in the same manner as a full-wave bridge configuration, and the scale indication will normally be equal to 1.11 times the applied dc-voltage value.

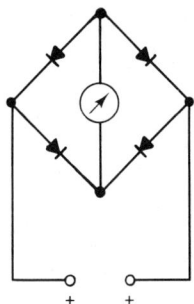

Figure 1-15 A full-wave bridge instrument-rectifier configuration.

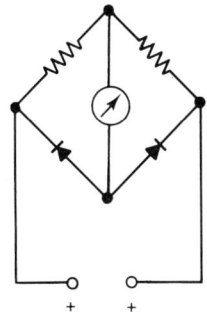

Figure 1-16. A half-bridge instrument-rectifier configuration.

Note that when ac voltages are measured with meters that use instrument rectifiers, the reading is subject to *waveform errors*. In other words, the scale is calibrated with reference to a pure sine-wave input. Any distortion of the input waveform will result in an incorrect scale indication. Most waveform distortions (although not all) are associated with harmonics. A sine waveform that is distorted by harmonics is described in terms of percentage of harmonic distortion. Figure 1-17 shows two extreme examples of harmonic distortion, viz., 30 percent and 100 percent third-harmonic content. Observe that the shape of a distorted sine wave depends upon the phase of a harmonic, in addition to the amplitude of the harmonic. As a rough rule of thumb, it can be stated that the error in scale indication may be as great as the percentage of harmonic content. For example, if a distorted sine wave contains a 20 percent second harmonic, and the ac voltmeter employs instrument rectifiers, the scale indication may be in error as much as ±20 percent.

1-8 Ohmmeter Checkout

Ohmmeter indication accuracy is checked with precision resistors. Basic tests are made with approximate half-scale values. For example, half-scale resistance values for the ohmmeter configuration depicted in Fig. 1-18 are 12, 1200, and 120,000 ohms. Corresponding commonly available ±1 percent resistor values are 12.1, 1210, and 121 kΩ. If a resistor decade box is used, such as that illustrated in Fig. 1-19, a comparatively wide range of test values can be obtained easily by inserting or removing various combinations of switch plugs. A less expensive resistor decade arrangement is shown in Fig. 1-20. It contains ±1 percent resistors, and is quite adequate for checking the ohmmeter function of a multimeter. However, if the ohmmeter function of a digital multimeter is to be checked closely, it is advisable to use high-precision standard resistors. A 1000-ohm high-precision standard resistor is illustrated in Fig. 1-21. The cost of this type of standard resistor is often prohibitive to the average electronics service shop.

Consider the accuracy that is provided on the ohmmeter function of a conventional multimeter. This accuracy is based to a considerable extent on the rated dc-voltage accuracy of the multimeter. For example, suppose that the dc-voltage function is rated at ±3 percent of full scale. On the 100-volt scale, this rating corresponds to an interval of 6 volts. In turn, when a precision resistor is checked on the ohmmeter function, the operator may refer to the 100-volt scale, to determine whether the pointer falls within a ±3 volts arc. This relation is depicted in Fig. 1-22. Here, the center-scale value is 12 ohms. If the 10-volt dc scale is used as reference, a ±3 percent tolerance corresponds to ±0.3 volt on this scale. In turn, the rated accuracy on the ohms scale cannot be better than the arc bounded by the two dotted lines in the diagram.

If the ohmmeter indication is found to be out of tolerance, the most likely cause is a weak battery. However, if a new battery does not correct the indication error, the trouble is most likely to be found in

1-8 Ohmmeter Checkout 25

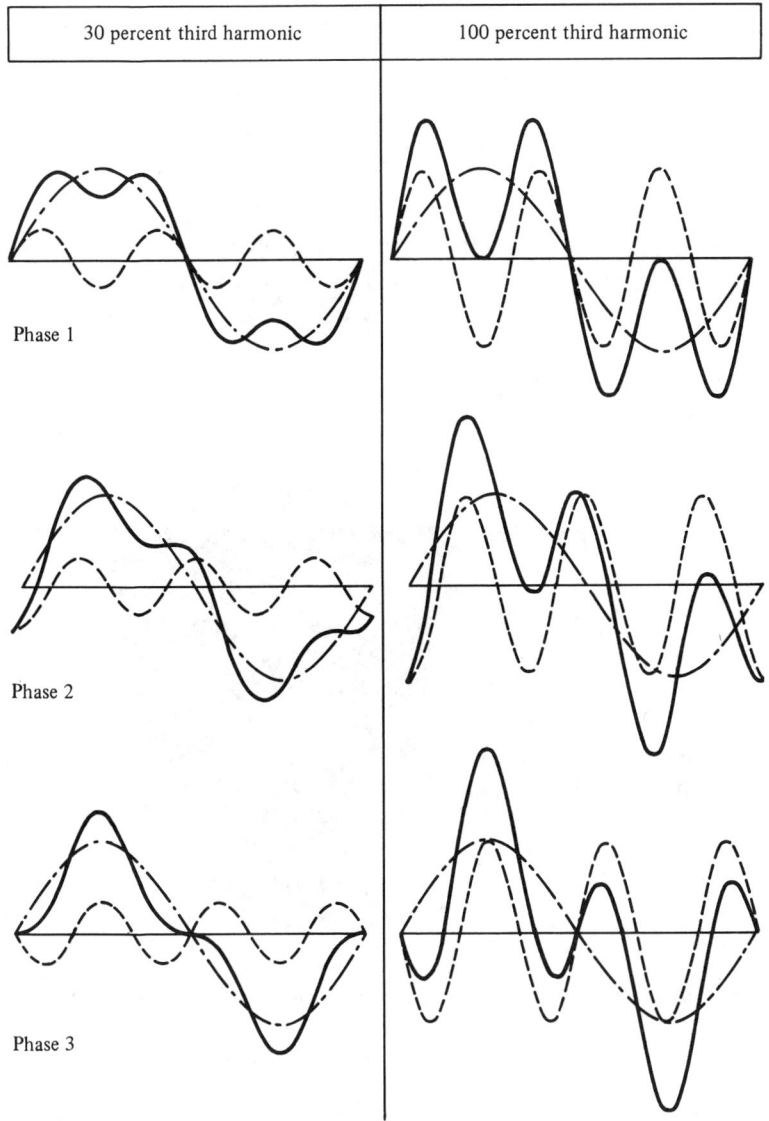

Figure 1-17 A fundamental and its third harmonic in different amplitudes and phases.

the multiplier resistor network. Presumably, the indication accuracy has been previously verified on the dc-voltage function of the multimeter, which eliminates the meter movement from suspicion. Multiplier resistors are common trouble sources. For example, if an ohmmeter is accidentally applied in a "live" ac or dc circuit, one or more of the

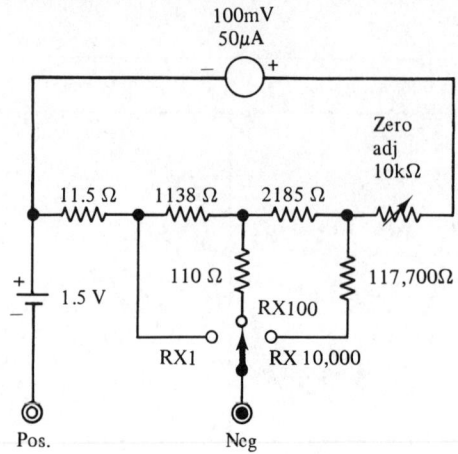

Figure 1-18 Configuration of a typical ohmmeter section in a multimeter.

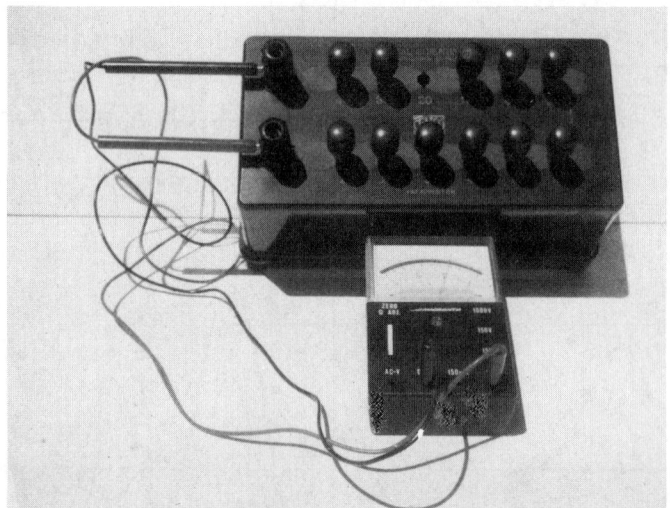

Figure 1-19 A resistor decade box provides convenience in ohmmeter checks.

Figure 1-20 An economical type of resistor decade.

1-8 Ohmmeter Checkout   27

Figure 1-21 A high-precision standard resistor.

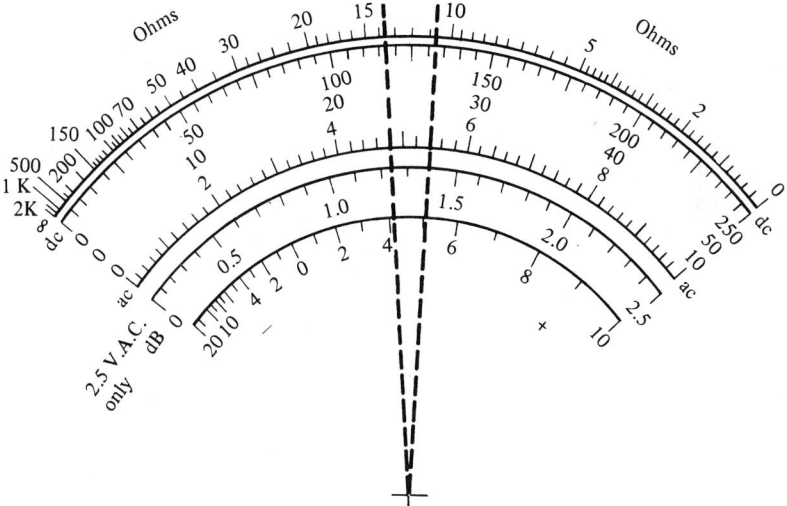

Figure 1-22 Example of indication accuracy.

resistors may become overheated and increased in value, or burned out completely. Since the meter movement in most modern multimeters is protected from overload damage by semiconductor diodes, the multiplier resistors are the components most likely to be affected. A characteristic for a typical germanium diode, or varistor, is shown in Fig. 1-23, with the meter-protective arrangement that is ordinarily used.

28 Performance Verification and Calibration

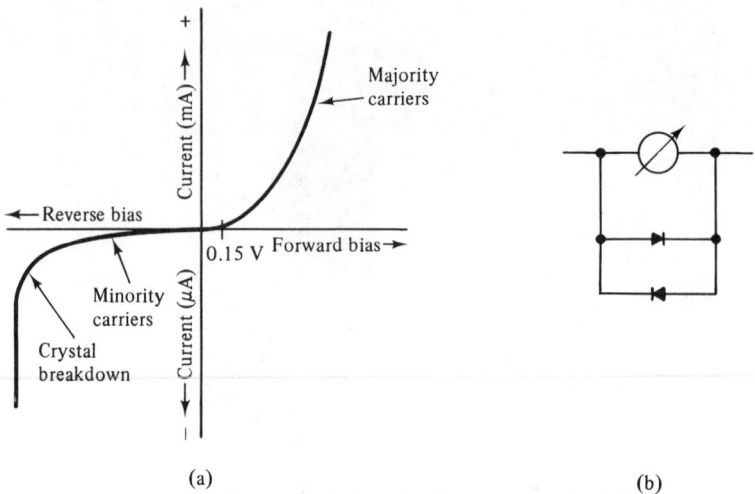

Figure 1-23 Meter protective arrangement characteristic: **(a)** germanium diode (varistor); **(b)** meter connections.

1-9 DC Current Calibration

A multimeter can be checked for indication accuracy on its dc-current ranges in somewhat the manner that was explained for dc-voltage indication. In other words, a known source voltage, as from a mercury battery, is applied to the multimeter through a precision resistor. In turn, the current flow can be calculated from Ohm's law, and this value is compared with the scale indication that is obtained on the dc-current function of the meter. A typical microammeter configuration is shown in Fig. 1-24. Observe that the instrument has appreciable input resistance, which must be taken into account when current flow in the test setup is calculated. It is evident that the input resistance of the exemplified microammeter is 2500 ohms. This configuration provides 100 μA full-scale indication.

Consider a test setup that utilizes a 1.25-volt mercury cell as a voltage source. According to Ohm's law, a current of 100 μA will flow through a resistance of 13,500 ohms. However, the input resistance of the microammeter must be subtracted from this calculated value, if a flow of 100 μA is to be obtained in a test circuit. Therefore, a series resistance of 11,000 ohms will be used to check the full-scale indication of the microammeter on its 100-μA range. Note that a mercury cell has a small value of internal resistance. However, if the cell is fresh, its internal resistance is negligible, particularly when the current drain

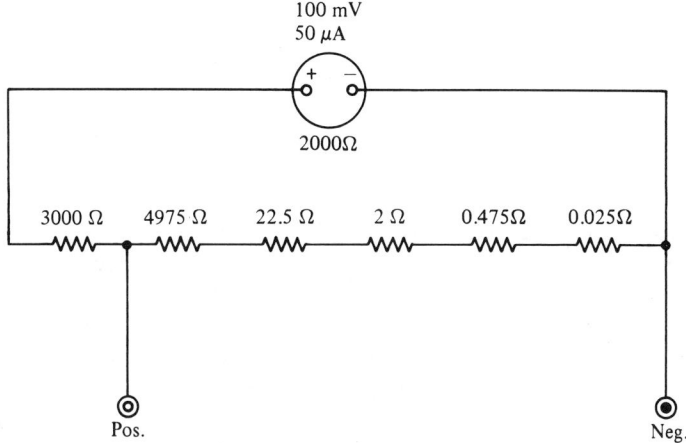

Figure 1-24 Typical configuration for a microammeter on its 100 μA range.

is small, as in the foregoing example. If the scale indication is found to be out of tolerance, it is reasonable to conclude that the trouble will be found in the ring shunt. This is a logical conclusion, because the accuracy of the meter movement has been established in the previous dc-voltage indication check.

Next, consider a check of indication accuracy on the 10-mA range of a 20,000 ohms-per-volt multimeter. A typical input-resistance value on this range is 25 ohms. If a 1.35-volt mercury cell is used as a source, the total resistance of the test circuit will be 135 ohms, according to Ohm's law. However, unless the input resistance of the meter is taken into account, the current flow in the test circuit will be 8.4 mA, instead of 10 mA. In other words, the series resistor in the test setup should have a value of 110 ohms. High-current ranges should be checked with a source that has very low internal resistance, such as a regulated power supply or a storage battery.

Resistors used in high-current calibration tests must have adequate power dissipation. For example, suppose that a 0.5-A current range is being tested from a 6-V storage-battery source. The total test-circuit resistance (which includes the meter input resistance) will be made equal to 12 ohms for a current flow of 0.5 A. A typical ammeter has an input resistance of 0.5 ohm on its 0.5-A range. If a highly accurate check is to be made, the resistance of the test leads must also be taken into account. On the other hand, if the resistance of the test leads is neglected, the check will be made with an 11.5-ohm precision power resistor. In accordance with Joule's law, this resistor must dissipate approximately 6 watts. However, it is good practice to use a resistor rated for 12 watts dissipation, or more, to ensure that it does not heat up substantially during the test, and tend to increase in value.

1-10 Waveform Errors

It is important to avoid being misled by apparent inaccuracy of multimeters on ac-voltage functions owing to waveform errors. A meter movement responds to the average value of the rectified current from the instrument rectifiers. This average value changes if the applied ac voltage (sine wave) contains harmonics. In other words, the rated accuracy of a multimeter on its ac-voltage function is stated with respect to a pure sine-wave input. The distinction between average and peak values of half-rectified and full-rectified sine waveforms is depicted in Fig. 1-25. Also shown is the difference between the average values of the waveforms and the rms value that is indicated on the meter scale.

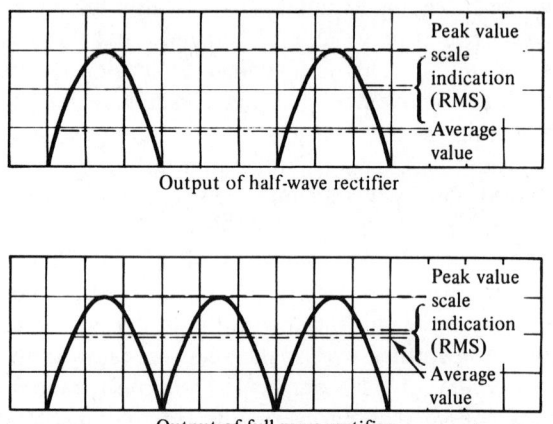

Figure 1-25 Meter movement responds to average value of rectified waveform; scale indicates rms value of input sine waveform.

As a rough rule of thumb, it can be stated that the indication error is in direct proportion to the harmonic percentage that is present in the applied waveform. For example, if the input sine waveform has a 2 percent harmonic content, the scale indication can be as much as 2 percent off value. Percentage of distortion is usually measured with a harmonic distortion meter, such as that illustrated in Fig. 1-26. This type of instrument is explained in greater detail in a following chapter.

Power-line waveforms usually have appreciable harmonic content

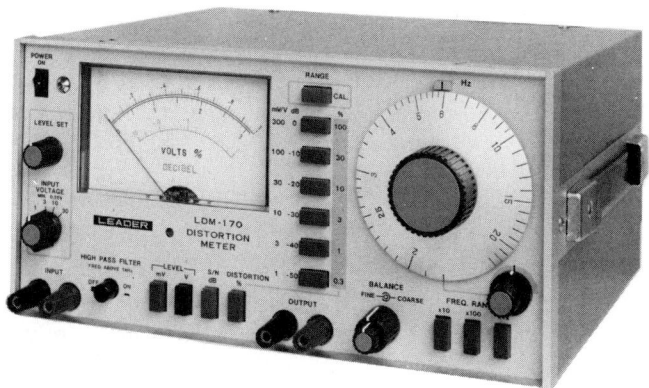

Figure 1-26 A lab-type harmonic distortion meter. (*Courtesy of Leader Electronics*)

—chiefly third harmonic, called the "iron third harmonic." This harmonic component is produced by the B-H curve of the iron in the distribution transformers. Therefore, it is advisable to use an audio oscillator with a low distortion rating for checking the ac-voltage ranges of a multimeter. It is interesting to note the rms values of some common complex waveforms, shown in Fig. 1-27. Although a multimeter or a transistor voltmeter (TVM) will not indicate the correct rms values of these waveforms, a true-rms voltmeter may be used for this purpose. It

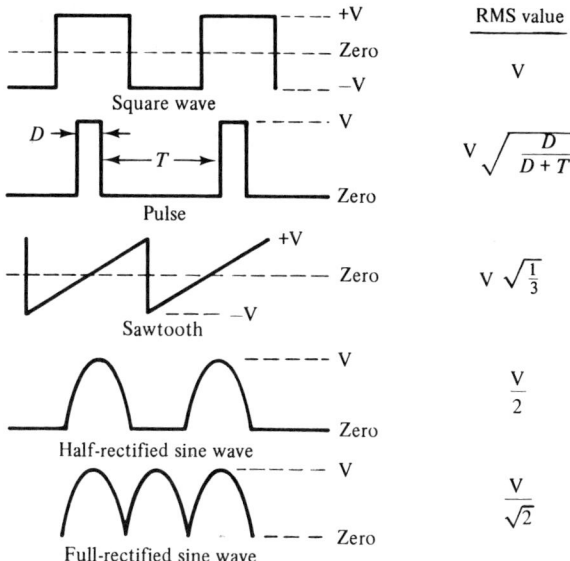

Figure 1-27 The rms values of some common complex waveforms.

is occasionally required to measure the rms values of complex waveforms, particularly in industrial-electronics troubleshooting. For example, a complex waveform from an inverter is used to power various units in some types of equipment. Inasmuch as the heating action (or true power) of a complex waveform is given by its rms value, this measurement is basic in such applications.

1-11 Calibration Procedure

Solid-state multimeters and digital voltmeters provide various calibration adjustments. For example, calibration adjustments for a typical solid-state multimeter are shown in Fig. 1-28. Calibration procedure is as follows:

Bias Adjustment. Potentiometer R44 is the bias adjustment. An external voltmeter is connected between the positive meter terminal on the instrument and the common ground point. The zero control is turned as required so that the pointer rests over the zero mark on the dc-volts scale. Then R44 is adjusted to make the external meter indicate 4.7 volts.

Balance Adjustment. Potentiometer R42 is the balance control. This control is adjusted so that when the front-panel zero control is turned fully clockwise and the function switch is in its DCV position, the meter will indicate between 0.55 and 0.65 V on the 0-to-1 volt scale. To adjust the balance control, the range switch is set to its 1-kV position, the function switch is set to one of its dc volts positions, and the front-panel zero control is turned fully clockwise. Then the balance control is adjusted to make the meter indicate 0.6 V on the 0-to-1 volt scale.

DC Volts Calibration. Potentiometer R18 is the dc volts calibration control. The mechanical and electrical zero adjustments are checked first. With the range switch set to its 0.1-volt position, and the function switch set to its +dc V position, the probe switch is set to its 100 kΩ position. A potential of precisely 0.1 volt dc is applied to the input terminals of the instrument, and the dc volts calibration control is adjusted to obtain a 1.0 indication on the 0-to-1 dc V/mA scale.

AC Volts Calibration. Potentiometer R47 is the ac volts calibration control. The mechanical and electrical zero adjustments are checked first. Then the range switch is set to its 0.1-volt position, and

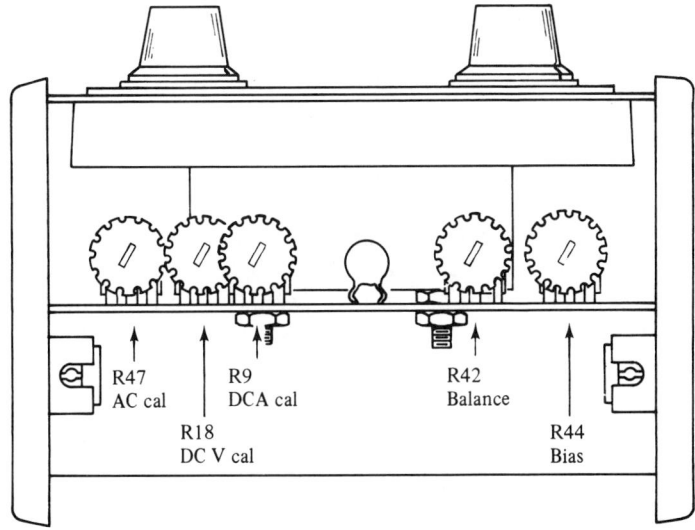

Figure 1-28 Calibration adjustments for a typical solid-state multimeter.

the function switch is set to its ac volts position. The test leads are short-circuited together and the meter is zeroed. A 0.1 rms sine-wave voltage with a frequency between 60 Hz and 1 kHz is applied to the meter, and the ac calibration potentiometer is adjusted to obtain an indication of 1 on the 0-to-1 ac volts rms scale.

DC Amperes Calibration. Potentiometer R9 is the DC amperes calibration control. The mechanical and electrical zero adjustments are checked first. Then the range switch is set to its 1-μA position, and the function switch is set to its dc amperes position. The probe switch is set to its direct setting. A 0.1-volt dc potential is applied to the meter, and the dc amperes calibration control is adjusted for an indication of 1 on the 0-to-1 dc V/mA scale.

1-12 Notes on Digital Voltmeters

A digital voltmeter displays a measured value in terms of discrete numerals, instead of a pointer deflection on a continuous scale. Note that most designs of digital voltmeters can be classified as: (1) ramp, (2) staircase ramp, (3) dual ramp integrating, (4) integrating, (5) integrating and potentiometric, (6) successive approximation, and (7) continuous balance. In each of these classifications, the basic function

34 *Performance Verification and Calibration*

that is performed is an *analog-to-digital* (A-to-D) conversion. Thus, a voltage value may be converted into a proportional time duration, which in turn starts and stops an accurate oscillator. The oscillator output is fed to an electronic counter, which drives a digital readout arrangement in terms of voltage values.

Several advantages are provided by digital readout, in comparison with analog indication of a pointer on a scale. An analog indication is subject to observational error, including parallax and estimation errors. Unless the scale is hand-drawn, an analog scale introduces an increasing indication error toward its low end. Digital readout eliminates these error sources. Simplified operation is also provided, inasmuch as a DVM has a minimum number of scales. Thus, the instrument is less confusing to inexperienced personnel. Moreover, the rated accuracy of most DVM's is much greater than that of a service-type analog voltmeter, and considerably greater than that of most laboratory-type analog voltmeters. Digital voltmeters are less costly in applications that require maximum accuracy of measurement. On the other hand, when only moderate accuracy is required, a digital voltmeter is more costly than comparable analog meters.

Digital voltmeters are classified according to the number of full digits displayed. An example of a three-digit DVM is seen in Fig. 1-29.

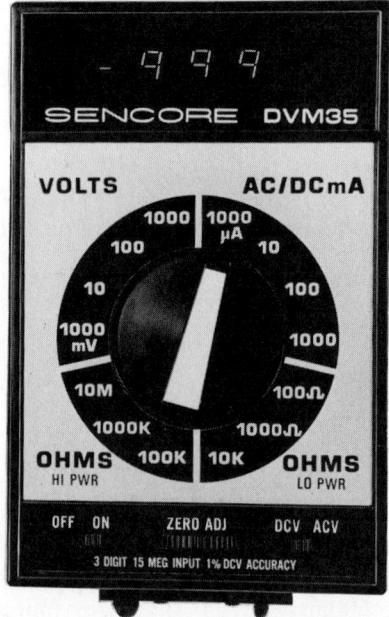

Figure 1-29 Three-digit DVM. (*Courtesy of Sencore, Inc.*)

An overrange digit is an extra digit added to allow the user to read values beyond full scale. An overrange digit is often called a "one-half" or a "partial" digit, since it cannot display all numbers through 9. Overranging extends the usefulness of a DVM by maintaining resolution up to, and beyond, full scale. For example, if a signal changes from 9.999 V to 10.012 V, a four-digit DVM without overranging could measure the first voltage value as "9.999 V," but would require a range change to make the second measurement with a resulting indication of "10.01 V." This 0.002-V change would not be observed. On the other hand, with overranging, the second measurement could be made as "10.012 V" with no loss of resolution.

Overranging is specified as a percentage. A four-digit DVM with 100 percent overranging would have a maximum display of "19999." A specification of 20 percent overranging would provide a maximum reading of "1199." Resolution denotes the maximum number of counts that can be displayed to the least number of counts. Full-scale resolution of a five-digit DVM is equal to 100,000 to 1, or to 0.001 percent. Overranging is usually ignored in resolution. Sensitivity denotes the smallest incremental voltage change that a DVM can detect. It is equal to the lowest full-scale range multiplied by the resolution of the DVM. The sensitivity of a five-digit DVM with resolution of 0.001 percent and a 100-mV lowest full-scale range is equal to 1 μV.

1-13 Notes on Analog Meter Circuit Tolerances

When resistors with a certain tolerance (such as ±1 percent) are connected in series, the tolerance on the combination remains the same as the tolerance on an individual resistor, as shown in Fig. 1-30. Thus, R1, R2, and R3 each have a tolerance of ±1 percent. In turn, the tolerance on their series combination is also ±1 percent. Observe carefully that although a meter movement has a certain tolerance, this is a mechanically based tolerance (not electrically based). In turn, the tolerance on a movement must be added to the tolerance on a multiplier. For example, in Fig. 1-30 the tolerance on the multiplier-movement configuration is ±3 percent. This method of calculating the tolerance on an ammeter instrument circuit is essentially the same. In other words, when resistors with a certain tolerance (such as ±1 percent) are connected in parallel to each other, or are connected in parallel with a meter movement, the tolerance on the shunt combination is also ±1 percent. However, if the movement in the ammeter configuration has a tolerance of ±2 percent, the tolerance on the instrument will be ±3 percent.

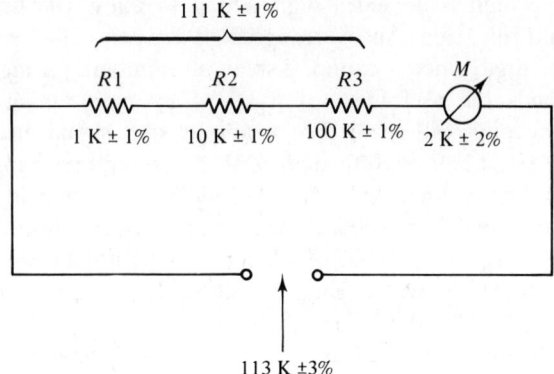

Figure 1-30 Example of tolerance on a voltmeter multiplier-movement configuration.

2
DC-Voltage Measuring Techniques

2-1 General Considerations

Multimeters and electronic meters have several scales for indication of electrical quantities such as dc voltage, ac voltage, resistance, dc current, and decibel values. A scale plate for a typical multimeter is illustrated in Fig. 2-1. Controls for selection of various functions and ranges are provided, as shown in Fig. 2-2. In this example, three controls are available. One control switches the instrument circuitry for indication of +dc, −dc, or ac voltage values. The other control selects five voltage ranges, five current ranges, and three resistance ranges. A zero-ohms control is also provided to set the pointer precisely at the end of the ohms scale when the instrument is operated on a resistance range.

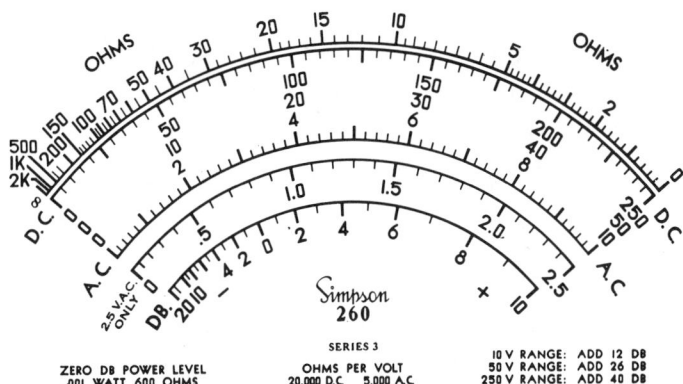

Figure 2-1 A scale plate for a standard multimeter. (Courtesy of Simpson Electric Co.)

38 DC-Voltage Measuring Techniques

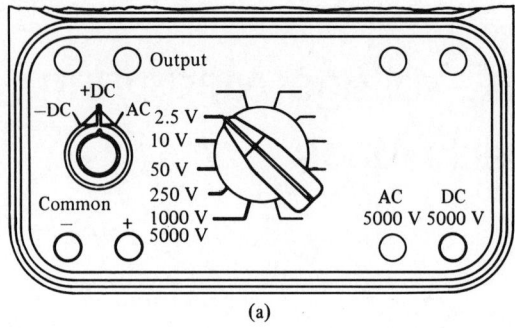

(a)

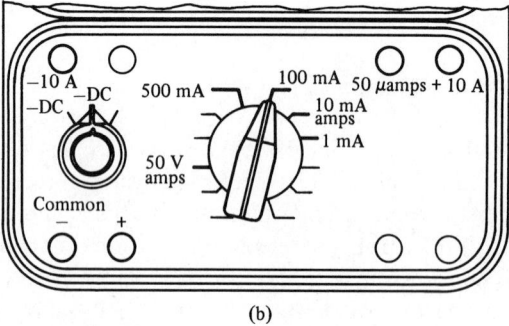

(b)

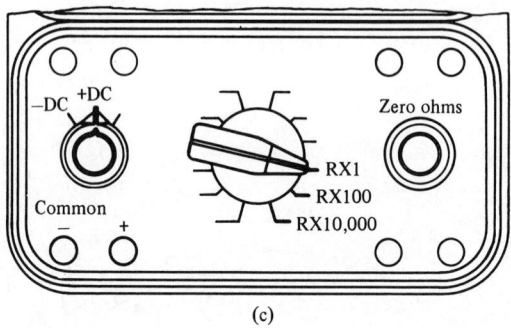

(c)

Figure 2-2 DC voltage, current, and resistance ranges associated with the scale plate shown in Fig. 2-1.

Multimeters are designed with various sensitivity ranges. For example, the three most common sensitivity ratings are 1000 ohms per volt, 20,000 ohms per volt, and 100,000 ohms per volt. These sensitivity ratings denote the input resistance of the multimeter when operated on its dc-voltage measuring function. Thus, a 1000 ohms-per-volt instrument will have an input resistance of 10,000 ohms on its 10-volt range; a 20,000 ohms-per-volt instrument will have an input resistance of 200,000 ohms on its 10-volt range; a 100,000 ohms-per-volt instrument will have an input resistance of 1 megohm on its 10-volt range. These relations are tabulated in Table 2-1.

TABLE 2-1

DC Voltmeter Input-Resistance Values

Meter Sensitivity	DC Voltmeter Input Resistance		
	1000 Ω/V	20,000 Ω/V	100,000 Ω/V
2.5-V Range	2500 Ω	50 KΩ	250 KΩ
10-V Range	10 KΩ	200 KΩ	1 MΩ
50-V Range	50 KΩ	1 MΩ	5 MΩ
250-V Range	250 KΩ	5 MΩ	25 MΩ
1000-V Range	1 MΩ	20 MΩ	100 MΩ
5000-V Range	5 MΩ	100 MΩ	500 MΩ

When voltages in high-resistance circuits are measured, the input resistance of a voltmeter is of great importance, as seen in Fig. 2-3. In this example, two 100-kΩ resistors are connected in series with a 10-volt battery. Before a voltmeter is connected into the circuit, there is a voltage drop of 5 volts across each of the 100-kΩ resistors, in accordance with Ohm's law. Next, if a 1000 ohms-per-volt meter is connected across one of the 100-kΩ resistors, the meter will not indicate 5 volts. As noted in the diagram, the meter will indicate only 0.45 volt, when operated on its 5-volt range. This is, of course, an extremely serious error that is caused by excessive loading of the circuit under test. In other words, the input resistance of the multimeter is 5000 ohms, and connection of the meter into the circuit shunts this 5000-ohm resistance across the 100-kΩ resistor. Effectively, the meter changes the circuit and disturbs the action of the unloaded circuit.

Next, if a 20,000 ohms-per-volt meter is utilized in the foregoing arrangement, the indication error will be much less. Although the meter does not indicate the correct value of 5 volts, it indicates 3.33 volts when operated on its 5-volt range. This is a 33 percent error—a

40 DC-Voltage Measuring Techniques

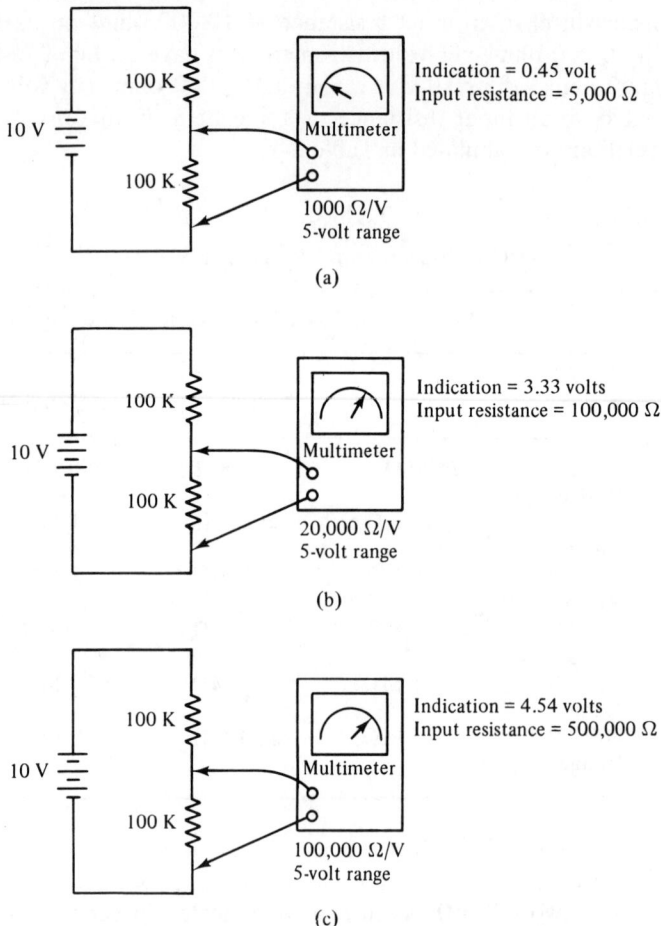

Figure 2-3 Indication errors owing to loading of circuit under test: **(a)** 1,000 ohms-per-volt meter indicates 0.45 volt; **(b)** 20,000 ohms-per-volt meter indicates 3.33 volts; **(c)** 100,000 ohms-per-volt meter indicates 4.54 volts.

greatly excessive error in practical troubleshooting procedures. Again, if a 100,000 ohms-per-volt meter is utilized in the foregoing arrangement, the indication error is reduced. An indication of 4.54 volts is obtained when the meter is operated on its 5-volt range. This is an error of 9 percent; although it is substantial, this error would be considered tolerable in various "rough" troubleshooting procedures. Finally, if an electronic voltmeter with an input resistance of 10 megohms is employed, the indication error in the foregoing arrangement will be only 0.6 percent—a negligible error for virtually all troubleshooting procedures.

2-2 Instrument Accuracy Ratings

In addition to circuit-loading indication errors, instrument accuracy ratings must be taken into consideration. No meter is 100 percent accurate; if a meter has been seriously overloaded or otherwise abused, it may be very inaccurate. As noted in the first chapter, service-type voltmeters are usually rated for some percentage of full-scale accuracy. For example, a lower-priced voltmeter may be rated for an accuracy of ± 3 percent of full-scale indication. It follows that if this meter is operated on its 250-volt range, the possible indication error at any point on the scale will be ± 7.5 volts. If a 125-volt potential is measured on the 250-volt range, the pointer may fall anywhere in the interval from 117.5 to 132.5 volts. Again, if a 5-volt potential is measured on the 10-volt range of the meter, the pointer may fall anywhere in the interval from 4.7 to 5.3 volts. On the other hand, if a ± 1.5 percent meter were employed, the rated indication range would be from 4.85 to 5.15 volts in the foregoing example.

Two practical conclusions can be drawn. First, it is good practice to use a voltmeter that has a comparatively high accuracy rating. Second, it is advisable to choose a range that provides indication on the upper half of the scale. For example, if a potential of 10 volts is being measured, a particular meter might provide a choice of operation on a 10-volt scale or on a 50-volt scale. If the accuracy rating of the instrument is ± 3 percent of full-scale indication, there is a possible accuracy error of ± 0.3 volt on the 10-volt range. On the other hand, there is a possible accuracy error of ± 1.5 volts on the 50-volt range—an error five times as great. This is a practical example of the advantage of operation on the higher portion of a meter scale. Another disadvantage of operation on the lower portion of a meter scale is the difficulty of reading the scale accurately, owing to the fact that a small change in pointer position corresponds to a large change of indicated voltage.

2-3 Polarity Considerations

Consider the use of a polarity-reversing switch in dc-voltage measurements with a multimeter or a solid-state electronic multimeter. With reference to Fig. 2-4, the dc voltage at the collector of a transistor might have either positive polarity or negative polarity, depending upon the type of transistor. If the test leads are applied with incorrect polarity, the pointer will swing off-scale to the left, instead of moving up-scale. To correct this test condition, either the test leads can be reversed, or the instrument's polarity reversing switch can be thrown to its other position. In most situations, it is convenient to throw the polarity reversing switch. Some voltmeters have a built-in polarity-reversing function, with a polarity indicator as exemplified in Fig. 2-4(c). As noted in Fig. 2-4, it is good practice to start by setting the range switch much higher than would seem to be needed, and then to reduce the setting as required. This procedure avoids the risk of meter overload, particularly in malfunctioning circuitry wherein high voltages may appear at unexpected circuit points.

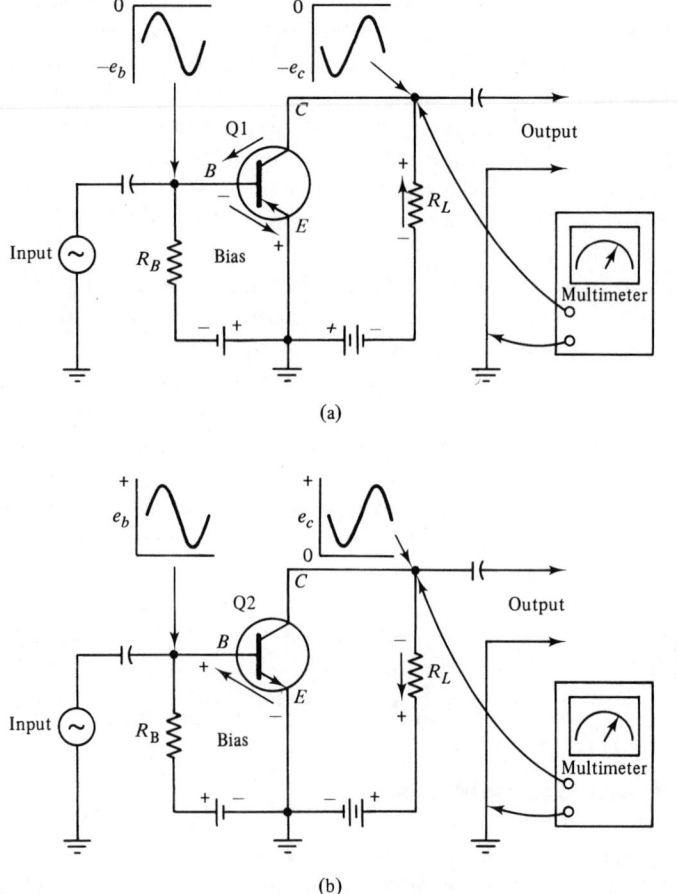

(a)

(b)

CAUTION

Before voltmeter is applied, set the range switch much higher than would seem to be needed. Then reduce the range-switch setting as required. This precaution will often prevent damage to the instrument.

Figure 2-4 Example of reverse polarities in similar circuits: **(a)** PNP transistor operates with negative collector voltage; **(b)** NPN transistor operates with positive collector voltage.

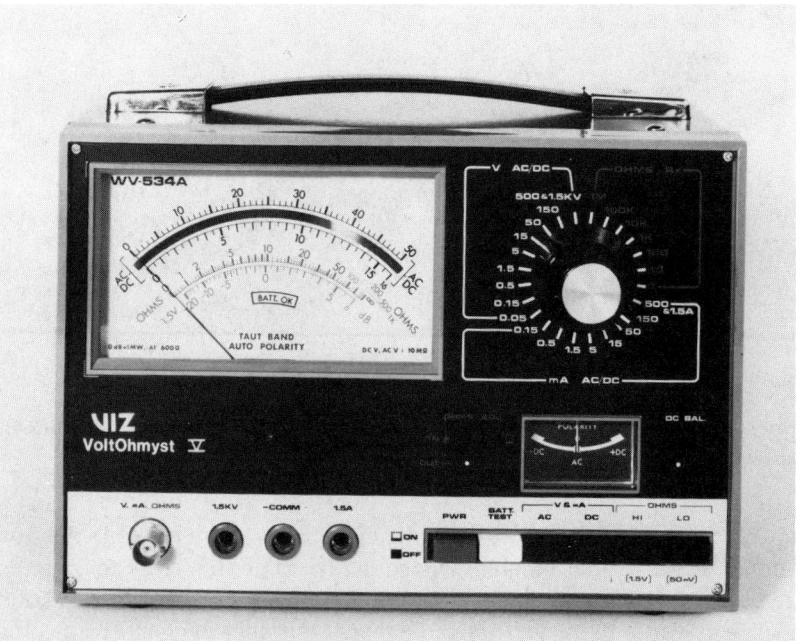

Figure 2-4(c) A voltmeter with a built-in automatic polarity reversal function. (*Courtesy of VIZ Instruments, Inc.*)

As noted in the first chapter, many modern multimeters are provided with meter-movement protection against overload. This protective means takes the form of semiconductor diode(s) connected across the terminals of the meter movement. In ordinary dc-voltage measurement procedures, protective diodes do not present a potential source of indication error. However, the technician should be on guard when measuring dc voltages that have substantial ac components. A practical example is a transistor configuration that works into an inductive load. In some cases, the load counter-emf produces an ac component in normal operation that exceeds the dc level. In turn, this ac component may be partially rectified by the protective diode(s), thereby developing an erroneous dc indication.

2-4 Measurement of High DC Voltages

High dc voltages are measured with a high-voltage dc probe, which functions as an external multiplier resistor for a multimeter or an electronic voltmeter (see Fig. 2-5). The value of the probe resistor depends

44 DC-Voltage Measuring Techniques

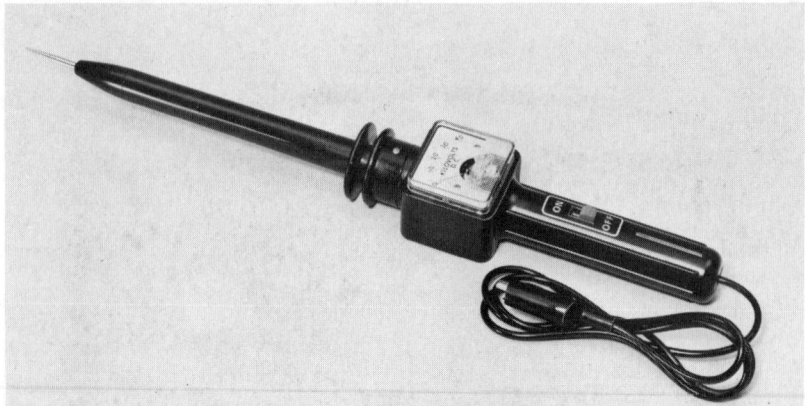

Figure 2-5 A high-voltage DC probe that has a built-in kilovoltmeter. (Courtesy of B & K Precision, Division of Dynascan Corp.)

upon the input resistance of the meter with which it is used. For example, suppose that a high-voltage probe is to be used to provide 25,000 volts full-scale indication on the 250-volt range of a 20,000 ohms-per-volt multimeter. Since the input resistance of the instrument is 5 megohms, the probe resistor must have a value of 495 megohms. In accordance with Joule's law, the probe resistor must dissipate 1.3 watts, approximately, at full-scale indication.

Note that if a high-voltage probe resistor has a tolerance of ±1 percent, and the multiplier resistors in a multimeter have tolerances of ±1 percent, the tolerance of the probe and multiplier combination is ±1 percent. In other words, these tolerances are not additive. As an illustration, if two resistors with a nominal value of 1000 ohms each are connected in series, the series combination has a nominal value of 2000 ohms. If each resistor has a tolerance of ±1 percent, the range of absolute resistance value on the combination is from 1980 ohms to 2020 ohms. This is a tolerance of ± 1 percent on the 2000-ohm nominal value. Again, suppose that the foregoing pair of resistors is connected in parallel. The nominal resistance of the combination is 500 ohms. If the tolerance on each resistor is ±1 percent, the absolute resistance value of the combination ranges from 495 ohms to 505 ohms. This range is a tolerance of ±1 percent on the parallel combination.

2-5 Measurement of Low DC Voltages

Modern multimeters generally provide a low dc-voltage range, such as 0-to-0.25 volt. However, old-style multimeters typically provide a

0-to-2.5 volt range for measurement of low voltages. In various applications, a low-voltage range is needed, which can be realized by using the microampere current range as a low-range voltmeter in an old-style multimeter. An example is shown in Fig. 2-6. This microammeter function has a range of 0 to 50 μA, with a 250-mV drop across the instrument terminals at full-scale indication. In turn, the "250" scale can be read in terms of millivolts when the instrument is operated on its 0-to-50-μA range. Note that the input resistance on this range is 2500 ohms. Accordingly, circuit loading can be a problem unless the circuit under test has a low internal resistance. If the base-emitter bias voltage of a transistor is being checked, an accurate indication will be obtained, inasmuch as a forward-biased base-emitter junction has a very low resistance.

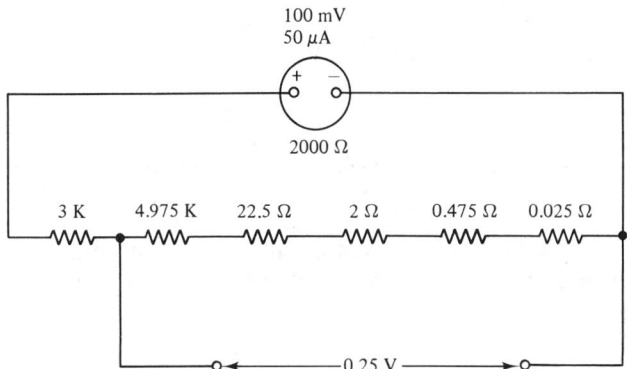

Figure 2-6 Example of a microammeter used as a low-range voltmeter.

2-6 Measurement of Difference Voltages

Troubleshooting procedures may involve difference voltages, as exemplified in Fig. 2-7. Here, the base voltage is -4.3 volts with respect to ground, and the emitter voltage is -4.5 volts with respect to ground. In turn, the transistor is forward-biased by 0.2 volt. Although it is possible to check the bias voltage by making a pair of measurements with respect to ground and calculating their difference, this is poor practice. In other words, small measurement errors can make a great difference in the answer obtained by calculation. Therefore, the preferred procedure is to apply the voltmeter between the base and emitter terminals, as shown in the diagram. The voltmeter then indicates the difference voltage directly. This is an example of an "above-ground" measurement. Since the emitter is more negative than the base, the base is

46 DC-Voltage Measuring Techniques

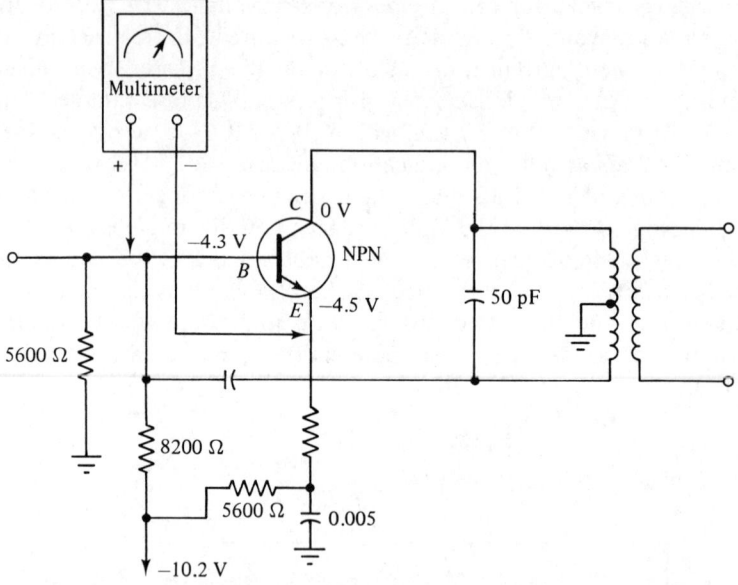

Figure 2-7 Example of difference voltage in transistor bias circuit.

positive with respect to the emitter. In turn, the positive lead of the voltmeter must be connected to the base in order to obtain an up-scale indication.

2-7 Use of Isolation Transformer

Even low-voltage dc measurements can impose a shock hazard to the technician when the equipment under test employs a transformerless power supply and the equipment chassis potential happens to be above earth ground potential. Therefore, it is good practice to use a line isolation transformer in this situation, as depicted in Fig. 2-8. Note in (a) that the chassis ground potential is 117 volts above earth ground potential. In turn, if the technician is standing on a damp floor or has any other contact with earth ground, he can receive a severe shock when he touches the equipment chassis. This hazard could be avoided by checking the chassis potential, and turning the power plug over, in case the chassis is "hot." However, it is preferable to utilize a 1-to-1 line-isolation transformer, as shown in (b). Then troubleshooting can be started at all times without bothering to remember to check the chassis to determine if it is "hot."

2-8 Measurement of DC Voltage Accompanied by High-voltage AC Pulses

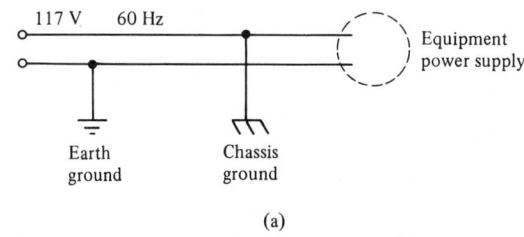

(a)

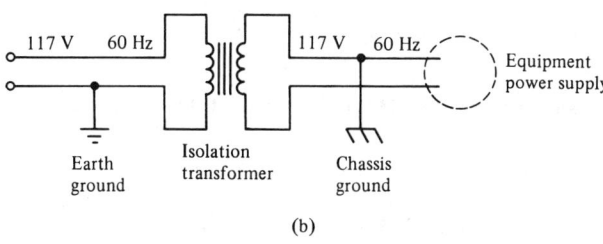

(b)

Figure 2-8 Elimination of "hot-chassis" hazard: **(a)** chassis ground may be above earth ground in transformerless power supply; **(b)** isolation transformer prevents chassis ground from being above earth ground potential.

2-8 Measurement of DC Voltage Accompanied by High-voltage AC Pulses

When a complex waveform generator works into an inductive load, the voltage under test may consist of a comparatively low dc voltage accompanied by high-voltage ac pulses. In such a case, attempted measurement of the dc voltage with a multimeter may result in damage to the meter. When this hazard exists, electronic service data usually note, "Do not measure" at the particular terminal. However, it is often helpful to measure this dc-voltage value in troubleshooting procedures, and this may be done if a suitable probe is used with the multimeter. A high-voltage dc probe will serve as a low-pass filter, owing to its stray capacitance, thereby greatly attenuating the ac pulses, and preventing damage to the multimeter. The probe used in this application should have a resistor that provides 10-to-1 dc-voltage attenuation when operated on a chosen range of the multimeter. For example, suppose that the low-pass probe is to be used to measure a nominal 350-volt potential on the 50-volt range of a 20,000 ohms-per-volt multimeter. In such a case, the input resistance of the instrument is 1 megohm, and the probe resistor will have a value of 9 megohms.

48 DC-Voltage Measuring Techniques

2-9 DC Voltage Indication in Corona Field

When dc voltages are measured in equipment that has high-voltage fields present, a dc voltmeter may indicate a substantial value of voltage, although the test probe is not connected to any terminal in the equipment. When a "floating" probe picks up dc voltage in this manner, it is the result of so-called "spark rectification" in a strong corona field. For example, if the probe is brought into the vicinity of a flyback and high-voltage transformer, the meter may indicate a substantial dc voltage, although the probe is not making contact with any metallic source of voltage.

2-10 Line-operated Voltmeters Must Be Properly Grounded

Inexperienced technicians sometimes suppose that, if the pointer deflects off-scale to the left on a line-operated TVM or VTVM, the test leads can be reversed to obtain up-scale deflection. This is a serious error, because the ground (black) lead of a line-operated voltmeter has a very low impedance to ground, and will load the circuit under test seriously. Moreover, when the "hot" lead of a VTVM is grounded to the chassis of the equipment, and the "ground" lead of the instrument is used as a probe, a very high level of hum voltage is fed into the meter, and highly erroneous indication results. All measurements should be made with respect to chassis ground when a line-operated VTVM is used, for example. Thus, if the cathode of a tube operates substantially above ground potential, it is poor practice to measure grid-cathode voltage by connecting the black lead of the VTVM to the cathode terminal of the tube. Substantially incorrect indication can result. The correct procedure is to make a measurement of grid voltage with respect to ground, and a measurement of cathode voltage with respect to ground. Then these values are subtracted to determine the grid-cathode voltage.

2-11 Quick Check for Circuit Loading with Multimeter

Since a multimeter has substantially different input resistance values on adjacent ranges, a range-switch test may be utilized for a quick check of possible circuit loading. First, a voltage measurement is made on a range that provides as nearly full-scale indication as possible. Then the meter is switched to its next higher range. If the indicated voltage value is substantially greater on the higher range, it is concluded that objectionable circuit loading is occurring, and that neither indicated value is correct. The proper procedure in this situation is to use a voltmeter that has considerably higher input resistance.

2-12 DC Voltmeter Response to Pulse Voltage

Beginners sometimes suppose that a dc voltmeter will respond to a pulse voltage such as that shown in Fig. 2-9. However, the meter will indicate zero. The pulse waveform in Fig. 2-9 is an ac waveform that has an average value of zero. Since a dc voltmeter responds to this average value, it indicates zero. Next, with reference to Fig. 2-10, a dc voltmeter responds to a dc pulse voltage. A dc pulse voltage has only one polarity—in this example, the waveform is a positive pulse. A dc voltmeter will respond to the average value of the pulse. This average value has the same level as the dc component in the waveform. A dc pulse is also called a pulsating dc waveform. Distinction is made between ac waveforms, pulsating dc waveforms, and ac waveforms with a dc component, as exemplified in Fig. 2-11. Note that a pulsating dc waveform does not cross the zero axis.

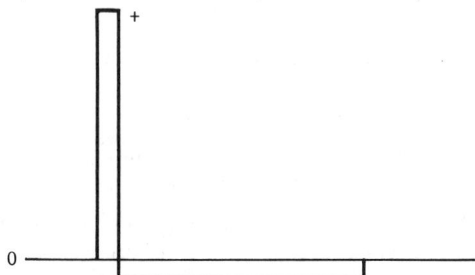

Figure 2-9 DC voltmeter indicates zero when energized by an ac pulse voltage.

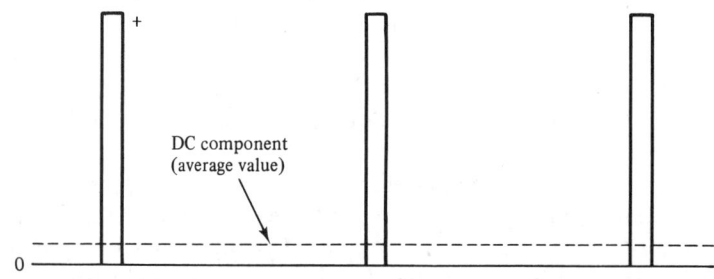

Figure 2-10 DC voltmeter responds to a dc pulse waveform.

50 DC-Voltage Measuring Techniques

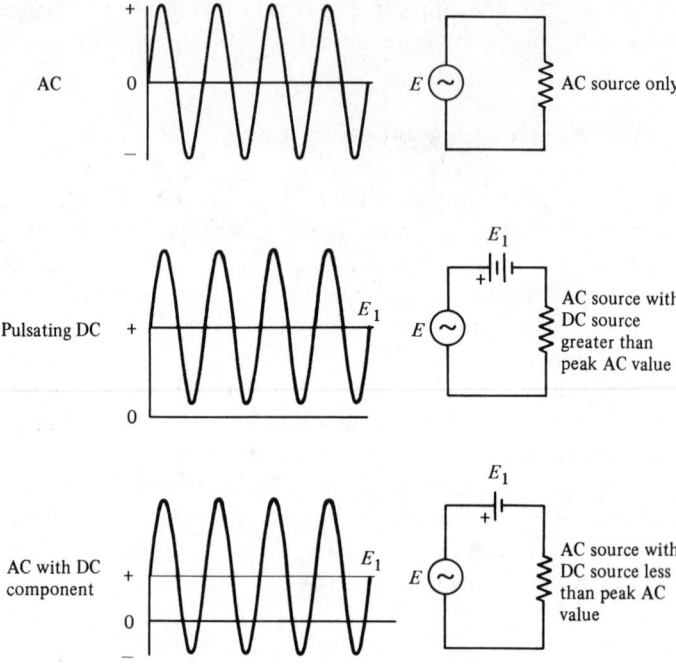

Figure 2-11 An ac waveform with three dc component levels.

2-13 DC Voltage Measurements in Nonlinear Circuits

DC voltage measurements can be made in linear circuits and in nonlinear circuits. However, the significance of the measurements is different in these two kinds of circuits. First, consider voltage measurement in a linear circuit, as exemplified in Fig. 2-12. Here, the dc voltage drop is being measured across a 2000-ohm resistor. Although the current flow is not being measured directly, it follows from Ohm's law that the current value is equal to $E/2000$, where E is the voltage value indicated by the TVM. For example, if the voltage measures 100 volts, it follows that the current value is 50 mA. This current-voltage relation is graphed in Fig. 2-13. It is a relation that is often used by electronic technicians to determine current flow through a resistor or a resistive component.

Next, consider the measurement of dc voltage across a nonlinear resistive component, such as a lamp filament. Although this measurement is easily made, as shown in Fig. 2-14, more data are required before the current flow through the filament can be determined. The lamp

2-13 DC Voltage Measurements in Nonlinear Circuits   51

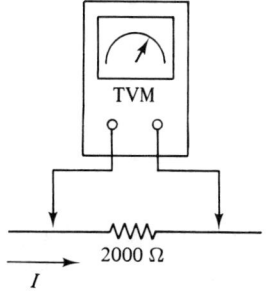

Figure 2-12   Current determination by voltage measurement.

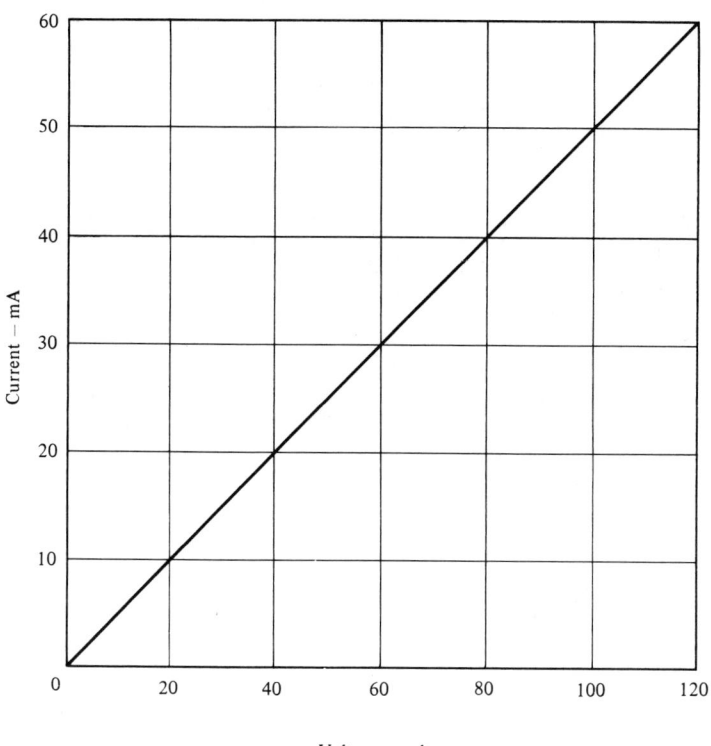

Figure 2-13   Voltage-current characteristic of a 2000-ohm resistor.

52 DC-Voltage Measuring Techniques

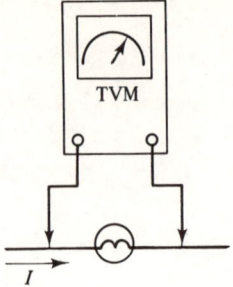

Figure 2-14 Voltage measurement across a lamp filament.

filament does not have a fixed resistance value; instead, the filament resistance increases as the voltage drop increases. This voltage-current relation is exemplified in Fig. 2-15. The graph is a curve—not a straight line. In turn, the resistance changes from point to point along the curve.

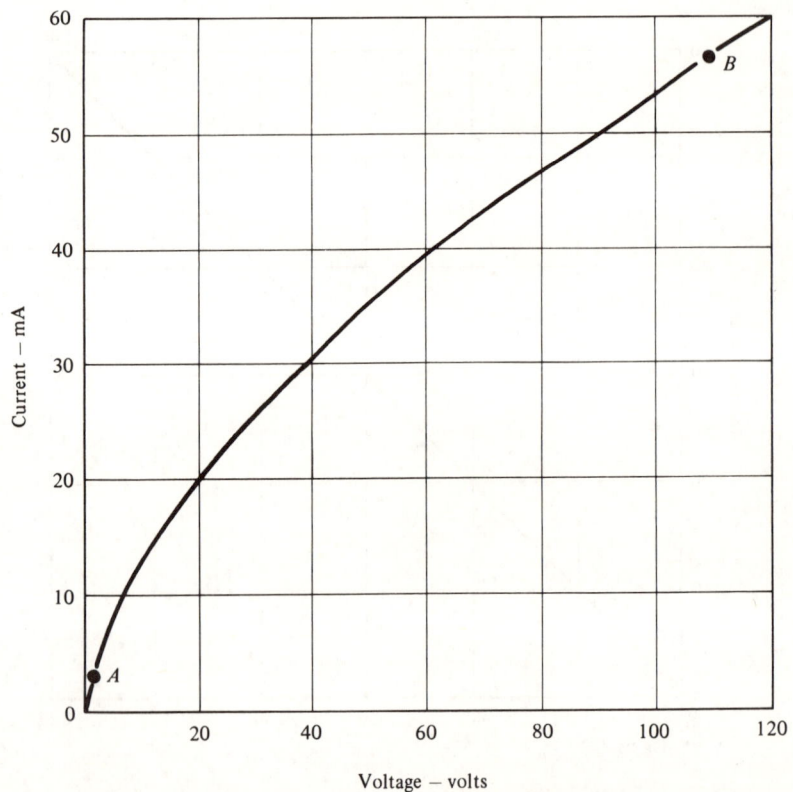

Figure 2-15 Voltage-current characteristic of a lamp filament.

Thus, the resistance at point A is approximately 300 ohms, whereas the resistance at point B is 2000 ohms. This is a resistance change of approximately 7 to 1 over the voltage interval from 1 volt to 100 volts. It follows that a voltage measurement in a nonlinear circuit does not provide full information unless the current or resistance variation versus voltage is also known, or measured.

2-14 Conduction in Reverse-biased Transistor

In general, a reverse-biased transistor such as that depicted in Fig. 2-16 is cut off, and no collector current is flowing. However, there are certain exceptions, as when a transistor is operated in class C. In such a case, the transistor appears to be cut off on the basis of dc-voltage measurements. On the other hand, it can be shown (as with an oscilloscope) that the base voltage is not pure dc, but instead is a form of pulsating dc that drives the transistor into conduction for brief intervals. In other words, the base is forward-biased with respect to the emitter for brief intervals, as shown in Fig. 2-16(b). On the average, however, the base is reverse-biased with respect to the emitter. Therefore, it should be kept in mind that dc-voltage measurements alone may sometimes be inconclusive when transistor circuit action is analyzed.

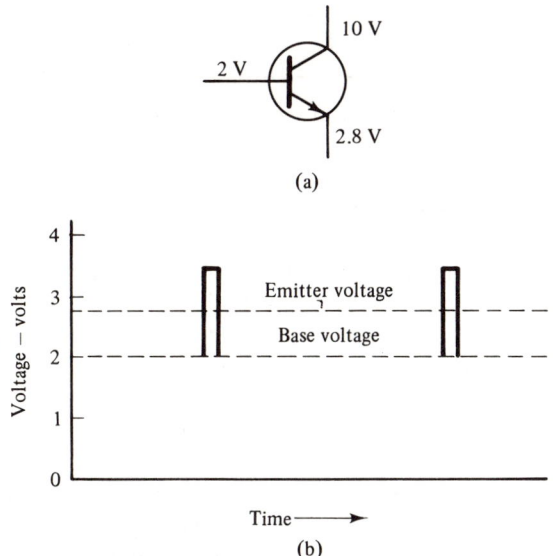

Figure 2-16 Conduction of reverse-biased transistor operated in class C: (a) representative terminal voltages; (b) base-voltage variation in time.

2-15 Measurement of Battery Voltage

Transistor radios and various other kinds of electronic equipment are battery operated. When the battery voltage falls below a certain critical value, the output from the radio becomes weak and distorted. Beginners sometimes make the error of measuring battery voltage under no-load conditions. This practice leads to false conclusions, because a battery can be nearly dead, and its open-circuit terminal voltage will be nearly normal. Therefore, a battery should be checked under normal load, as depicted in Fig. 2-17. In most applications, there should be no more than a 10 or 15 percent drop in battery voltage from no-load to full-load conditions. A greater drop indicates that the battery is becoming weak, or that the equipment is drawing excessive current from the battery.

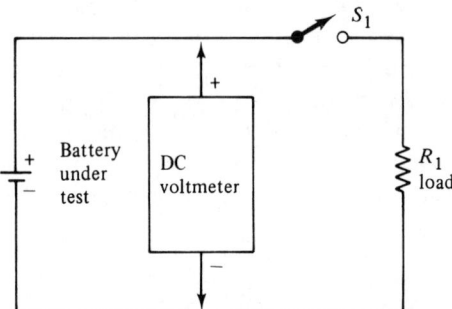

Figure 2-17 Battery voltage should be checked under load.

2-16 Open-circuit Testing Precaution

As a practical troubleshooting note, Fig. 2-18 shows how a multimeter can give misleading test results in open-circuit situations. An open-circuit condition can be caused by a defective solder joint, a cracked PC conductor, etc. When a multimeter is used to measure the source-to-ground voltage, the input resistance of the multimeter is shunted across the open circuit. In the case of a 1000 ohms-per-volt instrument operated on its 5-volt range, this input resistance is 5000 ohms. If a 20,000 ohms-per-volt instrument is operated on its 5-volt range, its input resistance is 100,000 ohms. In various types of electronic circuitry, the source voltage will measure approximately its normal value, and the malfunction will disappear while the voltage measurement is being made. This situation can be very puzzling to the beginning technician.

2-18 Application of Zero-center Indication   55

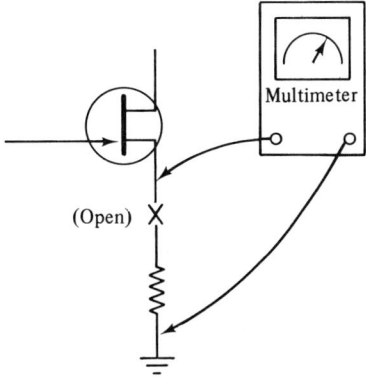

Figure 2-18   Internal resistance of the multimeter closes the open circuit.

2-17 Voltmeter Response to DC Pulses

A dc voltmeter responds to the average value of a train of dc pulses, although the indication becomes indistinct at low repetition rates. With reference to Fig. 2-19, the voltmeter operates in the usual manner until a certain low repetition rate is passed. For example, a typical multimeter responds smoothly down to 25 Hz, below which the pointer will vibrate more or less. At a repetition rate of 10 Hz, pointer vibration becomes quite large. Thus, it is impractical to use a dc voltmeter to measure the average value of a dc pulse train at very low repetition rates.

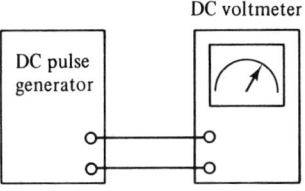

Figure 2-19   Checking voltmeter response to dc pulses.

2-18 Application of Zero-center Indication

Most VTVM's and TVM's provide for zero-center indication. Digital voltmeters (DVM's) usually display positive or negative signs in the readout to indicate whether the voltage under measurement is positive or negative. In any case, zero-center indication or its equivalent is occasionally needed in servicing procedures. As an illustration, when an FM detector is aligned, its output characteristic swings from positive

56 DC-Voltage Measuring Techniques

to negative through the frequency channel, as shown in Fig. 2-20. Therefore, a zero-center voltmeter indication is needed to check the detector response as the signal-generator frequency is varied from one end of the channel to the other.

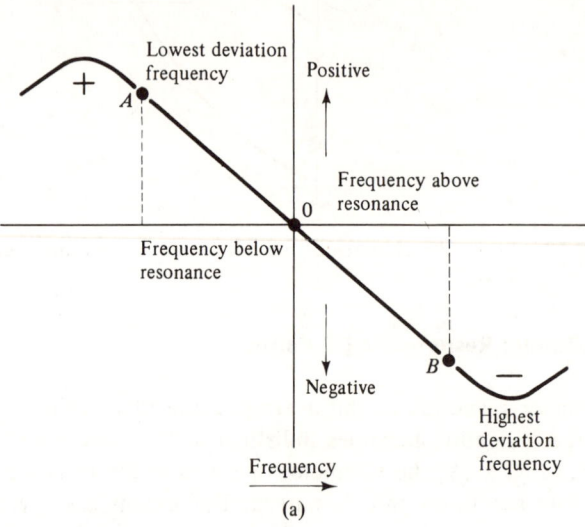

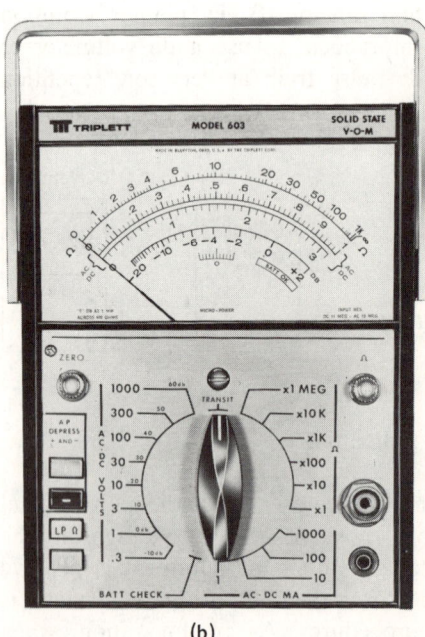

Figure 2-20 A typical S curve, or FM detector frequency response curve, and example of a zero-center scale on a solid-state VOM: (a) S curve specifications; (b) lowest scale on faceplate provides zero-center indication. (*Courtesy of Triplett Electrical Instrument Co.*)

2-19 Measuring Dwell Time of Contactor

A dc voltmeter provides a direct and precise measurement of the dwell time of a contactor. With reference to Fig. 2-21, the voltage waveform across the load is a rectangular wave with peak value of V volts. The voltmeter indicates the average value of the pulsating dc waveform. In turn, the ratio of the average value to the peak value is equal to the ratio of the "on" time to the "on" plus "off" time. In the example of Fig. 2-21, the "on" or dwell time is equal to one-third of the "on" plus "off" time. Accordingly, the indicated (average) value of voltage is equal to one-third of the peak voltage.

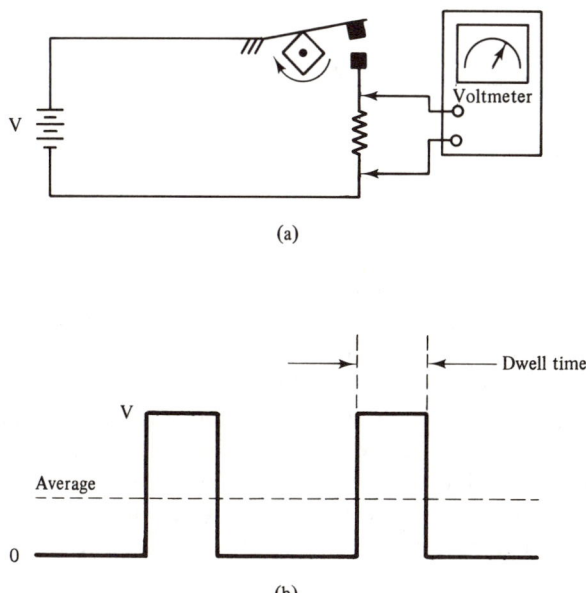

Figure 2-21 Measuring dwell time of contactor: **(a)** circuit arrangement; **(b)** rectangular operating waveform.

2-20 Photovoltaic Cell Output

A photovoltaic cell is checked with a multimeter, as shown in Fig. 2-22. When exposed to bright light, a typical cell generates 0.5 V with a current capability of 0.6 mA. When the cell works into a substantial load, its terminal voltage decreases accordingly. For example, if a cell works into a 100-kΩ load, and develops 0.25 volt under a given illumination level, it typically develops only 0.08 volt across a 1-kΩ load. Note that a cell tends to saturate when exposed to extremely bright light. In other words, it is not possible to increase the generated voltage indefinitely, simply by increasing the intensity of the incident light.

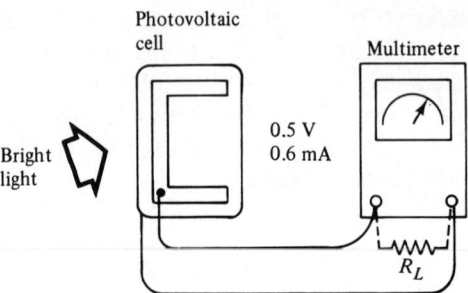

Figure 2-22 Checking a photovoltaic cell.

2-21 Zener Diode Checkout

To quick-check a Zener diode for open circuit, short circuit, or leakage, an ohmmeter may be connected to check the forward-resistance value, in the same manner as for a conventional diode. A more conclusive regulation test can be made with an adjustable regulated power supply, such as that illustrated in Fig. 2-23. Connect the output current from the power supply through a limiting resistor in series with the Zener diode under test. Slowly increase the applied voltage until the specified current is flowing through the Zener diode, using the arrangement shown in Fig. 2-24. Connect a dc voltmeter across the Zener diode to monitor the voltage drop. Vary the current above and below the specified Zener current. In turn, the voltage drop will remain constant if the Zener diode is operating normally.

2-22 Measurement of DC Voltage in HF Circuits

A shielded input cable, or open test leads to a voltmeter impose a capacitive load on the circuit under test. This capacitance does not disturb the action of a dc circuit. On the other hand, the action of high-frequency oscillator or narrow pulse circuits can be seriously disturbed; an oscillator can be "killed" in some cases by this input capacitance. In turn, the dc-voltage distribution in the circuit often becomes incorrect. This source of incorrect voltage indication is avoided by the

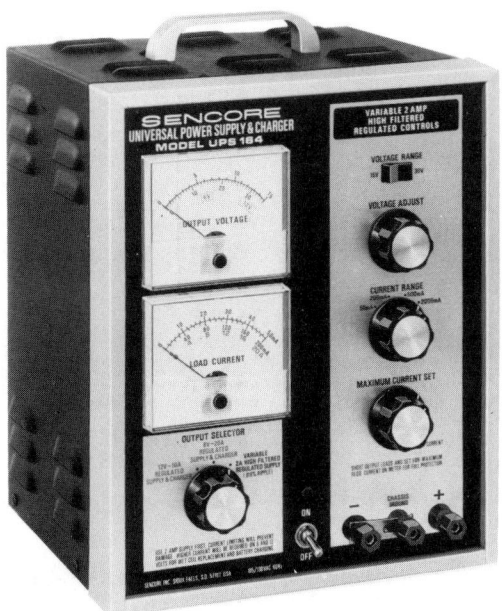

Figure 2-23 A regulated power supply for bench service. (*Courtesy of B & K Precision, Division of Dynascan Corp.*)

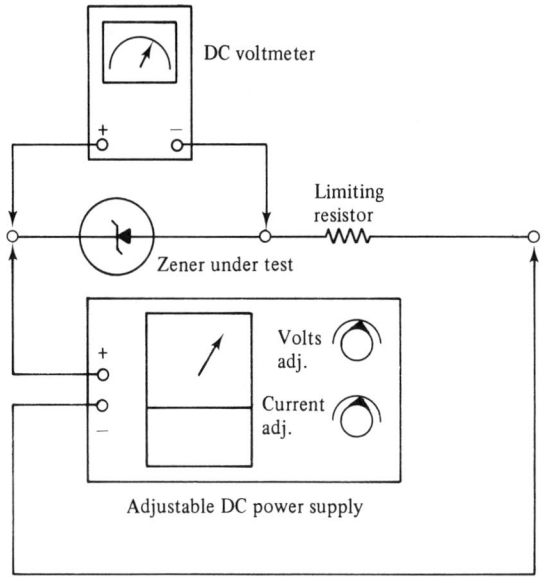

Figure 2-24 Test setup for checking Zener-diode operation.

60 DC-Voltage Measuring Techniques

use of an isolating probe, as depicted in Fig. 2-25. The probe contains an isolating resistor R_1 in its tip; this resistor has a sufficiently high value that the effective cable capacitance at the probe tip is minimized. As an illustration, a direct cable for an electronic multimeter has an input capacitance of 40 pF. On the other hand, when a 1-MΩ isolating resistor is utilized, the effective input capacitance is reduced to 1.5 pF. This small value of test capacitance can be tolerated even by very high frequency circuitry.

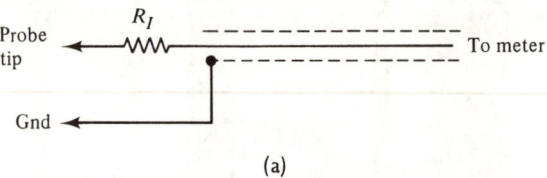

(a)

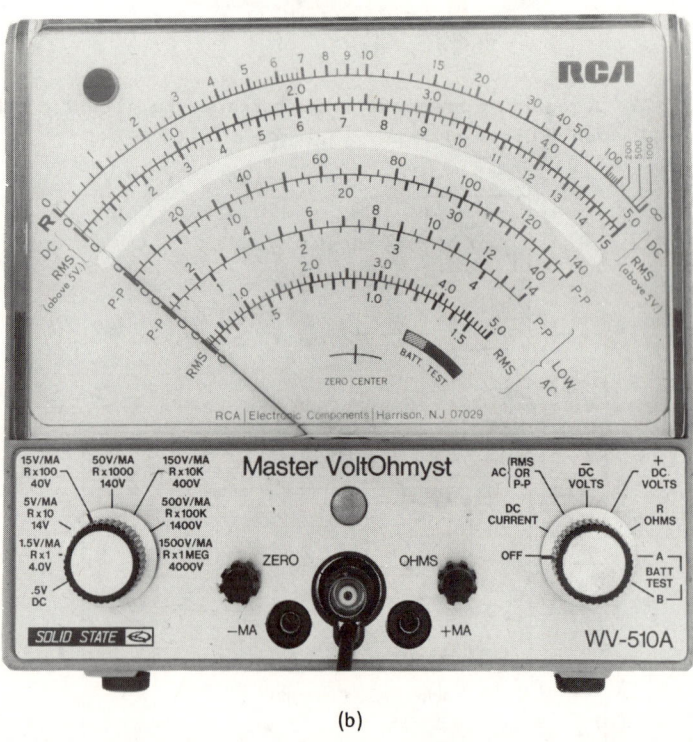

(b)

Figure 2-25 Isolating probe arrangement, and typical TVM that utilizes this type of probe: **(a)** probe configuration; **(b)** elaborate TVM designed for use with an isolating probe. (*Courtesy of RCA and VIZ, Inc.*)

2-22 Measurement of DC Voltage in HF Circuits

Isolating probes are not used with multimeters because of the comparatively low input resistance of these instruments. In other words, if the isolating resistance is made sufficiently great to serve its intended purpose, the sensitivity of the multimeter will be reduced to an impractical degree. On the other hand, an isolating probe does not impair the sensitivity of an electronic voltmeter because of the amplification that is available in the instrument circuitry. Also, the input resistance of an electronic voltmeter is typically in the range from 10 to 15 megohms, so that the addition of a 1-MΩ isolating resistance represents a comparatively small fraction of the inherent input resistance. In turn, most electronic multimeters are provided with isolating probes.

Note in passing that an electronic multimeter must be properly calibrated for use with an isolating probe. Otherwise, indication accuracy would be impaired. This requirement is seen in Fig. 2-26; it is

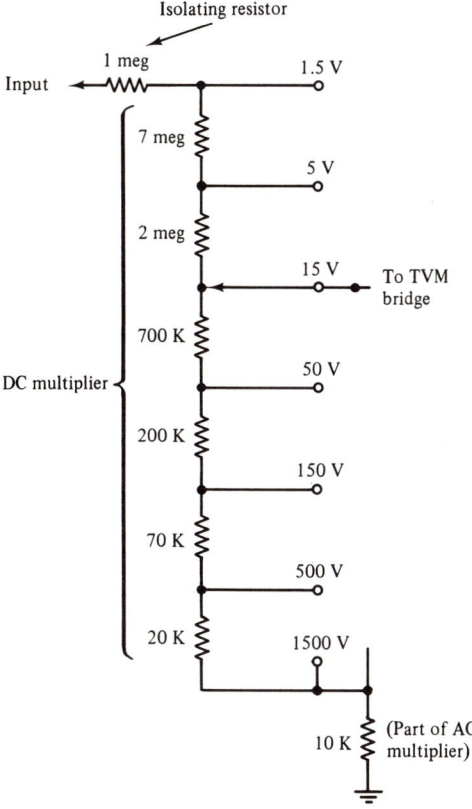

Figure 2-26 Example of a multiplier network for an electronic multimeter.

evident that the isolating resistor becomes a functioning portion of the multiplier's voltage-dividing action, and that its value must be taken into account when the instrument's dc function is being calibrated. Many electronic multimeters employ the same input cable for both dc-voltage and ac-voltage measurements. These instruments provide a switch with the isolating resistor in the probe tip, so that the resistor can be switched out (short-circuited) while ac-voltage measurements are made. This provision is required because an isolating resistor functions in combination with the input cable capacitance effectively as a low-pass filter. In turn, erroneous ac-voltage indication would be obtained (particularly at higher frequencies) if the isolating resistor were retained during ac-voltage measurements. Conversely, incorrect dc-voltage indication will result if the operator forgets to switch the isolating resistor into the measuring circuit during dc-voltage measurements.

3 Resistance Measuring Techniques

3-1 General Considerations

Resistance is a current/voltage ratio, as shown in Fig. 3-1. In turn, an ohmmeter must apply voltage to the component or device under test, and current must flow through the component or device. This is a practical consideration in troubleshooting procedures, because some ohmmeters may apply excessive voltage to various semiconductor devices or electrolytic capacitors with very low voltage ratings. Typical ohmmeter circuitry is shown in Fig. 3-2. This configuration applies up to 1.5 volts to the component under test when operated on its RX1 and RX10 ranges. It applies up to 7.5 volts to the component under test when operated on its RX10,000 range.

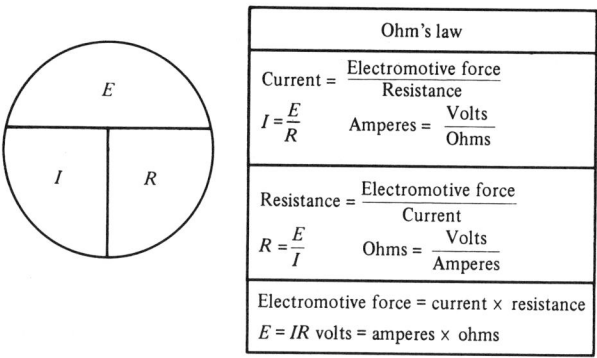

Figure 3-1 Ohm's law in diagram form.

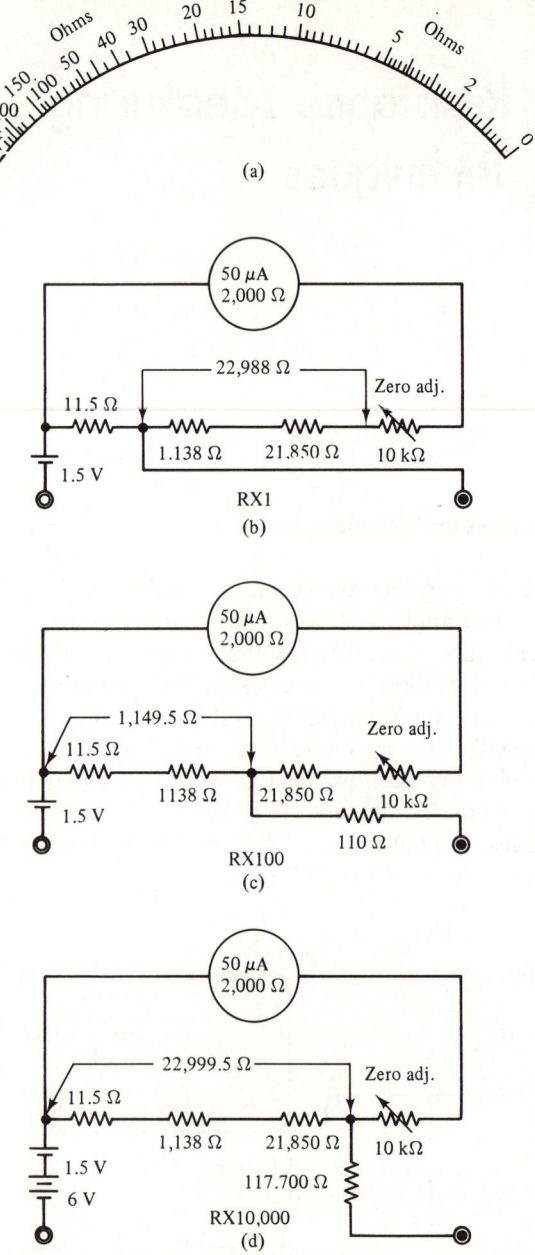

Figure 3-2 Typical ohmmeter configurations on various ranges: **(a)** scale used on all ranges; **(b)** meter circuitry on RX1 range; **(c)** meter circuitry on RX100 range; **(d)** meter circuitry on RX10,000 range.

Ohmmeter accuracy is related to the dc-voltage indication accuracy of a multimeter. An ohmmeter scale is cramped at its high-resistance end, as seen in Fig 3-2(a). Hence, the absolute error usually increases as the pointer moves toward the high-resistance end of the scale. The rated accuracy of an ohmmeter is usually given in terms of degrees of arc, as depicted in Fig. 3-3. A degrees-of-arc rating is based on the accuracy of the dc-voltage function. As an illustration, suppose that the dc-voltage indication accuracy is rated at ±2 percent. In turn, this denotes an interval of ±0.2 volt on a 10-volt scale. The arc thus defined is extended to the ohmmeter scale, and is considered its basic accuracy rating. However, it is possible for this accuracy indication to become impaired if the ohmmeter battery is weak, for example.

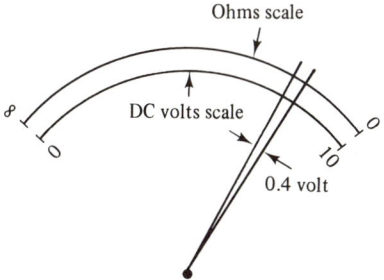

Figure 3-3 A dc-voltage function accuracy of ±2 percent corresponds to an interval of 0.4 volt on a 10-volt scale.

Consider a typical multimeter that has a dc-voltage indication accuracy rating of ±3 percent. This corresponds to an interval of ±7.5 volts on a 250-volt scale. With reference to Fig. 3-4, the center-scale indication on the ohms scale is 12 ohms, and is opposite the 125-volt point on the 250-volt scale. Since the arc of error is ±7.5 volts in this example, the interval from 117.5 volts to 132.5 volts corresponds to the rated accuracy with respect to a 125-volt potential. The corresponding interval on the ohms scale directly above defines the basic accuracy of the ohmmeter function. In other words, if a high-precision 12-ohm resistor were checked, and the pointer indicated some value in the range from 10.75 ohms to 13.5 ohms, this particular ohmmeter would be within its rated accuracy.

Many ohmmeters provide polarity reversing switches, and this facility is useful in some types of resistance measurements. As an illustration, a semiconductor diode is checked for front-to-back resistance ratio as shown in Fig. 3-5. As depicted in the diagram, the test leads can be reversed from (a) to (b). However, operating convenience is

66   Resistance Measuring Techniques

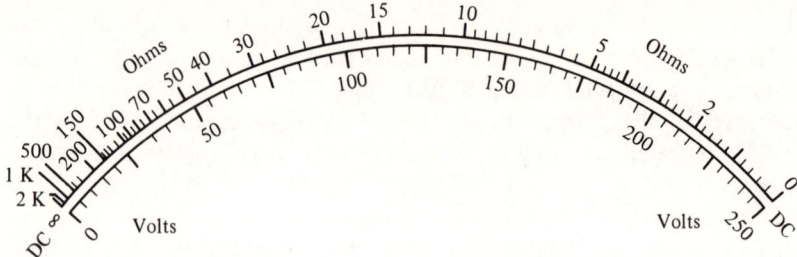

Figure 3-4 Comparison of a multimeter ohms scale and volts scale. (*Courtesy of Simpson Electric Co.*)

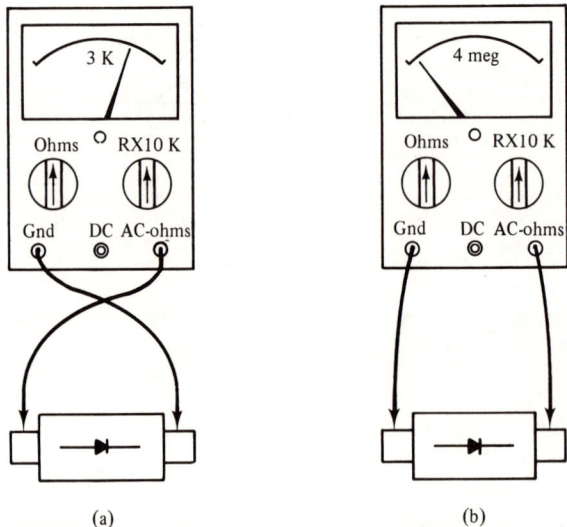

Figure 3-5 Measuring the forward and reverse resistance values of a semiconductor diode.

provided by a polarity switch that can be thrown from + to −. Similar facility is provided by a polarity reversing switch when the front-to-back resistance ratios of transistor junctions are tested, as depicted in Fig. 3-6. Each of the three determinations indicated in the diagram requires a pair of measurements; each second measurement is made in reversed polarity. If a polarity reversing switch is provided on the ohmmeter, the test procedure is facilitated.

Note that semiconductor junctions are examples of nonlinear resistance. In other words, the resistance of a junction is voltage-de-

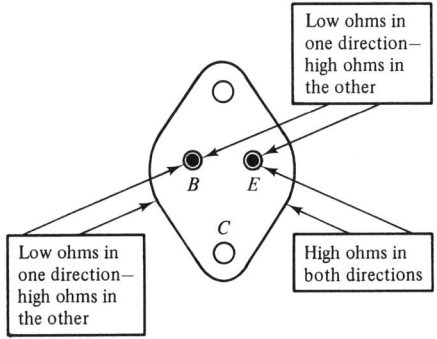

Figure 3-6 Ohmmeter test of a power transistor.

pendent. If an ohmmeter applies a low test voltage, the resistance indication will be comparatively high. On the other hand, if an ohmmeter applies a high test voltage, the resistance indication will be comparatively low. For this reason, the same ohmmeter will indicate a widely different value of junction resistance when switched from one range to another. It is customary to measure the forward resistance of a junction of the RX10 range, and to measure back resistance on a considerably higher range, such as RX1000. Table 3-1 exemplifies the effect of applied test voltage on ohmmeter readings in nonlinear situations.

TABLE 3-1

Typical Resistance Readings for a Semiconductor Diode

Ohmmeter operated on RX1 range:	Forward resistance 140 ohms Reverse resistance "infinity"
Ohmmeter operated on RX10 range:	Forward resistance 400 ohms Reverse resistance 200,000 ohms
Ohmmeter operated on RX10,000 range:	Forward resistance 2000 ohms Reverse resistance 200,000 ohms

Another aspect of this same principle is seen in Fig. 3-7. Here, progressive forward-resistance readings of stacked diodes are noted. When two similar diodes are connected in series, their total resistance (as measured by an ohmmeter) is more than twice the resistance of a single diode. Again, when three similar diodes are connected in series,

68 Resistance Measuring Techniques

their total resistance (as measured by an ohmmeter) is more than three times the resistance of a single diode. Observe that, in this example, three diodes connected in series have five times the resistance of a single diode, insofar as ohmmeter readings are concerned. In summary, when diodes are stacked, their total forward resistance increases out of proportion to the number of diodes in the stack, on the basis of ohmmeter measurements.

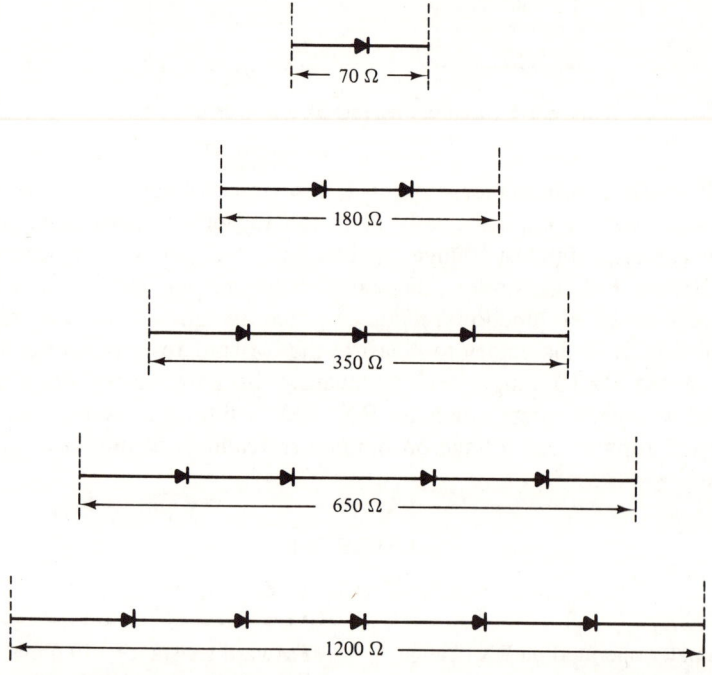

Figure 3-7 Ohmmeter forward-resistance readings of stacked diodes.

3-2 In-circuit Resistance Measurements

Considerable troubleshooting procedure involves in-circuit resistance measurements. Although in-circuit resistance measurements are not always possible, many circuit branches can be checked precisely, and approximate resistance measurements may be made in others. As a basic example, consider the measurement of emitter resistance in the arrangement of Fig. 3-8. An NPN transistor is present. If the emitter is biased positive with respect to the base, the transistor will not conduct

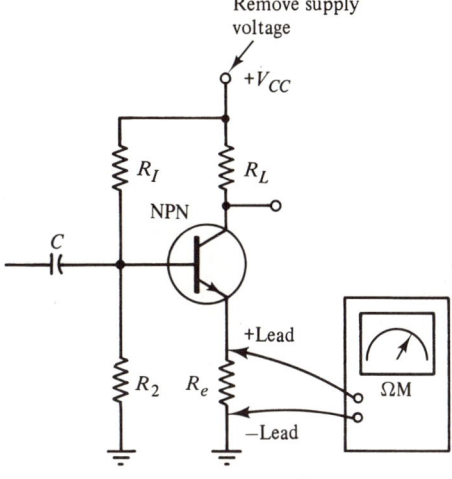

Figure 3-8 In-circuit measurement of emitter-resistor value.

test current under normal conditions. In turn, if the ohmmeter test leads are polarized as shown, the resistance of R_e can be measured precisely in-circuit under normal conditions. However, if the base-emitter junction happens to be leaky, the ohmmeter will indicate less than the actual resistance of R_e. Thus, if R_e measures its specified value, it can be dismissed from suspicion. On the other hand, if R_e measures low, further tests will be required to pinpoint the fault.

Note that most ohmmeters provide a red positive test lead and a black negative lead. However, this is not true of all instruments. Some multimeters have a red positive test lead for the dc-voltage function which becomes a negative test lead when the meter is operated on its ohmmeter function. Therefore, an ohmmeter should be checked for test-lead polarity if the operator is unfamiliar with the instrument. Test-lead polarity can be checked in various ways. Perhaps the most common method is to check the voltage output from the ohmmeter with a dc voltmeter. When the positive test lead of the ohmmeter is connected to the positive input terminal of the voltmeter, an up-scale indication will be obtained on the voltmeter.

Another example of polarity-dependent in-circuit resistance measurements is shown in Fig. 3-9. Three resistors are connected to the terminals of a bipolar transistor. Each of these resistors is shunted

70   Resistance Measuring Techniques

effectively by transistor junction resistance. In turn, the apparent value of the 10-kΩ resistor, when measured in-circuit, reads 26 ohms with emitter positive and base negative. On the other hand, the ohmmeter will read 10 kΩ with emitter negative and base positive. Next, consider the apparent value of the 1.2-megohm resistor when measured in-circuit. With collector positive and base negative, the ohmmeter will read 12-kΩ. On the other hand, with collector negative and base positive, the ohmmeter will read 1.2 megs. As noted previously, low-power ohmmeters are in general use, which apply less than 0.1 volt across the component or device under test. When this type of ohmmeter is used, the value of each of the resistors in Fig. 3-9 can be measured accurately in-circuit, regardless of the test-voltage polarity.

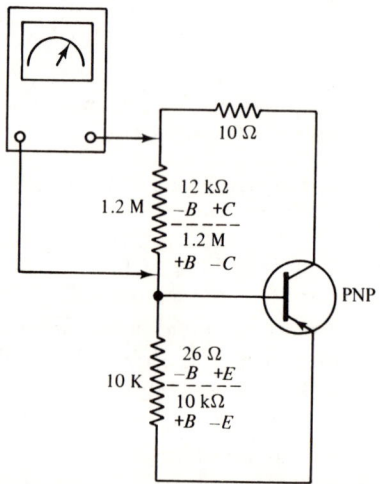

Figure 3-9   Example of polarity-dependent in-circuit resistance measurements.

Consider next the possibility of a defective transistor in the circuit of Fig. 3-9. For example, the transistor may be "shorted" (have defective junctions with very low forward and reverse resistance values). In such a case, a low-power ohmmeter indicates very low resistance values for the 10-kΩ and 1.2-meg resistors when measured in-circuit. The experienced troubleshooter evaluates these subnormal resistance values as evidence of a "shorted" transistor; it is very seldom that two resistors in a circuit both decrease greatly in value. Again, consider the possibility of an "open" transistor. This defect will also be indicated by an ohmmeter test—all three resistors in Fig. 3-8 will read correct values on a low-power ohmmeter, *and will also read correct values on a high-power ohmmeter.* In turn, the technician concludes that the transistor is "open."

A low-power ohmmeter is of no avail when the resistor under test is shunted by another resistor, or by a coil winding, for example. In such a case, the technician often unsolders one end of the resistor, to isolate it from its circuit. Instead of disconnecting one end of the resistor under test, it is also practical to make a razor slit in the associated printed-circuit conductor, as depicted in Fig. 3-10. This expedient temporarily isolates the resistor for test. After its value has been measured, the printed-circuit connection is restored by melting a small drop of solder across the slit.

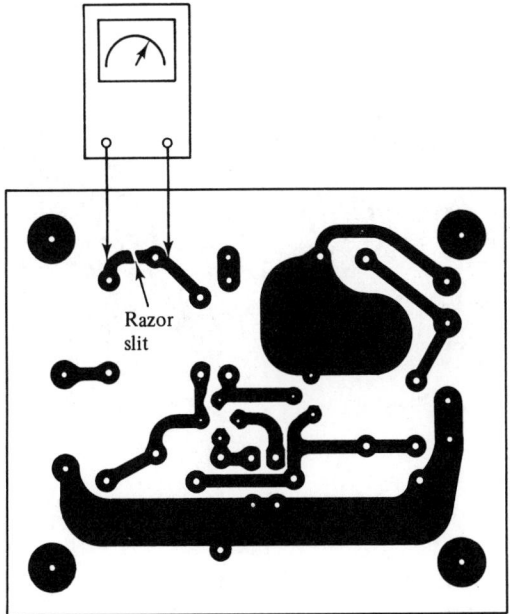

Figure 3-10 Printed-circuit conductor may be cut to make an in-circuit resistance measurement.

3-3 High-voltage Ohmmeter Application

Ohmmeters are used to locate poor insulation in some troubleshooting procedures. Ordinary ohmmeters may not show the presence of a leakage path in components or circuits that operate at comparatively high voltages. Therefore, it is good practice to use a high-voltage ohmmeter

72   Resistance Measuring Techniques

(also called a *megger*) in this type of test. A typical megger is illustrated in Fig. 3-11. It is basically similar to a conventional ohmmeter, except that the instrument circuit is powered from a 600-volt supply. Thus, a potential of 600 volts is applied across the component that is being tested. A single scale ranging from 0 to 100 megohms is used in this instrument. When insulation breakdown occurs, there is sometimes a small gap in the leakage path. In turn, a conventional ohmmeter will indicate an open circuit, whereas a megger will indicate practically a short circuit, or occasionally a resistive path, depending on the nature of the fault. A megger usually operates from self-contained batteries.

Figure 3-11   A 600-volt megohmmeter.

When an adapter is used with a multimeter to measure resistance values higher than can normally be measured with the instrument, a potential of 67.5 volts is typically applied between the points under test, as exemplified in Fig. 3-12. This adapter arrangement is used with the ohmmeter configuration shown in Fig. 3-2(d). The RX10,000 range is multiplied by 10 to provide an RX100,000 range. The adapter comprises a 1.1-MΩ 1 percent resistor and a 67.5-volt hearing-aid battery. It provides a 0-to-200-MΩ range. Not only does it permit measurement of comparatively high resistance values, but it improves the indication accuracy in the range from 1 to 20 megohms, because the readings are taken on the more expanded portion of the ohmmeter scale when the adapter is used.

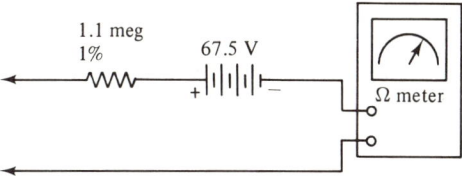

Figure 3-12 Configuration of a high-ohms adapter for a multimeter.

A high-ohms adapter for a VTVM or TVM is shown in Fig. 3-13. This arrangement is used with an instrument that has an input resistance of 10 megohms with a 1.5-volt ohmmeter battery. It multiplies the RX1-megohm range by 10; in turn, a 0-to-10,000-megohm range is provided. The 15-volt battery is a hearing-aid type, and the 1.5-volt cell is a penlight type. Note that the batteries are connected in opposite polarity, so that the adapter battery has an effective potential of 13.5 volts. When an ohmmeter is line-operated, it sometimes develops instability when used with a high-ohms probe. In other words, the pointer may tend to drift on the scale. In such a case, the ground terminal of the instrument should be connected to an earth ground, such as a cold-water pipe, as indicated in Fig. 3-13.

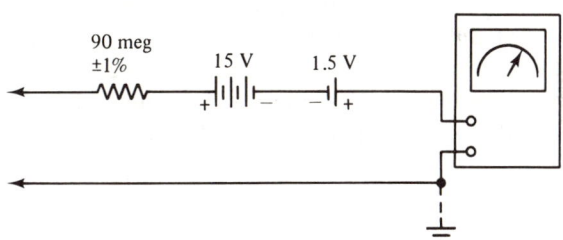

Figure 3-13 High-ohms adapter arrangement for a VTVM or TVM.

3-4 Low-ohms Adapter for Multimeter

A configuration for a low-ohms adapter is shown in Fig. 3-14. This arrangement is suitable for a multimeter that has a 100-μA current range with an input resistance of 2500 ohms, and an ohms scale with 12 ohms center-scale indication. The 1.15-ohm resistor should be wound from manganin wire, or similar resistive wire that has a very small temperature coefficient. This resistor must dissipate up to 1.7 watts. Note that the test leads must have a very low resistance, or indication error will result. However, the leads between the 1.15-ohm resistor and the multimeter carry a very small current, and may be

74 Resistance Measuring Techniques

of thin wire. The 1.5-volt battery must be sufficiently large to supply as much as 1.3 amperes while the adapter is in use. This adapter works into the 100-μA range of the multimeter. However, resistance values are indicated on the ohms scale. The adapter provides an RX0.1 range for the multimeter.

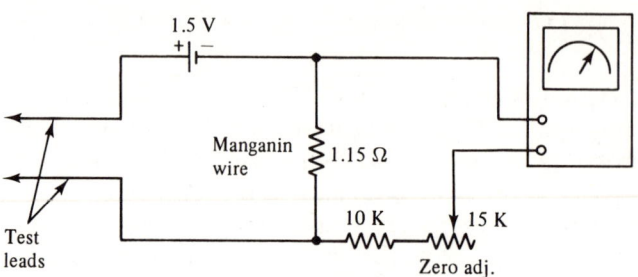

Figure 3-14 A low-ohms adapter for a multimeter.

A low-ohms adapter is helpful in running down short circuits in wiring systems or equipment in general. When a short circuit is present between a pair of conductors, for example, a resistance reading will be obtained that might be a fraction of an ohm. Then, as resistance values are measured at difference points along the conductors, the resistance reading will increase in one direction and decrease in the other direction. As the reading decreases, the short circuit is being approached. When the test leads are applied across the short circuit, a zero-ohms indication is obtained. A low-ohms adapter is also useful for checking cold-solder joints, switch-contact resistance, and the winding resistances of small coils.

3-5 Checking Special Types of Resistors

Light-dependent resistors (LDR's) have specified values in equipment service data. The specified resistance value is usually referenced to average room illumination. Note that the effective resistance of an LDR can vary greatly with changes in ambient light. For example, a typical LDR has a resistance that varies from 100 ohms in bright light to 0.5 megohm in complete darkness. With reference to Fig. 3-15, the measured resistance value is the same, regardless of test polarity. In other words, a CdS LDR is not a junction device. On the other hand, the technician will encounter diode-type and triode-type LDR's also. These devices employ junctions, and their measured resistance must be specified with respect to test polarity.

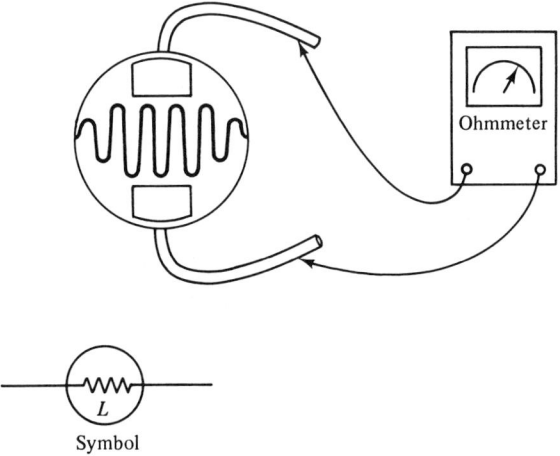

Figure 3-15 Resistance measurement of a CdS light-dependent resistor.

Thermistors provide another useful example of semiconductor resistance variation. A thermistor (Fig. 3-16) is basically a bead of certain metallic oxides. As its temperature is increased from 0° to 100°C, for example, the resistance of the device decreases typically from 1000 ohms to 100 ohms. If the temperature of the thermistor is further increased to the point of burnout, its resistance decreases typically to a value of 10 ohms. The resistance of a metallic-oxide thermistor is the same, regardless of ohmmeter polarity. On the other hand, junction diodes are also used in temperature-compensating circuits and function essentially as thermistors. In turn, the measured resistance of the diode device is polarity-dependent. Thermistor resistance is generally specified at average room temperature.

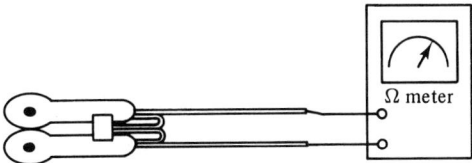

Figure 3-16 Resistance check of a thermistor.

Ballast resistors have a temperature characteristic that is the opposite of a thermistor. In other words, the resistance of an iron-wire/hydrogen ballast resistor increases rapidly as the temperature increases. A ballast resistor is employed typically in circuits that provide comparatively constant current flow under conditions of input voltage variation. A ballast resistor measures the same value regardless of ohmmeter polarity. Ballast resistance is generally specified at average room temperature. Note that "Globar" resistors used in radio receivers are sometimes called *ballast resistors*. However, a Globar resistor exhibits decreasing resistance with increasing temperature.

3-6 Resistance Charts

Receiver service data often provide resistance charts, as exemplified in Fig. 3-17. Resistance measurements are made from the specified test points to ground, unless otherwise noted. These charts facilitate troubleshooting procedures, because component or device defects are often associated with resistance changes. Although it is usually prohibitively difficult to calculate the normal resistance of an electronic network, a resistance chart provides the needed value for immediate reference. In the case of solid-state equipment, resistance charts are compiled on the basis of measurements with a low-power ohmmeter.

3-7 Measurement of Resistance of Meter Movement

When the resistance of a meter movement is measured, care must be taken to avoid overload. Most meter movements can be checked directly with a low-power ohmmeter that applies less than 0.08 volt across the points under test. For example, a typical meter movement for a 20,000 ohms-per-volt multimeter has a full-scale current rating of 50 μA, and a full-scale voltage drop of 100 MV. Therefore, this movement can be checked directly with a low-power ohmmeter. However, very sensitive movements can be overloaded or damaged, even by a potential of 0.08 volt. Therefore, in case of doubt, start the test with a series resistor of ample value, as shown in Fig. 3-18. It may be found possible to make the resistance measurement without any series resistor. However, if a series resistor is required to avoid driving the pointer off-scale, simply subtract the value of the series resistor from the ohmmeter reading to obtain the internal resistance of the movement.

3-8 Transistor Test with Ohmmeter

An ohmmeter can be used to determine whether a transistor provides gain, with the arrangement depicted in Fig. 3-19. Note that if an NPN transistor is under test, the ohmmeter polarity is reversed. The ohmmeter battery serves to energize the test circuit. A typical small-signal

Resistance measurements

Item	Tube	Pin 1	Pin 2	Pin 3	Pin 4	Pin 5	Pin 6	Pin 7	Pin 8	Pin 9
V1	3BZ6	325 K	47 Ω	11 Ω	9.5 Ω	†2000 Ω	†2000 Ω	0 Ω		
V2	3BZ6	0.1Ω	180 Ω	9.5 Ω	9 Ω	†2000 Ω	†2000 Ω	0 Ω		
V3	3BZ6	0.1Ω	180 Ω	9 Ω	8.5 Ω	†4200 Ω	†4200 Ω	0 Ω		
V4	12BY7A	27 Ω	3300 Ω	0 Ω	5 Ω	5 Ω	3.5 Ω	†7000 Ω	†1600 Ω	0 Ω
V5	5BR8	†50 K	200 K	†60 K	3.5 Ω	2.5 Ω	†56 K	†56 K	150 K	10 K
V6	5T8	1 NF	27 K	1 NF	1.5 Ω	0 Ω	550 K	0 Ω	10 meg	†470 K
V7	5AQ5	470 K	270 Ω	2.5 Ω	1.5 Ω	†970 Ω	†470 Ω	470 K		
V8	3CS6	†1 meg	0 Ω	7 Ω	6.5 Ω	†25 K	†7500 Ω	1.3 meg		
V9	10DE7	†200 Ω	•1.2 meg	•1.2 meg	15 Ω	13 Ω	•‡2 meg	•1.6 meg	0 Ω	270 Ω
V10	6CG7	†10 K	2.8 meg	1000 Ω	7 Ω	8.5 Ω	†47 K	120 K	1000 Ω	0 Ω
V11	12DQ6A	NC	18 Ω	TP	†12 K	1 meg	TP	21 Ω	0 Ω	Top cap †14 Ω
V12	12AX4GTA	NC	TP	¶850 K	NC	†40 Ω	NC	18 Ω	15 Ω	
V13	1B3GT				Pins 1 thru 8 have infinite resistance					Top cap ‡273 Ω
V14	17DKP4	6.5 Ω	35 K	‡750 K	•‡1.6 meg	NC	NC	•500 K	5 Ω	
V201	2BN4	0 Ω	670 K	11 Ω	11.5 Ω	†2500 Ω	0 Ω	670 K		
V202	5CG8	10 K	†15 K	0 Ω	13 Ω	11.5 Ω	†2700 Ω	†47 K	0 Ω	330 K
Item	Tube	Pin 1	Pin 2	Pin 3	Pin 4	Pin 5	Pin 6	Pin 7	Pin 8	Pin 9

¶This reading will vary depending upon the condition of the electrolytic in the circuit.
•This reading will vary. Control set for normal operation.
†Measured from 275 V source.

‡ Measured from pin 3 of V12.
NC no connection
TP tie point.

Figure 3-17 Typical resistance chart for in-circuit resistance measurements. (Courtesy of *Howard W. Sams & Co., Inc.*)

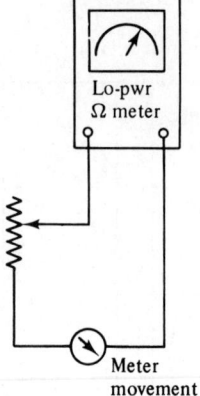

Figure 3-18 Measuring the internal resistance of a meter movement.

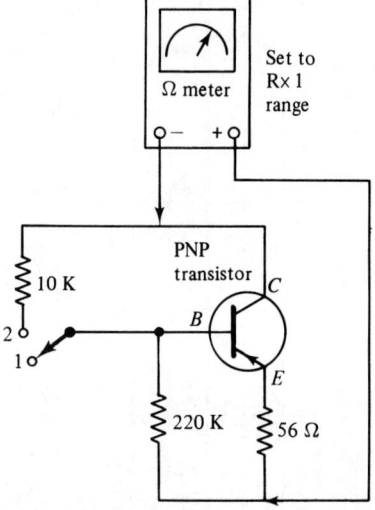

Figure 3-19 Bipolar transistor test with an ohmmeter.

transistor provides at least eight times as much deflection in position 2 of the switch, compared with position 1 of the switch. If less than eight times as much deflection is obtained, it is an indication that the transistor has a subnormal beta value.

An example of bipolar transistor junction resistance values is given in Fig. 3-20. A few bipolar transistors will show the same value of forward resistance for the emitter-base and collector-base junctions.

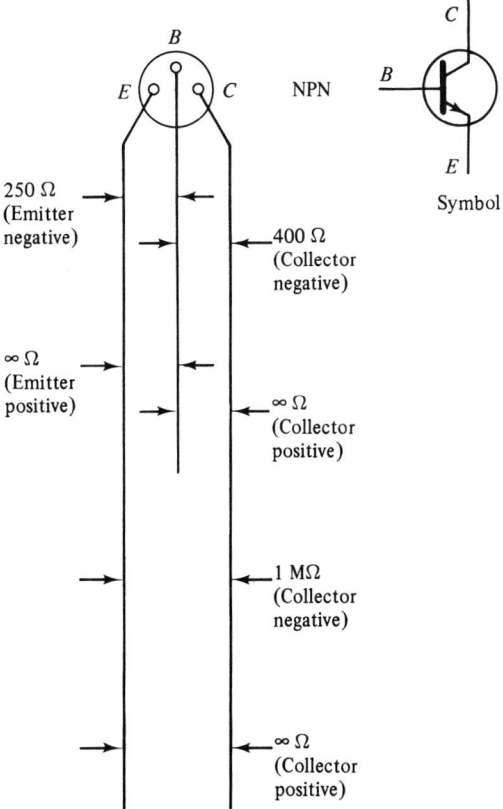

Figure 3-20 Example of bipolar transistor junction resistance values.

If the two junctions have unequal forward-resistance ratios, then the resistance from emitter to collector will depend upon the ohmmeter polarity. This is always a high resistance value, and the ohmmeter must be operated on a high range, such as RX1000. In the example of Fig. 3-20, the emitter-collector forward resistance has a value of 1 megohm. Note that if the transistor were a PNP type, all of the test polarities noted would be reversed.

Junction field-effect transistors show a different pattern of resistance readings, as depicted in Fig. 3-21. In other words, forward and reverse readings occur between gate and source, or between gate and drain only. The resistance reading from source to drain is the same, regardless of ohmmeter polarity. Note in passing that the resistance measured from source to drain is also the same, regardless of the ohmmeter range that is used. In other words, the source-drain resistance is

80 Resistance Measuring Techniques

linear. On the other hand, the junction resistance from gate to source, or from gate to drain, is nonlinear. The practical importance of this non-linearity is that the forward-resistance value that is measured across the junction will be different if the ohmmeter range is changed. Note also that ohmmeter polarity is not necessarily indicated correctly by the test-lead color coding. In other words, the red lead may be negative and the black lead positive. Therefore, it is good practice to check ohmmeter polarity with a dc voltmeter, in case of doubt.

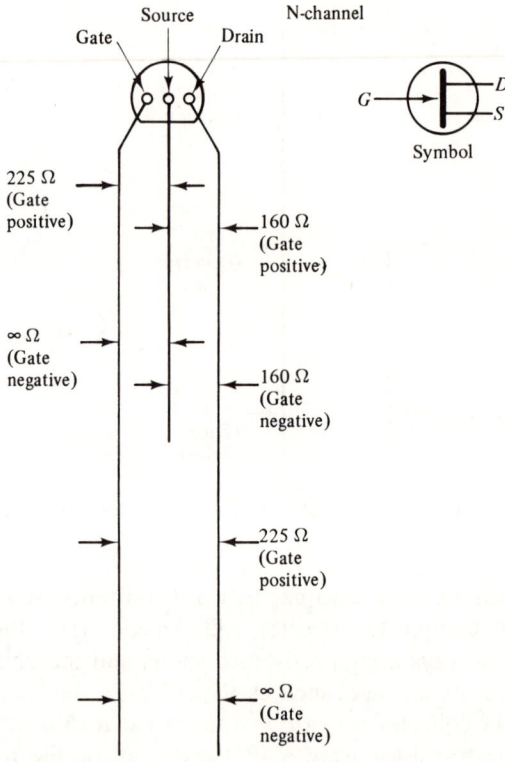

Figure 3-21 Example of junction-FET resistance values.

Insulated-gate field-effect transistors (IGFET's) are also called metallic-oxide substrate field-effect transistors (MOSFET's). They are different from junction field-effect transistors (JFET's) in that there is no junction between the gate and the semiconductor source-drain substance, as seen in Fig. 3-22. Note that MOSFET's are classified into

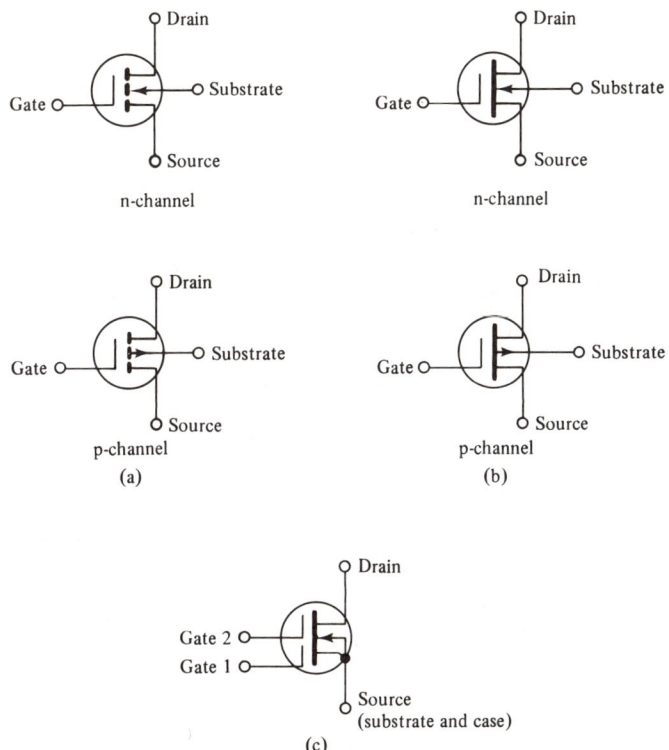

Figure 3-22 Basic types of IGFET's (MOSFET's): **(a)** enhancement-type metal-oxide semiconductor field-effect transistor (N-channel and P-channel types); **(b)** depletion-type metal-oxide semiconductor field-effect transistor (N-channel and P-channel types); **(c)** dual-gate N-channel depletion-type metal-oxide-semiconductor field-effect transistor.

enhancement and depletion types. These terms denote drain-source conduction characteristics as shown in Fig. 3-23. Because there is no junction between the gate and the semiconductor source-drain substance, a MOSFET shows infinite resistance between its gate and source, or between its gate and drain. However, a MOSFET shows a definite resistance value between source and drain terminals, in the same manner as a JFET. Note carefully that an ordinary MOSFET is very likely to be damaged if its gate terminal is disconnected from its source or drain terminal. This danger is present because of the extremely thin insulation provided under the gate electrode. A "floating" gate is likely to pick up enough static electricity to break down the insulation and ruin the MOSFET. Therefore, be sure to provide some kind of metallic

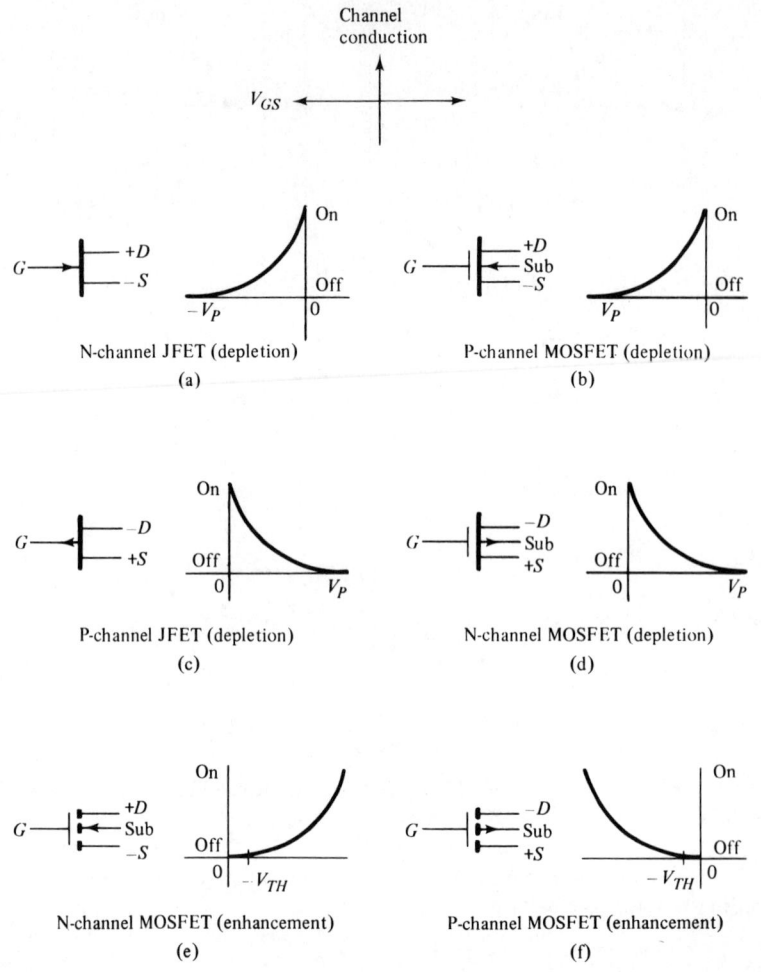

Figure 3-23 Drain-source conduction characteristics for depletion-type and enhancement-type FET's.

connection from the gate terminal to the source whenever a MOSFET is removed from its circuit, or when it is being connected into a circuit.

Dual-gate MOSFET's are basically the same as ordinary MOSFET's, except that two gate electrodes are provided. They are used in comparatively elaborate circuits. Insofar as handling is concerned during troubleshooting procedures, the same precaution applies as noted above. In other words, it is important to provide a short circuit between each gate terminal and the source terminal whenever a dual-gate MOS-

FET is not connected into its circuit—neither of the gate electrodes should ever be permitted to "float." However, this precaution does not apply to dual-gate-protected MOSFET's, such as that depicted in Fig. 3-24. In other words, this type of MOSFET contains built-in back-to-back zener diodes that bypass any excessive gate voltage to the source electrode. In turn, a dual-gate-protected MOSFET can be handled without any precautions against static electricity.

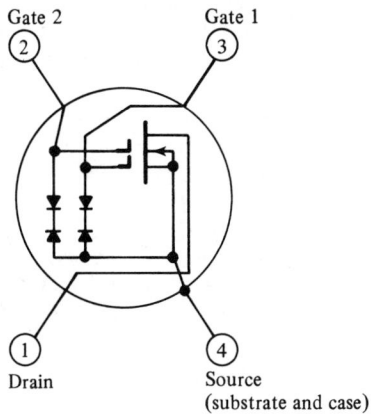

Figure 3-24 Internal arrangement of a dual-gate protected MOSFET.

When the gate-to-source resistance of a gate-protected MOSFET is checked with an ohmmeter, the resistance reading depends greatly upon the test voltage applied by the meter. Most ohmmeters (especially when operated on their lower ranges) do not apply sufficient test voltage to cause zener conduction, and such instruments will show an infinite-resistance reading for the transistor if it is in normal condition. On the other hand, an occasional ohmmeter, particularly if operated on a higher range, will apply sufficient test voltage to cause zener conduction. In such a case, the instrument will show a very low resistance reading. Quick checks of unipolar (FET) and bipolar transistors are made with a go/no-go type of tester, as in Fig. 3-25.

Sometimes the terminal identification on a bipolar transistor has been obliterated. In such a case, the ohmmeter checks depicted in Fig. 3-26 will identify the terminals of the device. The first step is to make resistance measurements between each pair of leads in both the forward-current and reverse-current (forward resistance and back resistance) directions. A resistance reading below 500 ohms indicates that the ohmmeter is forward-biasing a junction. Note that the highest for-

84   Resistance Measuring Techniques

Figure 3-25   A portable type of transistor tester. (*Courtesy of Sencore, Inc.*)

ward reading is obtained when the ohmmeter is applied between the emitter and collector leads. In turn, the third lead, which is not connected to the ohmmeter, is identified as the base lead. At this point it is still unknown whether the transistor is a PNP or an NPN type, and which of the two leads under test is the collector, and which the emitter lead.

In the second step, a resistance measurement is made between the base and one of the other leads. If forward resistance is indicated when the negative lead of the ohmmeter is connected to the base, it is established that the transistor is a PNP type. On the other hand, if forward resistance is indicated when the positive lead of the ohmmeter is connected to the base, it is evident that the transistor is an NPN type. Next, in the third step, it is determined which of the unknown leads is the collector, and which the emitter lead. Two resistance measurements are made between these leads, the ohmmeter polarity being reversed for the second measurement. Note the connections utilized in the lower resistance condition. For a PNP-type transistor, the negative lead of the ohmmeter will be connected to the collector lead. Conversely, for an NPN transistor, the positive lead of the ohmmeter will be applied to the collector lead.

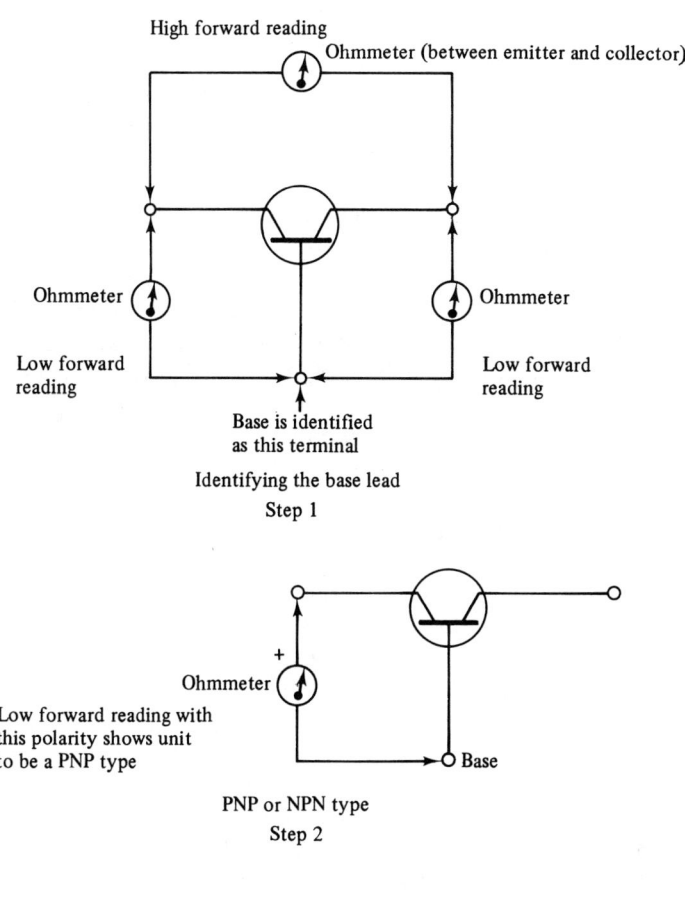

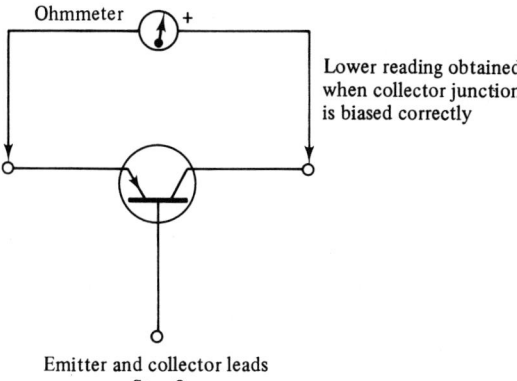

Figure 3-26 Identification of bipolar transistor type and terminals.

86 *Resistance Measuring Techniques*

To summarize the reasons for the foregoing test result, more current flows and a lower resistance reading is obtained when the ohmmeter applies test voltage in normal polarity (as in normal operation) to the emitter and collector of the transistor. The reasons are (1) The emitter doping is almost always heavier than the collector doping. This results in a slightly higher alpha value (current amplification) for the transistor when the emitter junction is forward-biased. (2) The collector junction has a larger area than the emitter junction. In turn, the collector junction has more leakage current than the emitter junction. (3) The total leakage current in the test situation is equal to the collector-junction leakage multiplied by the beta (current amplification) of the transistor. This product is greater when the ohmmeter applies test voltage in the polarity for normal operation.

3-9 Testing Integrated Circuits with an Ohmmeter

Various tests of the simpler types of integrated circuits can be made with conventional multimeters or TVM's. Additional test information can be obtained with a hi-lo FET meter. Consider the measurement of terminal resistance values in a basic integrated circuit. A hi-lo FET meter is connected to the IC terminals as shown in Fig. 3-27. With the meter controls set to measure lo-pwr ohms, resistance measurements are made between terminals 4-9, 4-10, and 5-10 in the example of Fig. 3-28. Because these resistance measurements were made on the low-power ohms function, the resulting equivalent circuit is as seen in Fig. 3-27. The 4-9 measurement is the resistance of R1 + R3, or 6 kΩ. The 4-10 measurement is the resistance of R2 + R3, or 6 kΩ. The 5-10 measurement is the resistance of R4, or 18 kΩ. If the measured values were found to be substantially incorrect, the technician would reject the IC.

Although other pairs of terminals could be used to measure resistance values in the example of Fig. 3-29, no further data would be obtained. In other words, there are four resistors in this circuit, each of which is included at least once in the 4-9, 4-10, and 5-10 measurements. Or each resistor could be measured alone. Consider checking the transistors in an IC for leakage. Proceed as before, and, with reference to Fig. 3-28, measure the resistance values between terminals 6-8, 1-2, 1-8, and 1-9. Since these are low-power ohms measurements, the equivalent circuit is as seen in Fig. 3-30. All junctions are normally open circuits in this test, and the meter should indicate infinity. If the reading(s) should be substantially less, the technician would reject the IC.

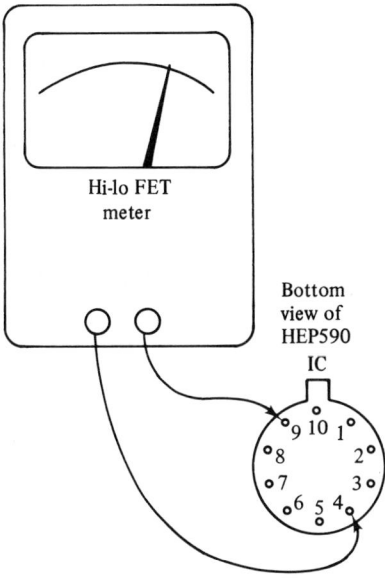

Figure 3-27 Test setup for checking the resistors in a Motorola HEP 590.

Next, consider checking a diode in an IC for leakage. Proceed as before, with the meter set to measure low-power ohms. With reference to Fig. 3-28, measure the resistance value between terminals 3-4. An infinite reading should be obtained; otherwise, the IC should be rejected. The substrate in an IC should also be checked for leakage, as follows. With the meter set to measure low-power ohms, and with reference to Fig. 3-28, measure the resistance value between terminal 7 and any other terminal of the IC. A reading of infinity should be obtained.

The front-to-back ratio of a transistor in an integrated circuit can be measured in various circumstances. With the meter set to measure high-power ohms, and with reference to Fig. 3-28, measure the resistance between terminals 2-1. Then, reverse the test leads and measure the resistance between terminals 2-1 again. Repeat the pair of measurements for terminals 6-8. These tests are essentially the same as out-of-circuit front-to-back ratio measurements. Typically, if the meter is operated on its RX100 range, the forward resistance will measure somewhat more than 1000 ohms, and the reverse resistance will measure infinity. A substantially incorrect front-to-back ratio indicates a defective IC. The diode in Fig. 3-28 can be tested similarly for front-to-back ratio between terminals 3-4. The foregoing transistor test yields the front-to-back ratio for the collector junction of Q2 and the emitter

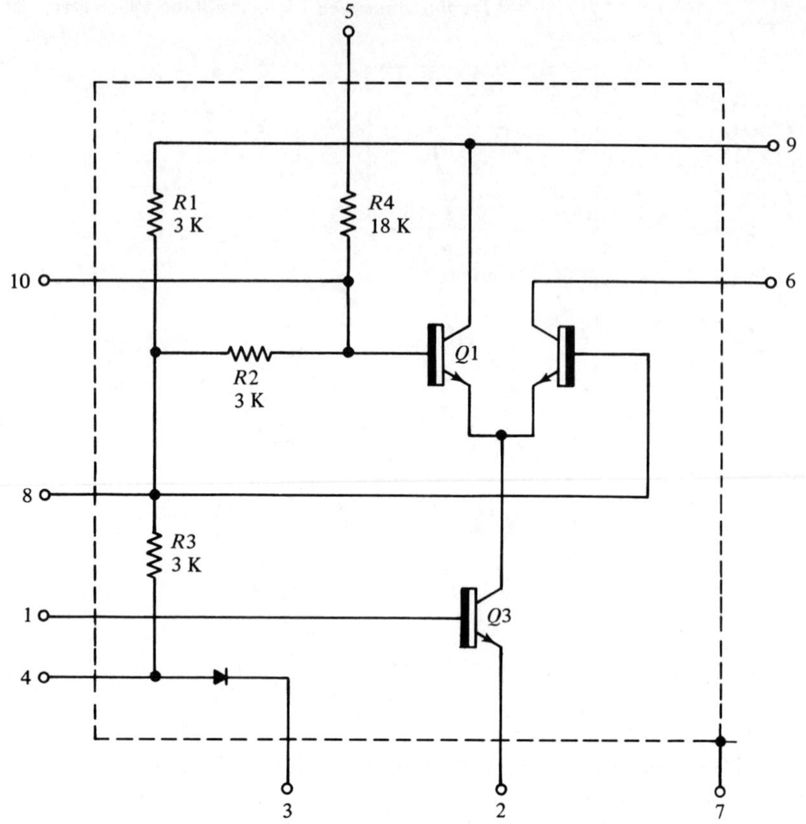

Figure 3-28 Complete circuit of the Motorola HEP 590 IC.

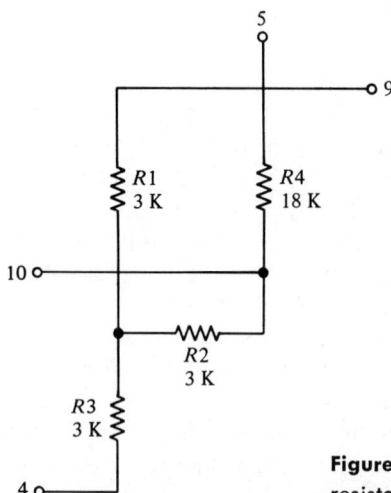

Figure 3-29 Equivalent circuit for low-power resistance measurements.

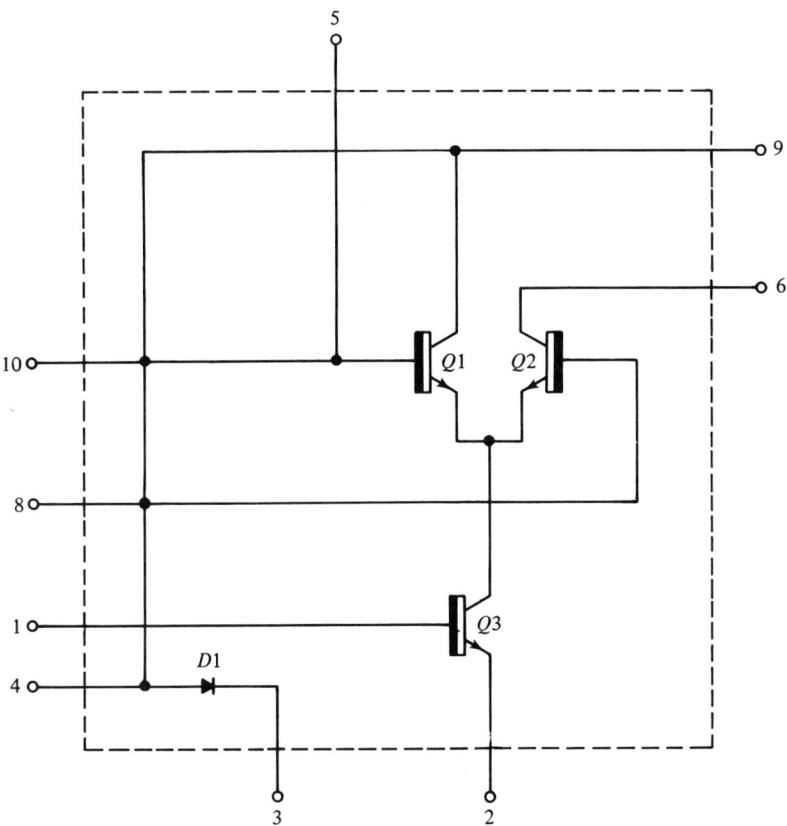

Figure 3-30 Equivalent circuit for transistor test, with all junctions as open circuits.

junction of Q3. No data are provided concerning the front-to-back ratio of the remaining junctions.

It is helpful to consider the checking of front-to-back ratio for a transistor shunted by resistance. With the meter controls set to measure high-power ohms, and with reference to Fig. 3-28, measure the resistance between terminals 9-10, with the RX100 range of the instrument. Then reverse the test leads and measure the resistance again. The equivalent circuit in this test is shown for all practical purposes in Fig. 3-31. Normally, the technician will measure somewhat more than 1000 ohms in the forward direction, and 6000 ohms in the reverse direction. Since 6000 ohms is considerably larger than the normal forward resistance of the collector-base junction, the practical measurement of forward resistance compares to that in the foregoing test. However, the

90 Resistance Measuring Techniques

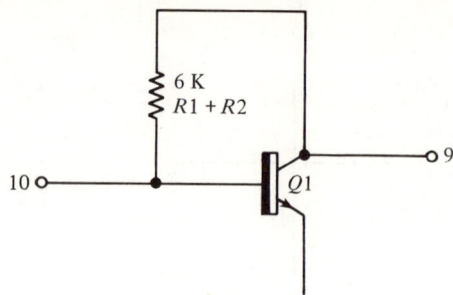

Figure 3-31 Front-to-back test of Q1's collector-base junction involves this equivalent circuit.

measurement of reverse resistance is established by the 6000-ohm shunt. Therefore, this is necessarily an inconclusive test.

Next, consider how to check the front-to-back ratio of a transistor with series resistance. With the meter controls set to measure high-power ohms, and with reference to Fig. 3-28, the front-to-back ratio of Q2 will be measured on the RX100 range of the instrument. Measure the resistance between terminals 4-6, and then repeat the measurement with the test leads reversed. The equivalent circuit in this test is shown in Fig. 3-30. Normally, the technician will measure approximately 3500 ohms in the forward direction, and infinity in the reverse direction. Since 3000 ohms is considerably greater than the normal junction resistance of Q2, it adds substantially to the junction-resistance measured value. On the other hand, 3000 ohms has no practical effect on the reverse-resistance measurement. (See Fig. 3-32.)

As was noted previously, a junction resistance is different from a fixed resistance, in that the effective value of a junction resistance depends on the value of the applied voltage. Therefore, junction resistances cannot be added and subtracted like fixed resistances. In the foregoing test, the effective value of the junction resistance is about 500 ohms, but if there were some other value of fixed resistance in this circuit, the effective resistance of the junction would not be the same. Observe that the foregoing test has provided no new data concerning the IC—it has been included because of its instructional content.

A control-action test of a transistor in the exemplified IC is made as follows: With the meter controls set to measure high-power ohms, and with reference to Fig. 3-28, apply the positive test lead to terminal 10, and apply the negative test lead to terminal 2. Note the scale reading, using the RX1-MΩ range. Then touch a finger to terminal 1, thereby driving Q3 with stray hum voltage. In turn, Q3 conducts on the positive peaks, and the ohmmeter reading normally decreases. If this response is not obtained, the IC should be rejected. Note that the exact

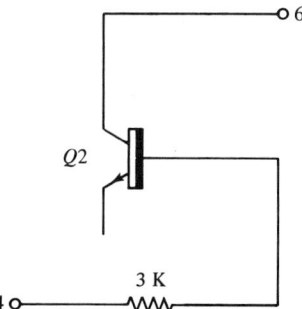

Figure 3-32 Series resistance is included in the equivalent circuit of Q2's collector-base junction.

amount of decrease in resistance reading will vary, depending on how much hum voltage is being injected. Therefore, this test is in the go/no-go class.

An IC may be checked for intermittents by tapping while any of the foregoing resistance measurements are being made. If the pointer "jumps" on the scale, the IC should be rejected. Thermal intermittents are checked by exposing the IC to moderate warmth, and then spraying with a coolant. The meter reading should "crawl" in typical situations, but it should never "jump." Note in passing that because IC resistors and semiconductors are thermally responsive, differential-amplifier configurations are widely used so that amplifier response will be based on resistance ratios instead of absolute resistance values.

3-10 Ohmmeter Test of Circuit Boards

Integrated circuits used in digital circuitry are usually arrayed on circuit boards or modules (plug-in circuit boards). Unlike faults in linear circuitry, most defects in digital circuitry are caused by deteriorated or nonoperative devices, broken wires or conductors, and short circuits. When a system has more than one fault, direct substitution of a new circuit board is not necessarily the best approach. If a circuit-board tester is available, comparison tests of suspected boards can be quickly made against new or known-good boards. As exemplified in Fig. 3-33, a circuit-board tester comprises a pair of circuit-board sockets for insertion of the suspected board and the reference board, together with switching facilities and an ohmmeter. Note that practically any circuit board can be given a comparative test with this arrangement. Various manufacturers sometimes provide different types of circuit-board sockets, and the tester must be constructed accordingly.

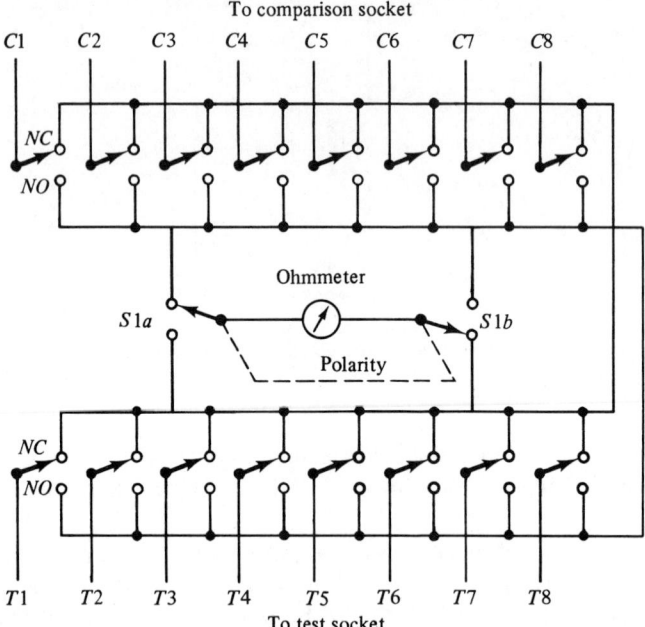

Figure 3-33 Typical circuit-board tester (for boards with eight terminals).

With reference to Fig. 3-33, when no pushbutton switch is pressed, all terminals of both sockets are connected together. On the other hand, if any one button is pressed, the socket terminal associated with that button is disconnected from the common bus and is connected to the other bus line. The ohmmeter is connected via polarity-reversing toggle switch S1 between these two bus lines. Thereby, the resistance at any terminal of the suspected circuit board can be compared with the resistance at the corresponding terminal of the reference circuit board. This resistance comparison can be made with both polarities of ohmmeter voltage. The technician need not concern himself with the internal circuitry of the board, since this is basically a go/no-go type of test.

4
DC Current Measurement Methods

4-1 General Considerations

DC current measurements are usually somewhat more involved than dc voltage measurements. In many instances, it is necessary to open the circuit under test, and to connect the current-meter test leads in series with the conductor. In the case of a printed circuit, a conductor can be opened by cutting a slit through it with a razor blade. Then the test prods can be applied on either side of the slit to connect the meter in series with the circuit. After the measurement is made, the PC conductor can then be repaired with a small drop of solder. In the case of a radio battery, the current demand can be easily measured as shown in Fig. 4-1. One end of the terminal strip is unplugged, and the meter is connected between the open terminals to complete the circuit.

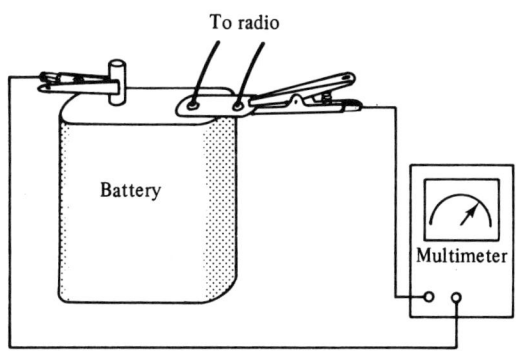

Figure 4-1 Measurement of current drain on a transistor radio battery.

94   DC Current Measurement Methods

Technicians frequently look for ways to measure current indirectly. As an illustration, if there is a resistor connected between the emitter of a transistor and ground, the emitter current can be measured indirectly with the voltmeter function of a multimeter or TVM. With reference to Fig. 4-2, the voltage drop is measured across the emitter resistor. Then the current flow is calculated from Ohm's law. A rough calculation can be made by observing the resistor's color code. A precise calculation can be made by measuring the value of the resistor. If a low-power ohmmeter is used, test polarity is not of concern. On the other hand, if a conventional ohmmeter is utilized, the negative test lead of the instrument must be connected to the emitter terminal of the transistor in the example of Fig. 4-2.

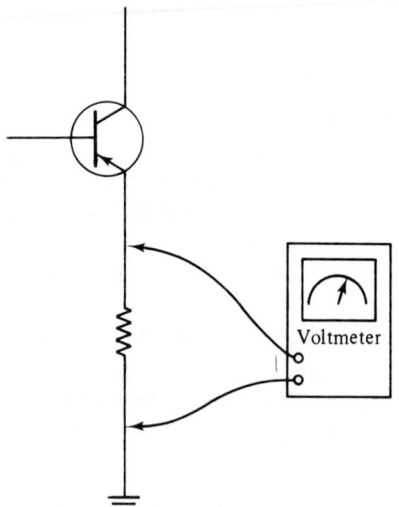

Figure 4-2   Indirect measurement of emitter current.

4-2 Measurement of High DC Current Values

Multimeters, TVM's, VTVM's, and DVM's are more or less limited in regard to dc current measurements. Multimeters typically measure up to 10 A; TVM's and DVM's usually measure up to 1 A; VTVM's may provide no dc current ranges. Accordingly, external shunts are used to measure high current values, as shown in Fig. 4-3. In this example, a 25-A shunt is utilized. This shunt provides a 0.1-volt drop at 25 A. In turn, the voltmeter is operated on its 0.1-volt range, and current values are indicated on the meter's "25" scale. Note that the voltmeter test leads are connected to the two inner terminals on the shunt. In

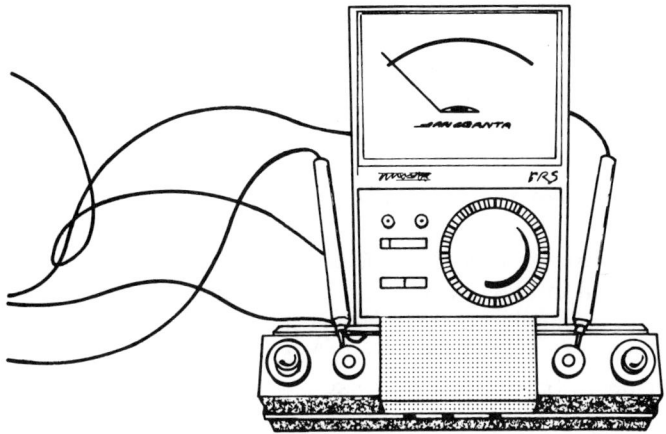

Figure 4-3 External shunts are used to measure high current values.

turn, the two outer terminals on the shunt are connected to the circuit under test. Shunts are available in a wide range of current values; for example, shunts rated for several hundred amperes of current may be used by industrial-electronics technicians. Heavy-current shunts are also used by automotive electric technicians to check the current capabilities of storage batteries.

4-3 Clip-around DC Current Probe

Current can be measured in a conductor without opening the circuit if a clip-around current probe is used. A typical dc current probe can be used with a TVM or a DVM to measure current values up to 10 A. This type of probe uses the Hall effect; a probe amplifier is employed that provides a sensitivity of 50 mV per mA. The clip-around dc current probe is not in extensive use outside of laboratories because of its comparatively high cost.

4-4 Ultrasensitive DC Microammeter

Measurement of dc current values as low as 0.0002 microampere is provided by an ultrasensitive dc microammeter. The top range of the instrument is 1 mA. This type of current meter is used chiefly in electronic research and development work. Although it is primarily a microammeter, it can also be used as a voltmeter with external multiplier resistors. When used with a 950-megohm multiplier, it provides a dc-voltmeter function with a sensitivity of 100 megohms per volt. The instrument has a 0.5-volt drop for full-scale deflection. It can be applied to measure resistance values up to billions of ohms.

4-5 Direct Current Galvanometer

A galvanometer is a current meter with zero-center scale indication, as exemplified in Fig. 4-4. In most cases, a galvanometer is a microammeter. For example, its range is typically ± 5 μA. Galvanometers are used chiefly as null indicators in Wheatstone bridges for very precise measurements of resistance values. A basic bridge configuration is shown in Fig. 4-5. The galvanometer indicates positive current flow by pointer deflection to the right of center, and indicates negative current flow by pointer deflection to the left of center. R_1 and R_2 are high-

Figure 4-4 A galvanometer scale.

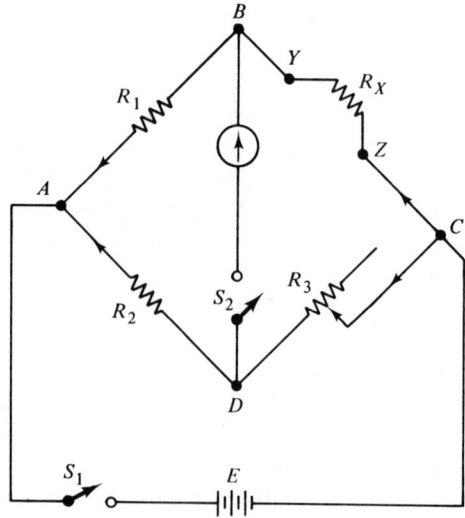

Figure 4-5 Basic Wheatstone bridge arrangement.

precision resistors that form two arms of the bridge. R_3 is a variable resistor, usually with a calibrated control dial to indicate resistance values. R_x is the unknown resistor whose value is to be measured. A battery is used to energize the bridge. When S_1 is closed, current flows in the direction of the arrows, and there is a voltage drop across all four resistors. The voltage drop across R_1 is equal to the voltage drop across R_2. R_3 is adjusted to bring the galvanometer pointer to zero, when S_2 is closed. Then R_3 is equal to R_x. At the null point, points B and D are at equal potential, and no current flows through the galvanometer.

4-6 Internal Resistance Considerations

Some test procedures that appear straightforward are objectionable because of circuit disturbance caused by the internal resistance of the microammeter that is utilized. An example is shown in Fig. 4-6. Here, the base circuit of a transistor has been opened, and a measurement of the base current made with the microammeter function of a 20,000 ohms-per-volt multimeter. In this example, the input (internal) resistance of the meter on its 250-μA range is 2500 ohms. When this resistance is connected in series with the base circuit, the current and

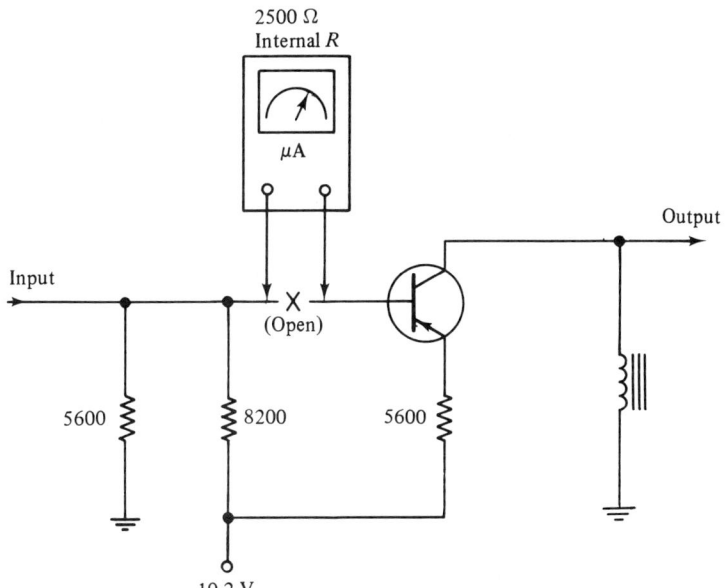

Figure 4-6 Internal resistance of microammeter seriously disturbs circuit action.

98 DC Current Measurement Methods

voltage distribution is upset substantially. In turn, a misleading current indication is obtained.

4-7 Current Measurement in Rectangular Waveform

Current measurement in a rectangular waveform of pulsating dc follows the same basic principles that are involved in dc voltage measurements. With reference to Fig. 4-7, a current meter responds to the average value of a rectangular current waveform. The ratio of average current value to peak current value is equal to the ratio of "on" time to the "on" plus "off" time. Thus, if the peak current can be measured, this value can be used with the average current value to determine the duty cycle of the current waveform. Note that the duty cycle is the ratio of "on" time to "off" time. Therefore, the duty cycle is equal to the ratio of the average value to the difference between the peak value and the average value.

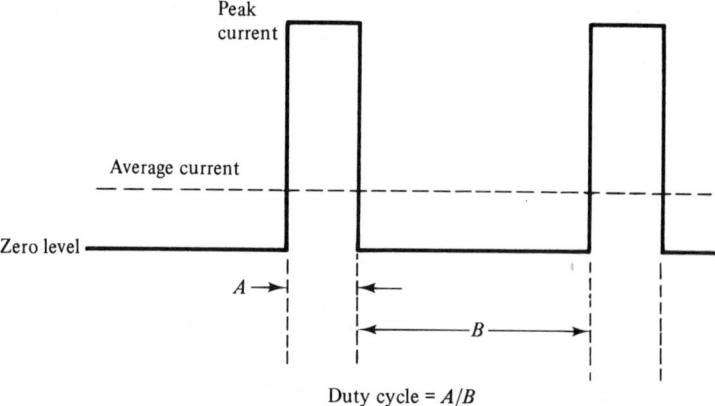

Figure 4-7 Meter responds to average current value in rectangular waveform.

Next, consider the determination of the rms value of a rectangular current waveform. This is the power equivalent-dc value of the waveform. It can be measured directly by a true-rms meter. The rms value of the waveform can be calculated from its average and peak current values as follows:

$$V_{rms} = V_{peak} \sqrt{\frac{V_{average}}{V_{peak}}}$$

4-8 Measuring True RMS Value of Pulsating DC Waveform

Unless pulsating dc has a known waveform (such as a rectangular waveshape), it is impossible to determine its rms value on the basis of average and peak voltage measurements. However, if a true rms current meter is used, such as that illustrated in Fig. 4-8, the rms value of a pulsating dc waveform can be measured directly. This type of instrument employs an electrodynamometer movement that responds to the rms value of a complex waveform. Note that this type of true rms meter is useful only at power repetition rate; in other words, the instrument is accurate only at 60 Hz and associated harmonics thereof. If a pulsating dc waveform has a repetition rate in the audio-frequency range, for example, its rms value is generally measured with a thermocouple-type current meter, such as that illustrated in Fig. 4-9.

Figure 4-8 An electrodynamometer type of current meter. *(Courtesy of Simpson Electric Co.)*

Consider the power developed by a pulsating dc current. If the current flows through a resistive load, the power in watts is equal to the square of the rms current multiplied by the resistance of the load. When a resistive load is utilized, the voltage drop across the load also has a

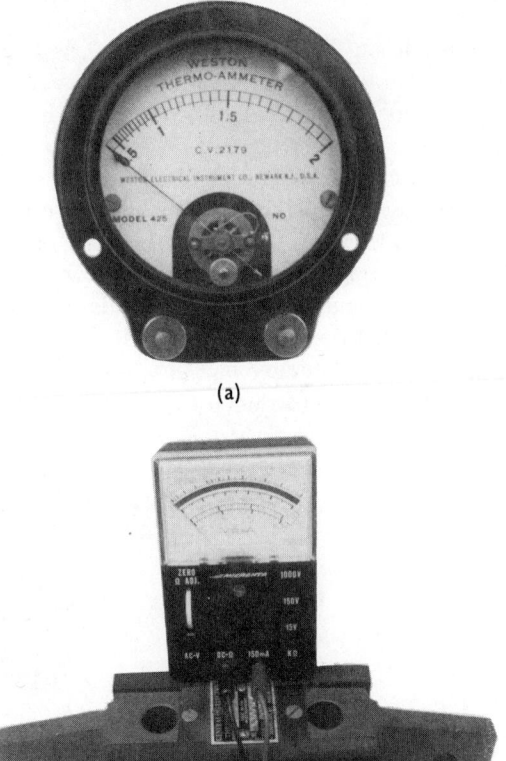

Figure 4-9 Thermocouple meter arrangements: **(a)** current meter with built-in thermocouple; **(b)** external thermocouple.

pulsating dc waveform. In other words, the load voltage is a pulsating dc voltage. In turn, the power in the load is equal to the product of the rms values of the pulsating dc voltage and the pulsating dc current. On the other hand, suppose that the load for the pulsating dc current is a capacitor. Then the voltage drop across the load does not have the same waveshape as the current. Moreover, there is no real power developed in the load; all of the load power is reactive power.

With reference to Fig. 4-8, electrodynamometer meters are used to measure voltage and power, in addition to current. This type of instrument indicates rms voltage, rms current, and real power. It does not indicate reactive power. Accordingly, if a pulsating dc current flows into and out of a capacitor load, an electrodynamometer meter will indicate zero power. Note that this type of instrument is equally useful for indicating pure dc voltage and current values. If a pure dc voltage is multiplied by a pure dc current, the product is in watts of real power. Similarly, if an rms pulsating dc voltage is multiplied by an rms pulsating dc current, the product is in watts of real power. By way of comparison, reactive power is measured in volt-amperes-reactive (VARS). Reactive power is measured with a VAR meter.

4-9 Application of Kirchhoff's Current Law

Kirchhoff's current law states that the sum of the currents entering a branch point is equal to the sum of the currents leaving that branch point. This current law can eliminate the necessity for making a current measurement in some situations. As a simple example, Fig. 4-10 shows a power supply connected to a bleeder resistor and to a load. The current from the power supply is necessarily equal to the bleeder current plus the load current. In turn, after any two of these values have been measured, the third value can be easily calculated.

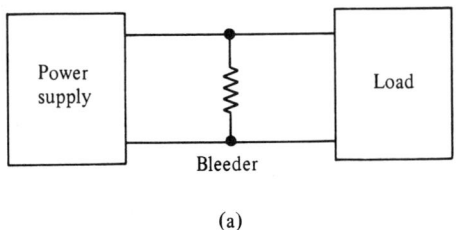

(a)

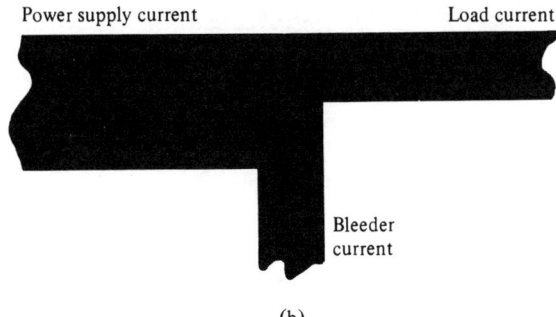

(b)

Figure 4-10 Example of current distribution: **(a)** power supply, bleeder, and load; **(b)** supply current equals bleeder current plus load current.

4-10 Constant-current Principle

A source of electrical power can sometimes be regarded as a constant-current source (or current source). A practical example is shown in Fig. 4-11. A silicon transistor is operated in the CE mode, with a variable collector voltage, and a current meter in the collector circuit. As seen from the family of collector characteristics, the collector current remains virtually constant as the collector voltage varies from 0.5 volt to 5 volts. In turn, the collector of the transistor is regarded as a current source. This principle finds extensive application in semiconductor

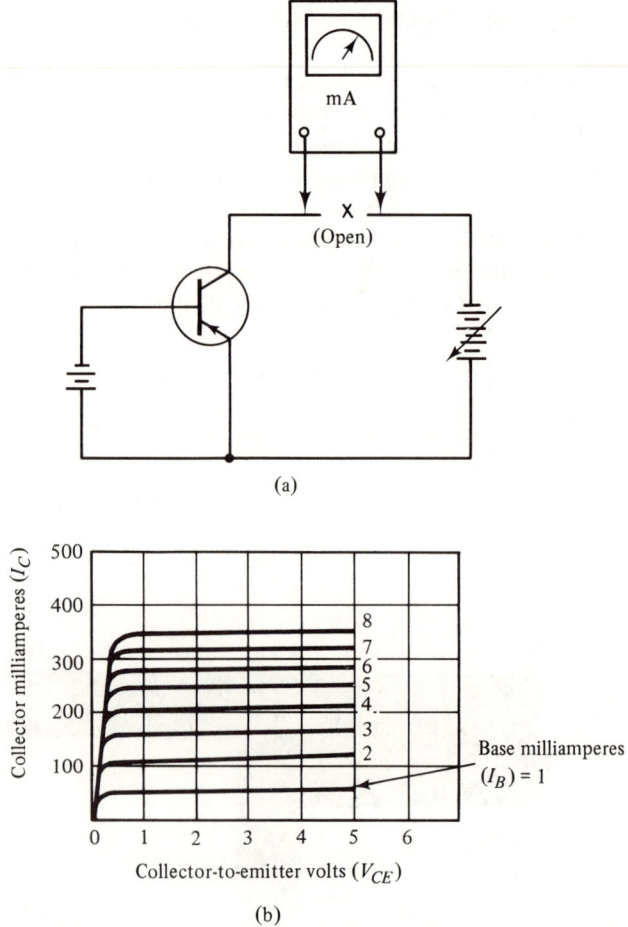

Figure 4-11 Example of a constant-current source: **(a)** silicon transistor in CE configuration; **(b)** collector current is virtually constant as collector voltage varies.

technology. By way of comparison, a zener diode is regarded as a voltage source, or constant-voltage source. In other words, the voltage drop across a zener diode remains virtually constant as the diode current varies over a substantial range. Observe in Fig. 4-11 that the collector output resistance (current source resistance) is not constant, but changes as the collector voltage changes. This change in source resistance is essentially the precise amount required to maintain a constant output current as the collector voltage is varied. Note in passing that a zener diode dissipates constant power if it is connected in series with the collector circuit of a silicon transistor. In other words, the transistor maintains a constant value of collector current as the source voltage varies (with the base-emitter bias held constant). In turn, since the voltage drop across the zener diode remains constant as the source voltage varies, the diode dissipates a constant value of power.

4-11 Checking Current Balance

Some maintenance and troubleshooting procedures require checks of current balance. For example, the output transistors in a high-fidelity amplifier normally draw equal emitter currents. Emitter currents can be individually measured and compared, or the output stage can be checked for balance with a single test, as depicted in Fig. 4-12. If the output stage has normal dc balance, there will be zero volts from the emitter of Q1 to the emitter of Q2. It is advantageous to operate the

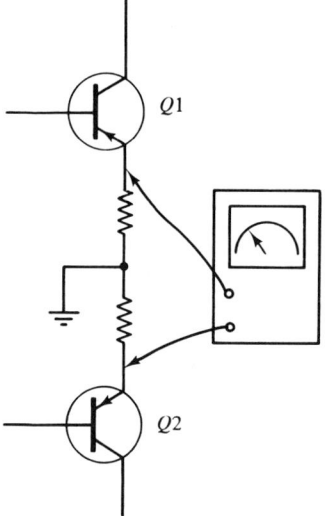

Figure 4-12 Typical check of current balance.

multimeter or TVM on its lowest voltage range, to obtain a clear indication of any small unbalance. However, the initial test should be made on a comparatively high voltage range, to avoid possible overload of the meter in case of stage malfunction.

Another example of a current-balance check is shown in Fig. 4-13. This is an example of a bias-current-balanced inverting amplifier (operational amplifier). In normal operation, the bias currents I_1 and I_2 cancel, so there is no offset bias current. An offset current is equal to the difference between the bias currents at the inverting input and at the noninverting input. In practice, R3 is adjusted as required to obtain virtually zero offset bias current. This check can be easily made in one test by connecting a dc voltmeter between the inverting and the noninverting inputs. A zero indication should be obtained when the instrument is operated on its most sensitive voltage range. Note that bias-current balance is generally obtained when the value of R3 is equal to the parallel value of R1 and R2.

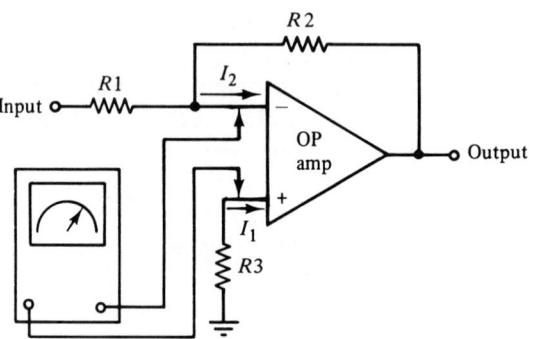

Figure 4-13   Current balance check of an op amp.

4-12 Transistor Electrode Currents

Troubleshooting procedures are facilitated by an understanding of transistor electrode currents under no-signal conditions and signal conditions. With reference to Fig. 4-14, the exemplified transistor has a forward bias current of 2 μA under no-signal conditions, with a collector current of 98 μA and an emitter current of 100 μA. Next, when a 1-μA signal current is applied to the base in the forward direction, the total base current becomes 3 μA. In turn, the collector current becomes 147 μA, and the emitter current becomes 150 μA. Since a change of 1 μA in the base circuit causes a change of 49 μA in the collector circuit, the beta value for this transistor is 49. Observe that the emitter current is equal to the collector current plus the base current.

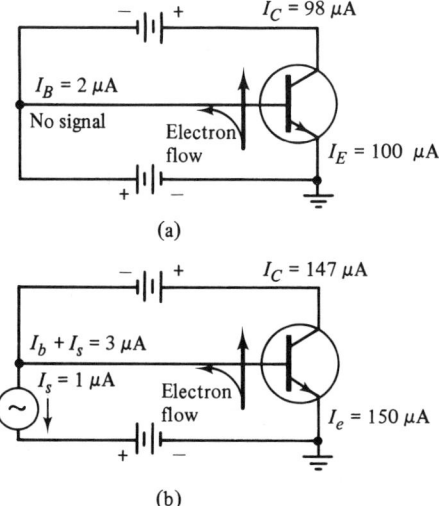

Figure 4-14 Bipolar transistor electrode currents: **(a)** currents under no-signal conditions; **(b)** currents under signal conditions.

4-13 Transistor Current Measurements

Basic transistor current measurements are made by transistor testers. I_{CBO}, commonly called I_{CO}, is the dc collector current that flows when a specified voltage V_{CBO} is applied from collector to base, the emitter being left open (unconnected). The polarity of the applied voltage is such that the collector-base junction is biased in a reverse direction. (Collector is negative with respect to the base for a PNP transistor). This test is made as shown in Fig. 4-15.

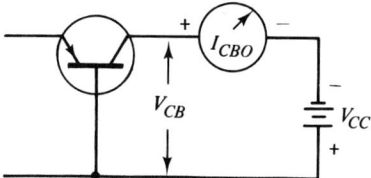

Figure 4-15 Measurement of I_{CBO}.

The dc current that flows when a specified voltage is applied from emitter to base is commonly called I_{EBO} or I_{EO}, the collector being left open (unconnected). The polarity of the applied voltage is such that the emitter-base junction is biased in a reverse direction. (The emitter is negative with respect to the base for a PNP transistor.) This test is made as shown in Fig. 4-16.

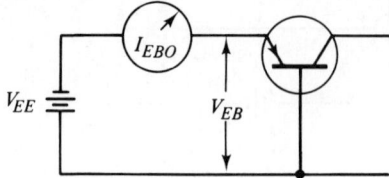

Figure 4-16 Measurement of I_{EBO}.

The dc collector current that flows when a specified voltage is applied from collector to emitter is called I_{CEO}, the base being left open (unconnected). The polarity of the applied voltage is such that the collector-base junction is biased in a reverse direction. (The collector is negative with respect to the emitter for a PNP transistor.) This test is made as shown in Fig. 4-17.

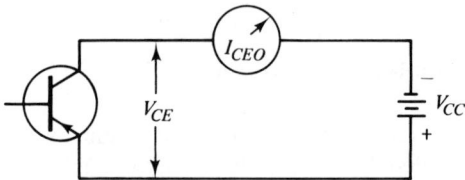

Figure 4-17 Measurement of I_{CEO}.

The collector current that flows when a specified voltage is applied from collector to emitter is called I_{CES}, the base being short-circuited to the emitter. The polarity of the applied voltage is such that the collector-base junction is biased in a reverse direction. (The collector is negative with respect to the emitter for a PNP transistor.) This test is made as shown in Fig. 4-18.

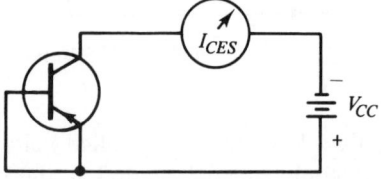

Figure 4-18 Measurement of I_{CES}.

The dc emitter current that flows when a specified voltage is applied from emitter to collector is called I_{ECS}, the base being short-circuited to the collector. The polarity of the applied voltage is such that the emitter-base junction is biased in a reverse direction. (The emitter is negative with respect to the collector for a PNP transistor.) This test is made as shown in Fig. 4-19.

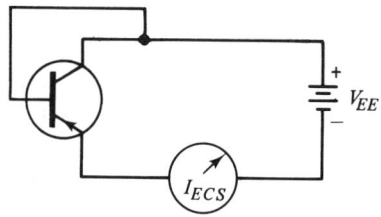

Figure 4-19 Measurement of I_{ECS}.

A bipolar/unipolar field-effect transistor (FET) in-and-out-of-circuit transistor tester is illustrated in Fig. 4-20. It combines a quick, good-bad in-circuit test, and out-of-circuit parameter test for servicing situations in which precise data concerning gain and leakage are needed.

Figure 4-20 A comprehensive in-and-out-of-circuit transistor tester. (Courtesy of Sencore, Inc.)

When this type of instrument is used, there is no need to know the basing diagram, polarity, or even if the device under test is a regular transistor or an FET, unless parameter testing is desired. Generally, the technician needs to know only if a transistor is good or bad, since in most cases a transistor will fail completely or develop leakage that causes it to perform improperly. However, there are times when transistors must be selected for beta current values; FET's must sometimes be selected for transconductance (Gm) values, or for zero-bias drain current (IDSS).

The test provided by this instrument depends upon the capability of a bipolar transistor or FET to develop a signal polarity reversal from input to output when operating in the common-emitter or common-source configuration. A 2-kHz square wave is generated by an integrated circuit (IC1 in Fig. 4-21), and is coupled to the base or gate of the device under test by permutator switches. Permutator switches also connect V_c from the power supply to the collector or drain and ground of the emitter or source terminals. If the device under test is good, the collector or drain signal developed in the test and coupled via C106 to IC2 (Fig. 4-21) will be 180 deg out of phase with the reference signal coupled from IC1 to the base of TR104.

The collector signal is amplified by IC2 and is coupled to a NAND gate comprised of TR103 and TR104. The NAND gate will provide an output when the base and collector signals of the device under test are 180 deg out of phase. This output signal then provides both audio and visual indication of a good transistor or FET. With a collector signal in phase with the base signal, as in the case of a short-circuited transistor, or no collector signal in the case of an open-circuited transistor, there will be no output signal from the NAND gate and thus, no audio or visual indication.

The out-of-circuit gain test provides a reading of ac beta for regular transistors and mutual conductance for FET's. With the permutator test button indicating proper connection depressed and the Gain button depressed, a 2-kHz square wave calibrated signal is coupled to the base (or gate) of the device under test. The resulting signal from the collector (or drain) is amplified by IC3 and IC4 (Fig. 4-21), detected by the meter bridge circuit, and read on the meter directly in beta or microhms. The gain test can be activated only when a permutator button giving a good indication (proper connections) is depressed (Fig. 4-21). The inverted collector (or drain) signal, fed through the NAND gate TR103 and TR104, turns on TR106, which activates the reed relay, L1. L1 then provides dc voltage for the gain test. L1 is self-latching until the depressed permutator button is released or another button is depressed. This action provides protection for the device under test by not applying voltage should the device be improperly connected through the permutator by pushing an incorrect test button. Test-button functions are as follows:

Button 1A: I_{CBO} is the leakage current that flows in a bipolar transistor when a voltage is applied between the collector and base, with the emitter open and the collector-base junction reverse-biased. (The collector is positive with respect to base for an NPN transistor.) In an FET, this leakage is called I_{GDO}, and its effect upon the dc bias of the circuit is even more deleterious than for I_{CBO}. Larger silicon and smaller germanium transistors may safely indicate up to 50 μA of leakage, while some special high-power germanium transistors may indicate up to 3000 μA and still be within rated tolerance.

Button 2A: I_{BEO} in bipolar transistors is the current that flows through the forward-biased base-emitter junction. (The base is positive with respect to the emitter for an NPN transistor.) This button should produce a full-scale indication for bipolar transistors. For FET's, this leakage is called I_{sgo} and indicates full scale for junction FET's and zero for MOS or IG FET's.

Button 3A: I_{ECO} is the leakage current that flows in a bipolar transistor when a voltage is applied between emitter and collector with the base open-circuited. (The emitter is positive with respect to collector for an NPN transistor.) I_{ECO} is a measure of the transistor's ability to block reverse voltage, such as would be encountered in circuits with an inductive load in the collector circuit. In FET's this current would be called I_{DSO} and should indicate full scale because of the normal conduction of the low-resistance drain-source channel.

Button 1B: I_{EBO} is the leakage current that flows in a bipolar transistor when a voltage is applied between emitter and base, with the collector open-circuited and the emitter-base junction reverse-biased. (The emitter is positive with respect to base for an NPN transistor.) I_{EBO} is most important in pulse circuits, where the base is driven deep into reverse bias, and the leakage current could influence the pulse-shaping circuits. In an FET, this leakage is called I_{GSO} and it is a measurement of leakage current that flows from gate to source, with the gate-source junction reverse-biased. Larger silicon and small germanium transistors may safely indicate up to 50 μA of leakage, while some special high-power germanium transistors may indicate up to 3000 μA and still be within tolerance.

Button 2B: I_{CEO} is the leakage current that flows in a bipolar transistor when a voltage is applied between its collector and emitter, with the base open-circuited. (The collector is positive with respect to emitter for an NPN transistor.) Excessive I_{CEO} will cause a transistor to operate unreliably in any circuit; however, the transistors most prone to this type of leakage are high-power types such as those used in audio output circuits and power supply regulators. In FET's this current is

110 DC Current Measurement Methods

called I_{SDO} and should indicate full scale because of the normal conduction of the low-resistance source-drain channel.

Button 3B: I_{CBO} in bipolar transistors is the current that flows through the forward-biased base-collector junction. (The base is positive with respect to collector for an NPN transistor.) This button should produce a full-scale indication for bipolar transistors. For FET's this leakage would be called I_{DGO} and would indicate full scale for junction FET's and zero for MOS and IG FET's.

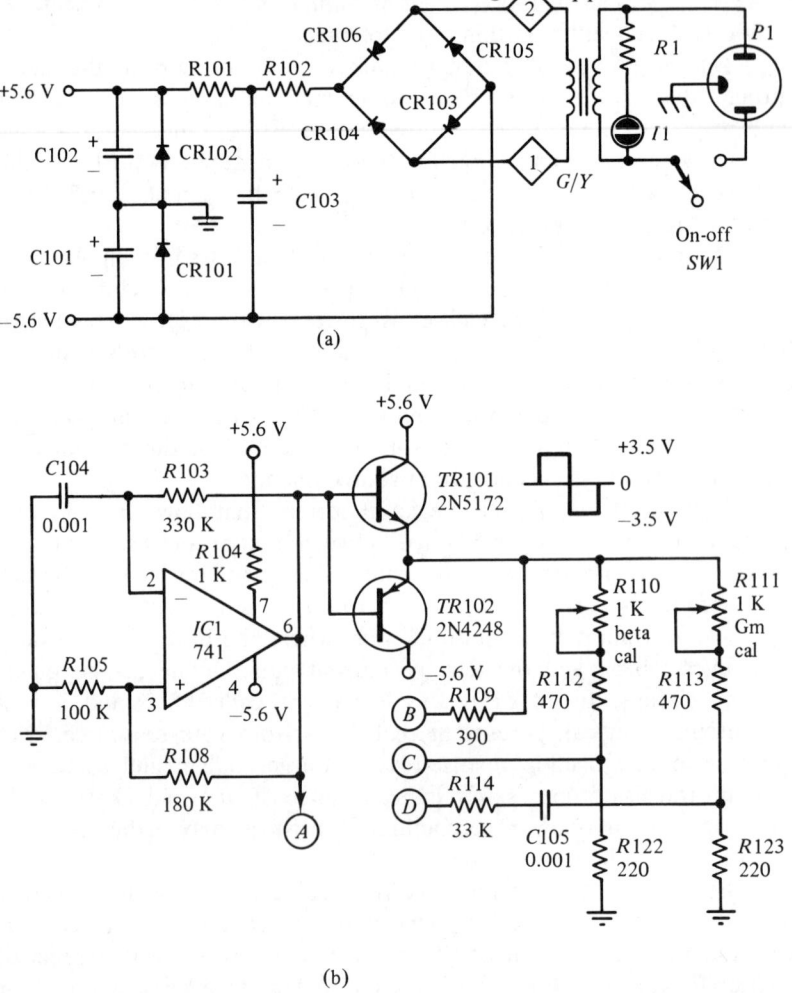

Figure 4-21 Transistor tester circuitry: (a) power supply; (b) 2-kHz oscillator and emitter followers.

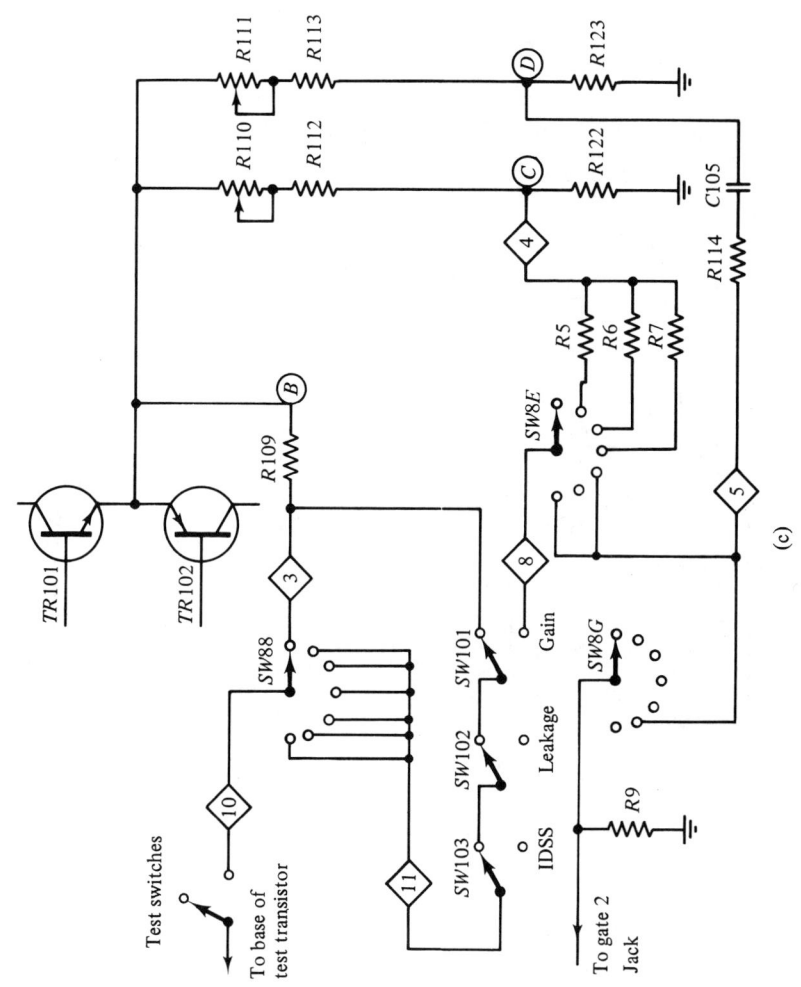

Figure 4-21(c) Test signal distribution.

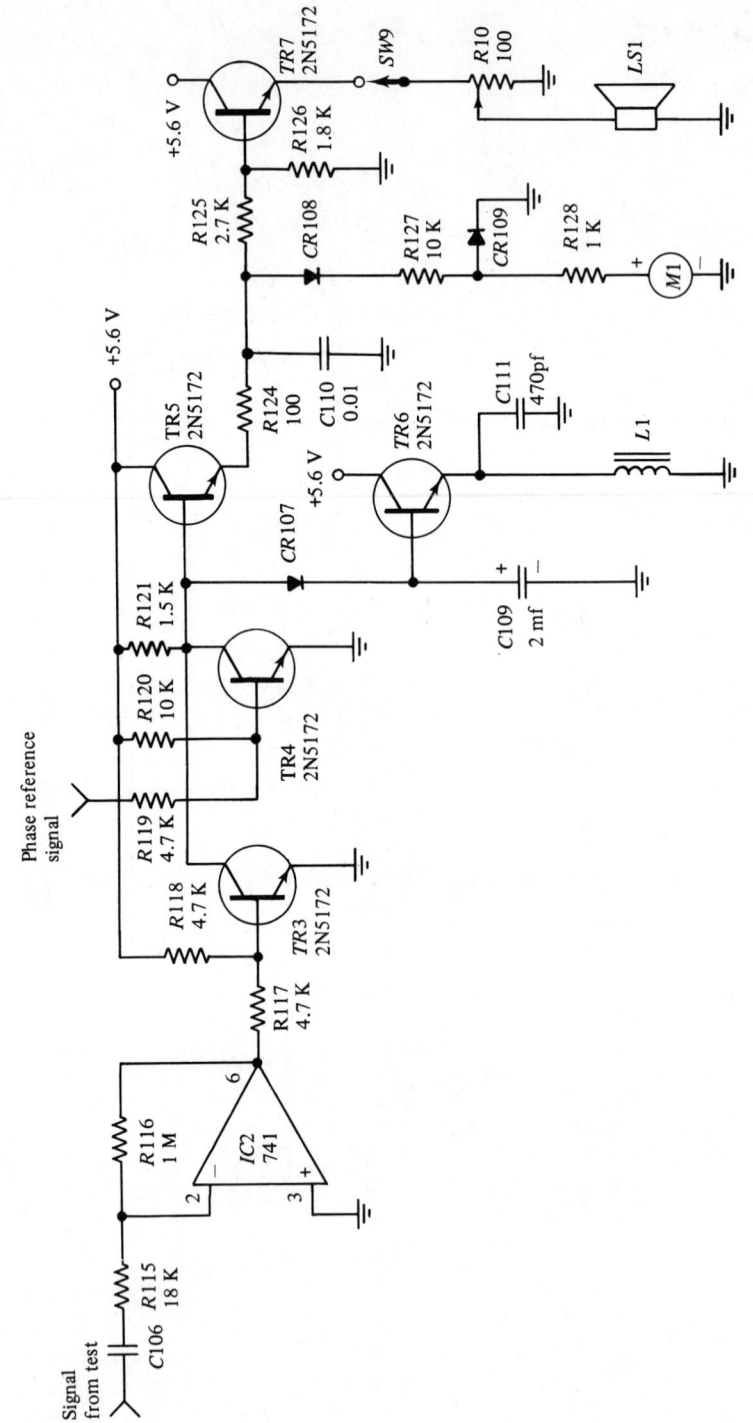

Figure 4-21(d) Phase comparison and indicator circuitry.

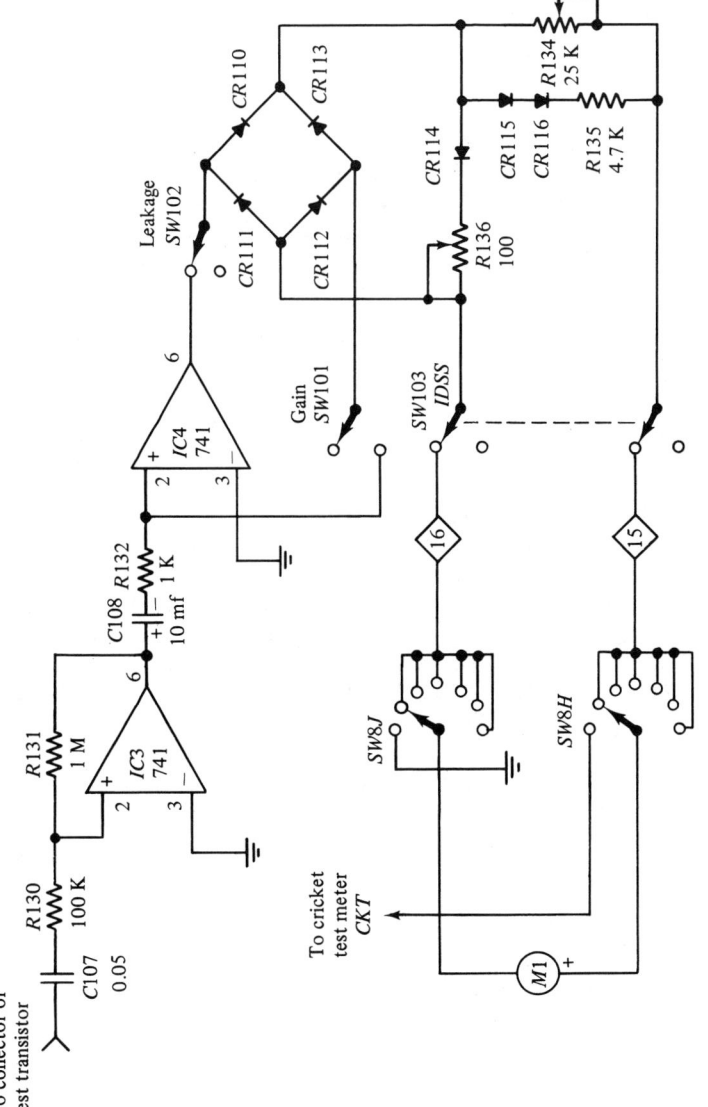

Figure 4-21(e) Gain test and meter circuit.

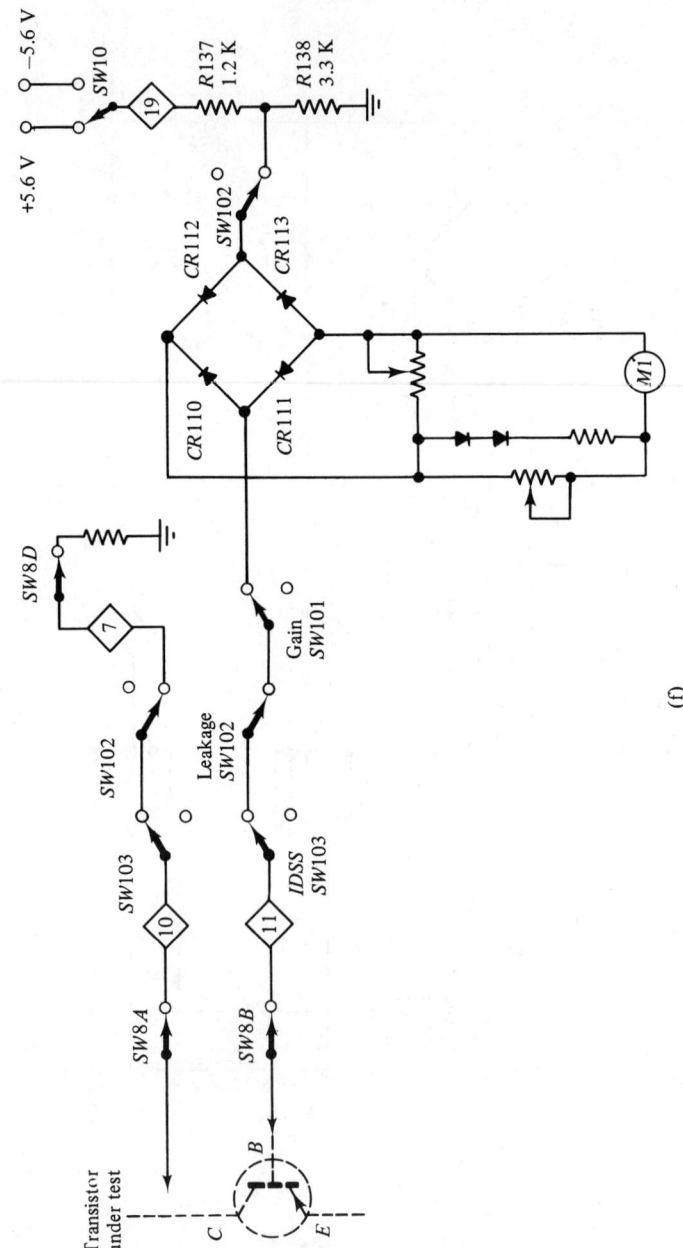

Figure 4-21(f) Leakage test meter circuit.

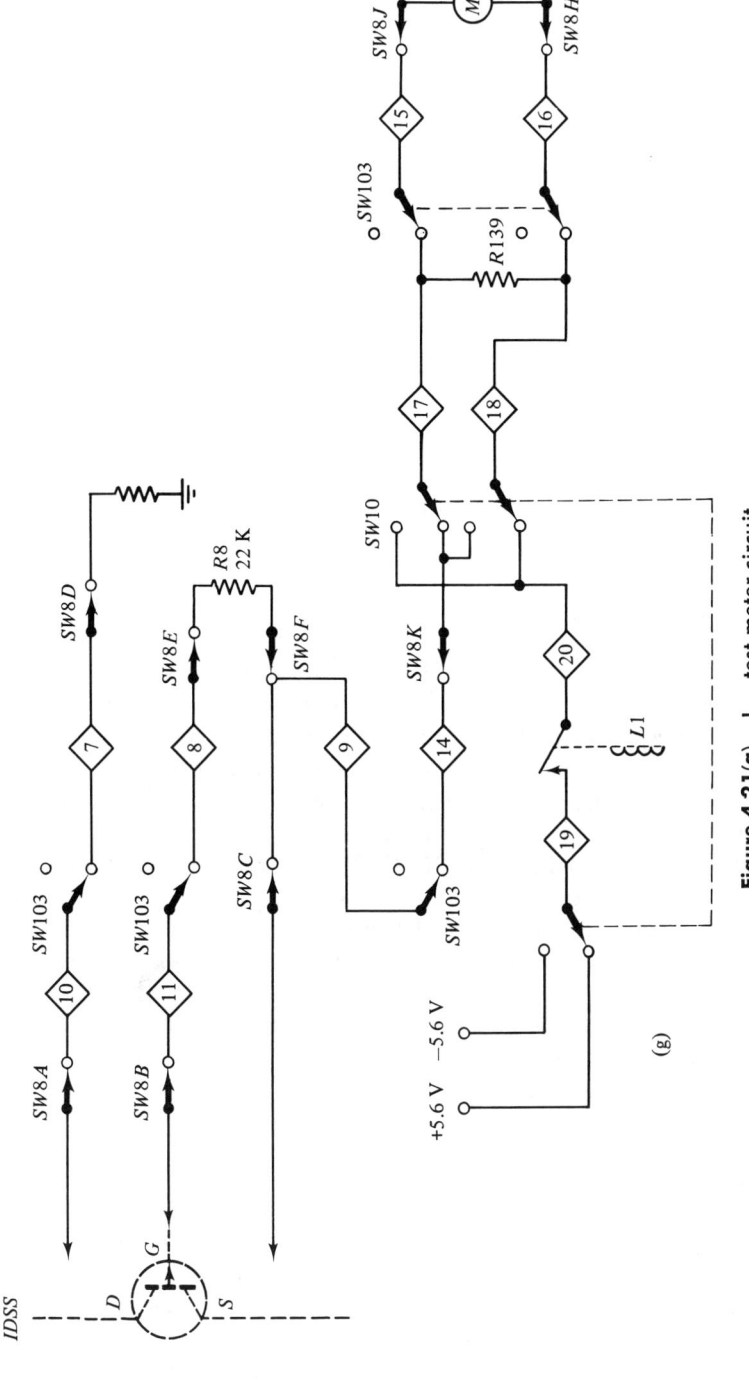

Figure 4-21(g) I_{dss} test meter circuit.

5
AC-Voltage Measurement Procedures

5-1 General Considerations

A multimeter has a lower sensitivity rating on its ac-voltage function than on its dc-voltage function. For example, a 20,000 ohms-per-volt instrument may have a sensitivity of 5000 ohms per volt on its ac-voltage function. Another 20,000 ohms-per-volt multimeter may have a sensitivity of 10,000 ohms per volt on ac operation. In turn, a multimeter generally loads an ac circuit more than it loads a dc circuit. A solid-state electronic multimeter usually provides considerably higher sensitivity on its ac-voltage function. As an illustration, a typical transistor volt-ohm-milliammeter (TVOM) has an input resistance of 15 megohms on its dc-voltage function, and 10 megohms input resistance on its ac-voltage function. This 10 megohms is shunted by approximately 100 pF of input capacitance. The input capacitance of an ac voltmeter is a significant factor in circuit loading at higher operating frequencies. A digital multimeter (DVM) has a sensitivity rating on its ac-voltage function similar to that of a TVOM. A vacuum-tube voltmeter (VTVM) has a typical ac-voltage sensitivity rating of 1 megohm shunted by 35 pF.

Another important consideration is the useful frequency range of an ac voltmeter. For example, Fig. 5-1 shows a typical frequency-response curve for a 20,000 ohms-per-volt multimeter. Its indication accuracy becomes impaired at 20 kHz, and the instrument indication error is excessive at higher frequencies. On the other hand, a solid-state electronic multimeter is typically rated for accurate ac-voltage indication to 140 kHz. A typical digital multimeter is rated for accurate ac-voltage indication to 1 MHz. Similarly, service-type VTVM's are generally rated for accurate ac-voltage indication to 1 MHz. The lowest fre-

5-1 General Considerations 117

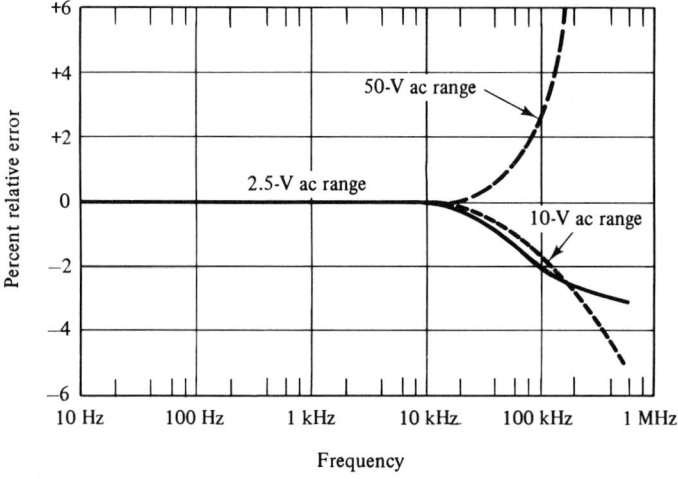

Figure 5-1 Typical frequency response of a VOM on its ac-voltage function.

quency that can be processed satisfactorily by an ac voltmeter is determined by the damping of the meter movement. In other words, the pointer begins to vibrate excessively at frequencies below 25 Hz, in most cases. This vibration makes it difficult to read the scale indication accurately.

A multimeter provides an accurate ac-voltage indication only when there is no dc component present. For example, an ac voltmeter will indicate an incorrect value of amplifier output voltage in Fig. 5-2 because the sine waveform is accompanied by a dc component. Therefore, a multimeter is operated on its output function to measure the sine-wave voltage. The output function is the same as the ac-voltage function, except that a series blocking capacitor is included in the instrument circuit.

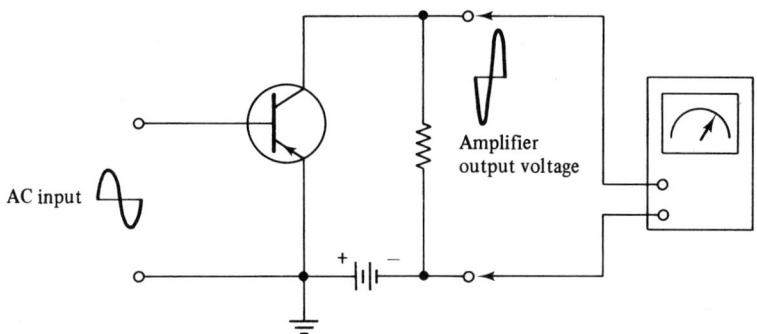

Figure 5-2 The ac output voltage has a dc component.

In turn, the low-frequency response of the instrument becomes somewhat limited on its output function. Also, the reactance of the blocking capacitor reduces the accuracy of ac-voltage indication to some extent, as exemplified in Fig. 5-3. Observe that if the meter is operated on its 2.5-volt ac range, frequencies below 500 Hz will not be processed accurately.

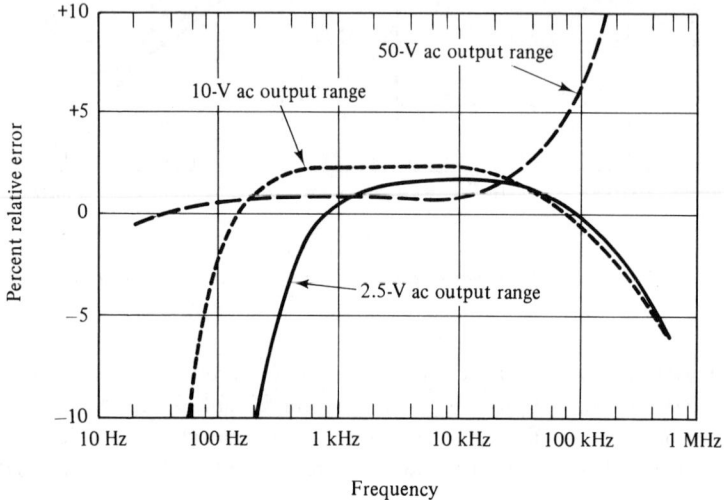

Figure 5-3 Typical frequency response of a VOM on its output function.

5-2 Measurement of AC Voltages at High Frequencies

Useful checks of ac voltages at high frequencies can be made with a multimeter, provided that the test leads are replaced with an RF probe, as shown in Fig. 5-4. The multimeter is operated on its dc-voltage function. This type of probe permits checks of audio-frequency, radio-frequency, and very-high-frequency ac voltages up to 200 MHz from low-impedance sources. The probe has marginal utility in high-impedance circuits because of substantial circuit loading. Its chief usefulness is to indicate the presence or absence of signal voltages; calibration of the probe response is not very practical. Since a germanium diode may be damaged by application of more than 20 peak volts, the probe should be applied only in low-level circuits.

On the other hand, an RF probe can be used with a TVM, VTVM, or DVM to make accurate ac-voltage measurements at frequencies up to 200 MHz. This difference in utility is based on the

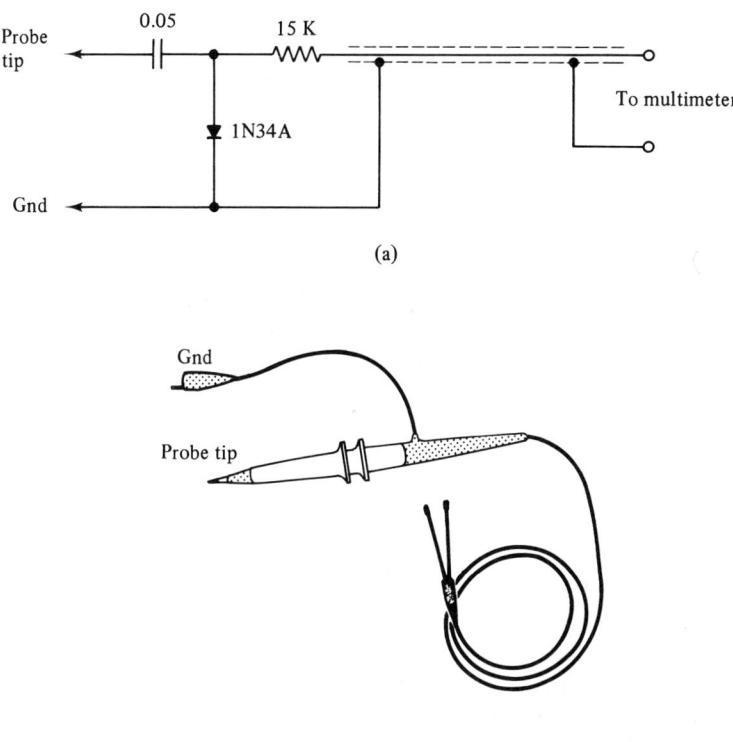

Figure 5-4 RF probe for a multimeter: **(a)** circuit; **(b)** appearance.

comparatively small current demand of the foregoing instruments. With reference to Fig. 5-4, an RF probe used with a TVM generally utilizes a 100-kΩ isolating resistor, instead of a 15-kΩ isolating resistor. If an RF probe is employed with a VTVM, a 1-MΩ isolating resistor is usually chosen. In turn, peak-voltage indication is obtained when the voltmeter is operated on its dc-voltage function. Application of more than 20 peak volts input signal should be avoided, or the diode is likely to be damaged.

5-3 Decibel Measurements with AC Voltmeters

Since the response of the eye and ear to signal levels is logarithmic, audio and video gains and losses are measured to advantage in decibel units. Although most ac voltmeters used in service shops have decibel

120 AC-Voltage Measurement Procedures

scales, it is important to note that the decibel is a power ratio, and not a voltage ratio. Voltage ratios correspond to power ratios, if the voltages are measured across equal load resistances. However, if the load resistances are unequal, voltage values do not correspond directly to power values. Decibel units are additive and subtractive. For example, if there is a 20-dB loss in a volume control followed by a gain of 30 dB in an amplifier, the over-all gain is 10 dB. In terms of loudness units at moderate sound levels, a gain of 10 dB is generally judged to be approximately twice as loud as the original level, and a loss of 10 dB is generally judged to be about half as loud as the original level.

All dB values are referenced to some chosen power value, to which the value 0 dB is assigned. Some dB scales are referenced to 6 mW in 500 ohms, and other dB scales are referenced to 1 mW in 600 ohms. (See Fig. 5-5.) Still other reference levels will be encountered on occasion. The dB scales on an ac voltmeter read directly only when the measurements are made across a value of load resistance for which the scales have been referenced. However, calculations can always be made to correct dB values both for nonstandard load resistances and for some other 0-dB reference level. A simple example of a dB check across 600-ohm loads with ac voltmeters designed to indicate dB values across 600-ohm loads is shown in Fig. 5-6. Note that a good sine waveform must be utilized, or the measured values will be subject to waveform error. In this example, the audio oscillator supplies a signal level of 2 dB to the 600-ohm volume control, and the amplifier increases the signal level 12 dB. In turn, the over-all gain is the difference between these two values, or 10 dB.

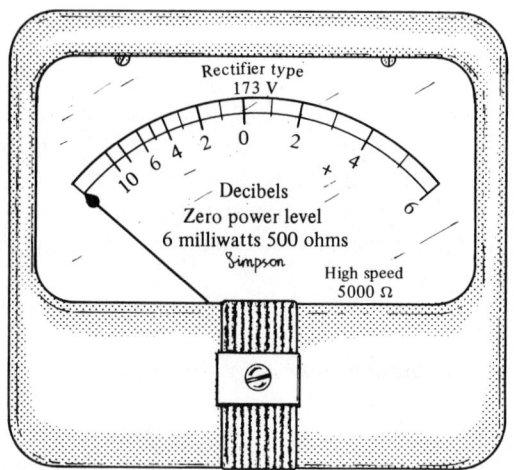

Figure 5-5 A dB meter scale referenced to 6 mW in 500 ohms. (Courtesy of Simpson Electric Co.)

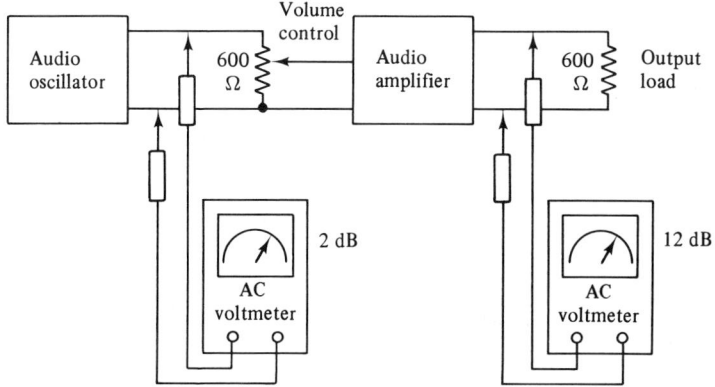

Figure 5-6 Simple example of dB measurements across 600-ohm resistive loads.

Note that many ac meters have only one dB scale, which is usually applicable to the lowest ac-voltage range of the instrument. In turn, when the meter is operated on a higher ac-voltage range, it is necessary to add a suitable number of dB to the scale reading. This scale factor will be printed on the meter scale plate, or in the instruction manual for the instrument. As a practical note, although two meters are shown in Fig. 5-6, only one meter is needed; in practice, the test leads are simply moved from the input terminals to the output terminals of the unit or system under test. The basic relation between dB values, power ratios, and voltage ratios can be seen in Table 5-1. Note that dB measurements are valid across resistive loads only; because an impedance load develops both real power and reactive power, dB indications across impedances are meaningless. In other words, the power ratio corresponding to a dB value is understood to be a ratio of two real-power values. No decibel table can be completely comprehensive, and some dB values must necessarily be calculated. Such calculations in the range from 0.01 to 990 ratios of power can be made with minimum effort by referring to Table 5-2.

Two basic situations are encountered in the measurement of dB values across resistive loads that have different values from that for which the dB meter scale has been calibrated. In the first situation, the load resistances are different from the reference value for the dB meter, but these load resistances are equal in value, as exemplified in Fig. 5-7(a). In this situation, the technician can determine the dB gain or loss of the system by merely taking the difference between the two readings. For example, the reference value for the dB meter might be 600 ohms. If the input and output resistances of an amplifier should

TABLE 5-1
Power Ratios, Voltage Ratios, and dB Values
(Voltage ratios based on equal load resistances)

Power Ratio	Voltage Ratio	db −← +→	Voltage Ratio	Power Ratio
1.000	1.0000	0	1.000	1.000
0.9772	0.9886	0.1	1.012	1.023
0.9550	0.9772	0.2	1.023	1.047
0.9333	0.9661	0.3	1.035	1.072
0.9120	0.9550	0.4	1.047	1.096
0.8913	0.9441	0.5	1.059	1.122
0.8710	0.9333	0.6	1.072	1.148
0.8511	0.9226	0.7	1.084	1.175
0.8318	0.9120	0.8	1.096	1.202
0.8128	0.9016	0.9	1.109	1.230
0.7943	0.8913	1.0	1.122	1.259
0.6310	0.7943	2.0	1.259	1.585
0.5012	0.7079	3.0	1.413	1.995
0.3981	0.6310	4.0	1.585	2.512
0.3162	0.5623	5.0	1.778	3.162
0.2512	0.5012	6.0	1.995	3.981
0.1995	0.4467	7.0	2.239	5.012
0.1585	0.3981	8.0	2.512	6.310
0.1259	0.3548	9.0	2.818	7.943
0.10000	0.3162	10.0	3.162	10.00
0.07943	0.2818	11.0	3.548	12.59
0.06310	0.2512	12.0	3.981	15.85
0.05012	0.2293	13.0	4.467	19.95
0.03981	0.1995	14.0	5.012	25.12
0.03162	0.1778	15.0	5.623	31.62
0.02512	0.1585	16.0	6.310	39.81
0.01995	0.1413	17.0	7.079	50.12
0.01585	0.1259	18.0	7.943	63.10
0.01259	0.1122	19.0	8.913	79.43
0.01000	0.1000	20.0	10.000	100.00
10^{-3}	3.162×10^{-2}	30.0	3.162×10	10^3
10^{-4}	10^{-2}	40.0	10^2	10^4
10^{-5}	3.162×10^{-3}	50.0	3.162×10^2	10^5
10^{-6}	10^{-3}	60.0	10^3	10^6
10^{-7}	3.162×10^{-4}	70.0	3.162×10^3	10^7
10^{-8}	10^{-4}	80.0	10^4	10^8
10^{-9}	3.162×10^{-5}	90.0	3.162×10^4	10^9
10^{-10}	10^{-5}	100.0	10^5	10^{10}

TABLE 5-2

DB Calculations for Power Ratios from 0.01 to 990

Power ratios expressed in + dB

Ratio Power	0.0	0.1	0.2	0.3	0.4	0.5	0.6	0.7	0.8	0.9
1	0.000	0.414	0.792	1.139	1.461	1.761	2.041	2.304	2.553	2.788
2	3.010	3.222	3.424	3.617	3.802	3.979	4.150	4.314	4.472	4.624
3	4.771	4.914	5.051	5.185	5.315	5.441	5.563	5.682	5.798	5.911
4	6.021	6.128	6.232	6.335	6.435	6.532	6.628	6.721	6.812	6.902
5	6.990	7.076	7.160	7.243	7.324	7.404	7.482	7.559	7.634	7.709
6	7.782	7.853	7.924	7.993	8.062	8.129	8.195	8.261	8.325	8.388
7	8.451	8.513	8.573	8.633	8.692	8.751	8.808	8.865	8.921	8.976
8	9.031	9.085	9.138	9.191	9.243	9.294	9.345	9.395	9.445	9.494
9	9.542	9.590	9.638	9.685	9.731	9.777	9.823	9.868	9.912	9.956

For power ratios between 0.01 and 0.099, use above table to find dB for 100 times the ratio and subtract 20 dB.

For power ratios between 0.1 and 0.99, use above table to find dB for 10 times the ratio and subtract 10 dB.

For power ratios between 1 and 9.9, use above table directly.

For power ratios between 10 and 99, use above table to find dB for 1/10th of the ratio and add 10 dB.

For power ratios between 100 and 990, use above table to find dB for 1/100th of the ratio and add 20 dB.

both be 75 ohms, the technician can still use the dB meter that has a 600-ohm reference value. With reference to Fig. 5-7(a), neither the 3-dB nor the 13-dB reading is correct in itself. On the other hand, the difference between these two readings, or 10 dB, is correct inasmuch as these readings were made across equal load resistances. Thus, the gain of the system in this example is 10 dB.

With reference to Fig. 5-7(b), the dB measurements are made across unequal load resistances. Neither of these resistance values corresponds to the 600-ohm reference value of the dB meter. In turn, neither the 15-dB reading nor the 3-dB reading is correct in itself, nor is the difference between these two readings correct. It may be noted that this system provides gain, although the dB meter readings seem to indicate that there is a loss. In order to evaluate these dB readings, they must be converted to suitable resistive reference values. In other words, a corrective factor must be applied to determine the true dB gain or loss. The chart shown in Chart 5-1 is suitable for this purpose. For example, if the first measurement is made across 100 kΩ, and the second measurement is made across 1000 ohms, this resistance ratio is 100. In

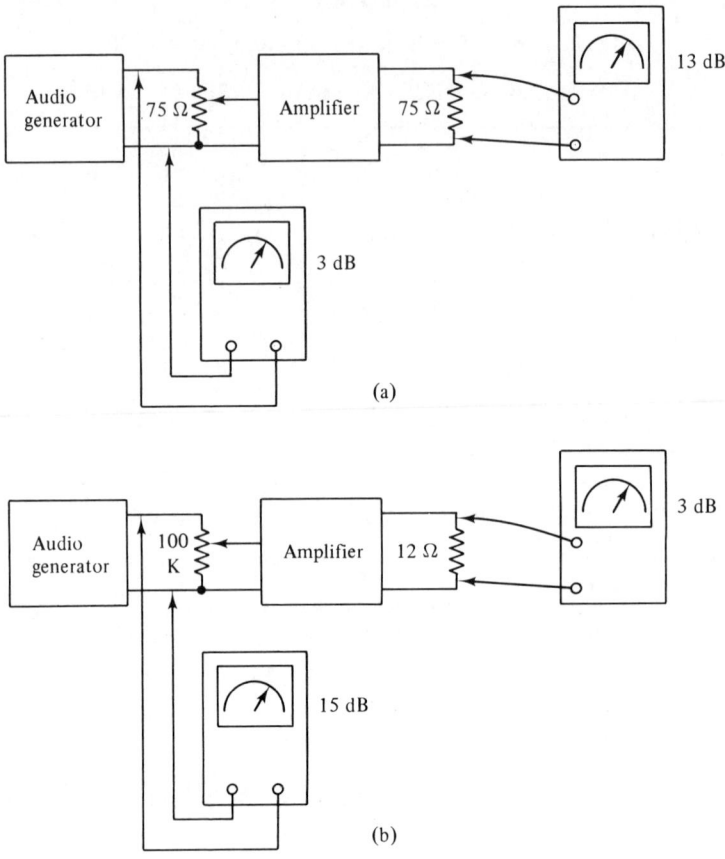

Figure 5-7 Examples of dB measurements: **(a)** across equal resistances; **(b)** across unequal resistances.

turn, the chart shows that 20 dB must be added or subtracted to the apparent gain or loss that is indicated by the measured values.

Consider the distinction between relative and absolute dB levels. Any dB scale has a certain power ratio assigned as zero dB. In turn, all dB values below this reference power ratio are negative dB values, and all above zero dB are positive values. Red and black numerals are often utilized on the scale to call attention to this difference in sign. Positive and negative signs must be taken into account when one is comparing dB levels on either side of zero. Thus, the difference between −5 dB and +3 dB is 8 dB. When dB values are measured across a resistance that has the reference value for the meter, these values can be evaluated directly as power values by reference to a suitable table (or by

Chart 5-1 Decibel correction chart.

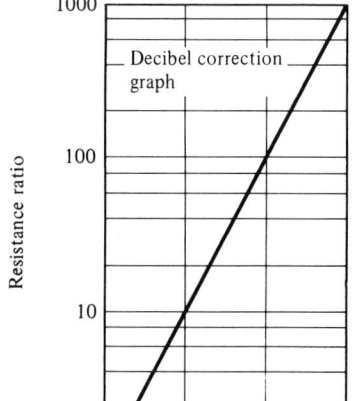

calculation). On the other hand, when dB values are measured across a resistance that has some value other than the meter's reference value, these values are only indirectly related to power values. As noted previously, if the technician takes a pair of dB readings across 1000-ohm loads with a 600-ohm dB meter, the difference between these readings will be the true difference in dB between the two levels, although neither reading is correct in itself. These readings are called relative dB values.

Consider the dBm unit, and its measurement. This is an abbreviation for decibels above (or below) a power level of 1 mW. In turn, a dBm value corresponds to a quantity of power expressed in terms of its ratio to 1 mW. It is standard practice to reference dBm values to 1 mW in 600 ohms. The graph in Fig. 5-8 can be used conveniently to relate dBm values to rms ac values. Although dBm measurements are commonly made at 1000 Hz, as are other basic audio measurements, the test frequency is arbitrary. Zero dBm indicates a power level of 1 mW in 600 ohms; 10 dBm indicates a power level of 10 mW, and so on. If an ac voltmeter with a dBm scale is used to measure dBm values across a resistive load other than 600 ohms, the correction factors given in Fig. 5-9 must be algebraically added to a dBm value found from the chart in Fig. 5-8. As noted in Fig. 5-8, zero dBm corresponds to 0.775 volt rms across 600 ohms. Similarly, 20 dBm corresponds to 7.75 volts rms.

126 AC-Voltage Measurement Procedures

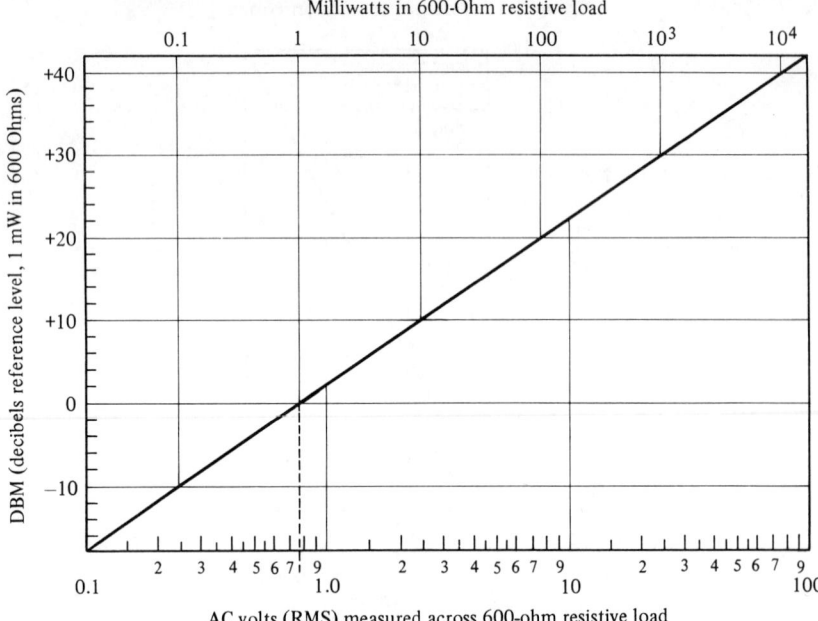

Figure 5-8 Graph for conversion of rms voltages to dBm values.

Resistive load	DBM*
600	0
500	+ 0.8
300	+ 3.0
250	+ 3.8
150	+ 6.0
50	+10.8
15	+16.0
8	+18.8
3.2	+22.7

*DBM is the increment to be added algebraically to the DBM value read from the graph.

Figure 5-9 List of dBm correction factors for Fig. 5-8.

5-4 Measurement of Small AC Voltages

Small ac voltages are measured with an audio voltmeter, such as that illustrated in Fig. 5-10. This instrument has 10 ranges from 0.01 volt to 300 volts rms. These instruments are basically ac voltmeters of the

5-4 Measurement of Small AC Voltages 127

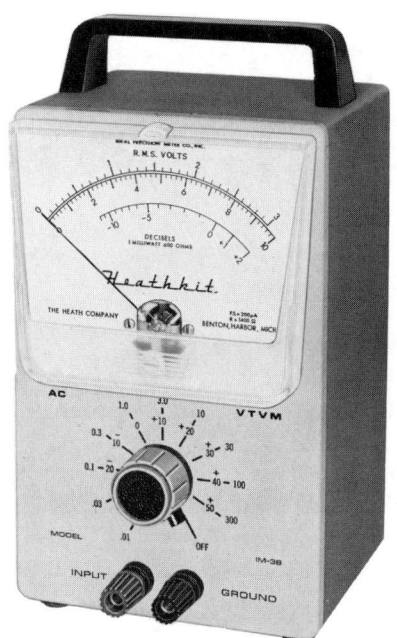

Figure 5-10 Audio voltmeter.
(Courtesy of Heath Co.)

rectifier type with high sensitivity provided by a preamplifier. Thus, the indication is subject to waveform error, and measurements should be made with a pure sine waveform. All audio voltmeters have a flat frequency response through the audio-frequency range from 20 Hz to 20 kHz, and many of these instruments provide extended high-frequency response to 0.5 MHz, or more. A typical instrument has an input resistance of 10 megohms shunted by 20 pF of capacitance. If a shielded input cable is utilized, the effective input capacitance is increased accordingly.

Most audio voltmeters use a dB scale reference of 1 mW in 600 ohms, and indicate dBm directly on the first range. A scale factor must be employed on the second range to measure dBm, and a larger scale factor must be utilized on the third range. Audio voltmeters are essential for troubleshooting low-level circuitry, such as preamplifiers driven by tape heads, magnetic phono cartridges, and so on. For example, the output from a magnetic cartridge is normally from 5 to 10 mV. An audio voltmeter responds to noise voltages; this response is most noticeable on the low ranges. Noise voltages cause the pointer to bounce up and down erratically. The reason for this random response is seen from the irregular peaks in a noise waveform, as exemplified in Fig. 5-11. To measure the noise level of a preamplifier (or other amplifier), the technician must estimate the average scale indication.

128 AC-Voltage Measurement Procedures

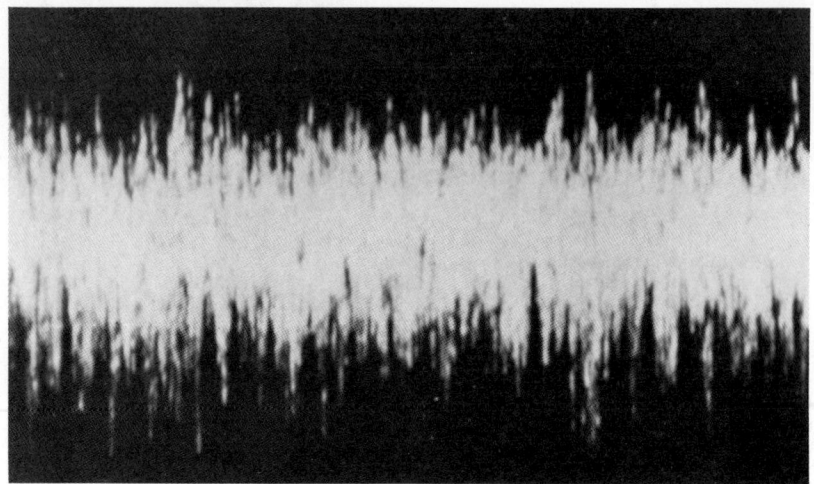

Figure 5-11   A typical noise waveform.

5-5 Volume Unit Measurements

A volume unit (VU) is a power ratio that indicates the level of a complex wave in dB above a reference volume. Thus, a VU measurement denotes a complex-wave power reading on a VU meter. Volume units are not used to indicate the power level of a sine-wave signal; dBm units are employed for this purpose. If the power level of a sine-wave signal is measured with a VU meter, a reading will be obtained in dBm units. A VU measurement implies a voice or music waveform, which is characterized by high peaks. As a rule of thumb, it is commonly assumed that the average peak value in a program waveform is 10 dB above a sine-wave peak level. In practice, an audio system operating at a level of +12 VU will be tested for percentage of distortion at a sine-wave level of +22 dBm.

In VU measurements, the reference volume is specified as a strength of program wave that produces a reading of 0 VU on a meter as described above. This type of meter has specified damping, and is calibrated to indicate 0 VU on a 1-kHz sine wave with a power of 1 mW in 600 ohms. Thus, reference volume is not a precise concept, and cannot be defined in fundamental terms. Volume unit measurements are of basic importance in the monitoring of audio systems in radio broadcast operations.

5-6 Tuned AC Voltmeters

Tuned ac voltmeters respond to a selected band of frequencies, or reject a chosen frequency. As an illustration, an intermodulation analyzer such as that illustrated in Fig. 5-12 is an ac voltmeter with associated high-pass and low-pass filters, as depicted in Fig. 5-13. The signal channel passes higher audio frequencies from approximately 2 kHz to 20 kHz. After demodulation, the high-pass filter output is passed through a filter that has an upper frequency limit of about 700 Hz. A two-tone test signal is utilized in operation of an IM analyzer. The meter employs instrument-type rectifiers, and the scale is calibrated in percentage of intermodulation distortion. Application of this type of tuned ac voltmeter is explained in the following chapter.

Figure 5-12 An intermodulation distortion analyzer. (*Courtesy of Heath Co.*)

Consider next the basic characteristics of a harmonic distortion meter, such as that illustrated in Fig. 5-14. This is an example of a tuned ac voltmeter that rejects a chosen frequency. A basic block diagram for the instrument is shown in Fig. 5-15. A sine-wave test frequency, usually 1 kHz, is utilized in operation of a harmonic distortion meter. When harmonic distortion occurs, second and/or third harmonics of the test frequency are produced. Sometimes even higher harmonics will be produced. The HD meter operates by filtering out the test frequency (fundamental frequency), and passing the harmonic frequencies into the meter circuit. The meter utilizes instrument-type rectifiers, and the scale is calibrated in percentage of harmonic distortion. For example, if an amplifier under test happens to develop a second har-

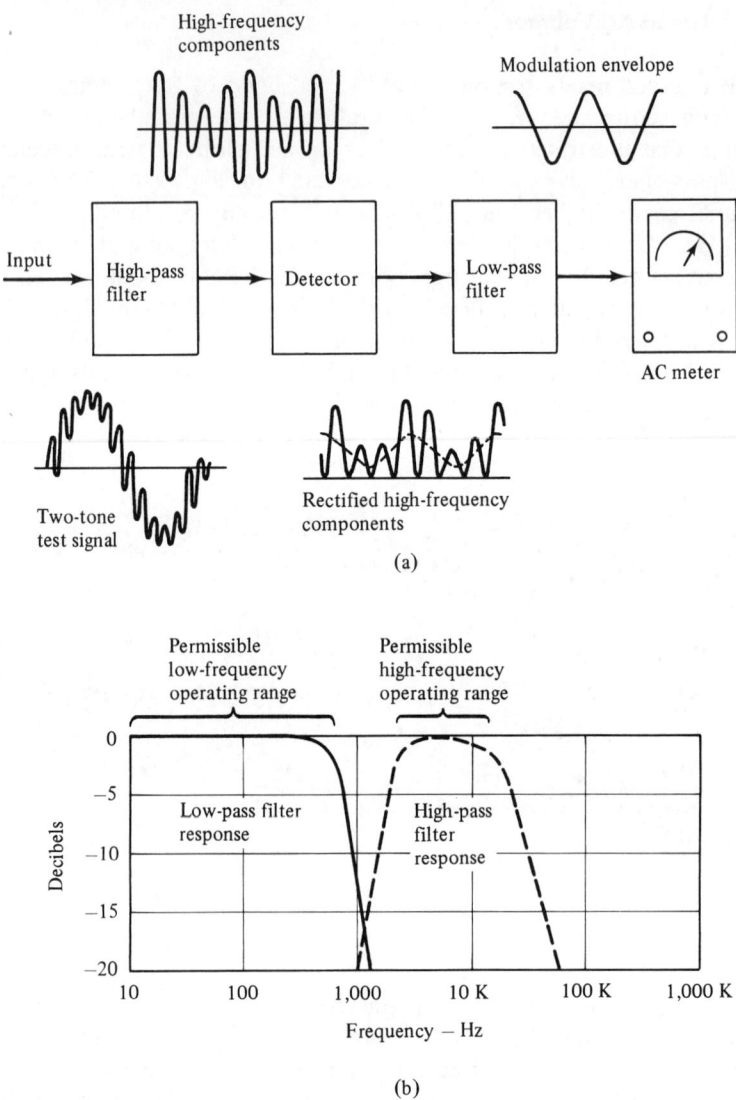

Figure 5-13 Intermodulation distortion analyzer features: **(a)** basic arrangement; **(b)** filter frequency response curves.

monic that has an amplitude equal to 2 percent of the test-frequency amplitude, the HD meter will indicate 2 percent on its scale.

Another example of a tuned ac voltmeter is seen in Fig. 5-16. This is a field-strength meter that indicates the amplitudes of VHF and

5-6 Tuned AC Voltmeters   131

Figure 5-14 Harmonic distortion meter. (*Courtesy of Heath Co.*)

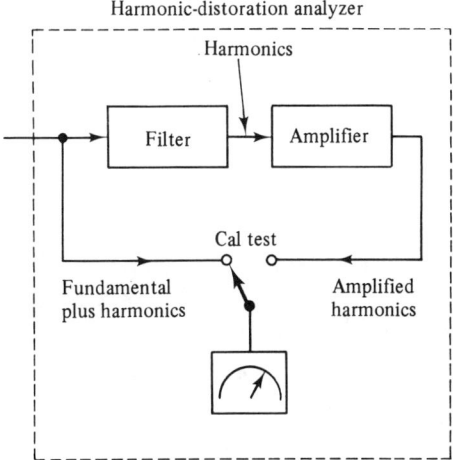

Figure 5-15 Basic block diagram for a harmonic distortion meter.

UHF signals in microvolt units. It is basically a front end (tuner) such as that utilized in a television receiver, with a rectifier-type ac voltmeter. The tuner passes one channel signal at a time, as depicted in Fig. 5-17. Thus, if the FS meter is tuned to a color-TV broadcast, the meter indicates the combined amplitudes of the picture, color, and sound signals. A built-in speaker is often provided in the instrument, to alert the operator to possible interference in the TV signal, which would cause an abnormally high scale indication.

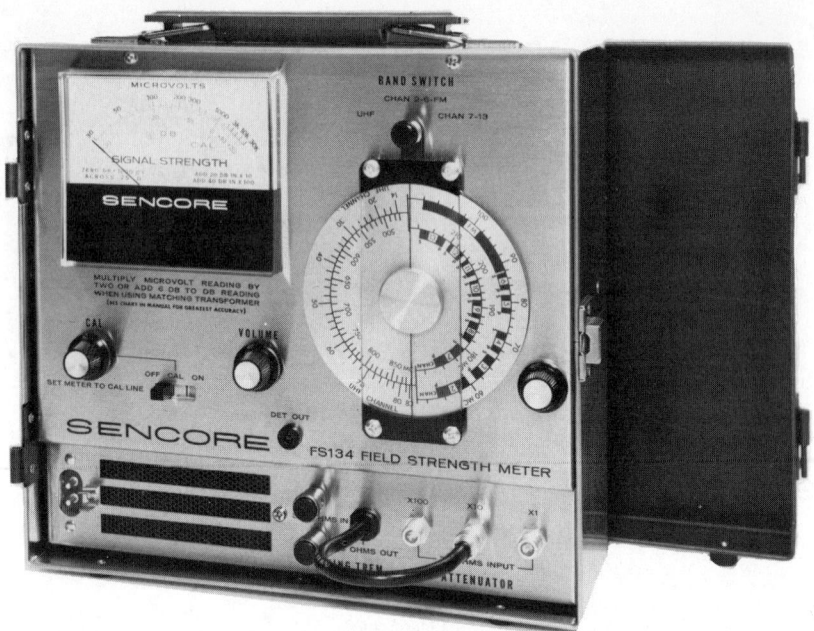

Figure 5-16 A television field-strength meter. (*Courtesy of Sencore, Inc.*)

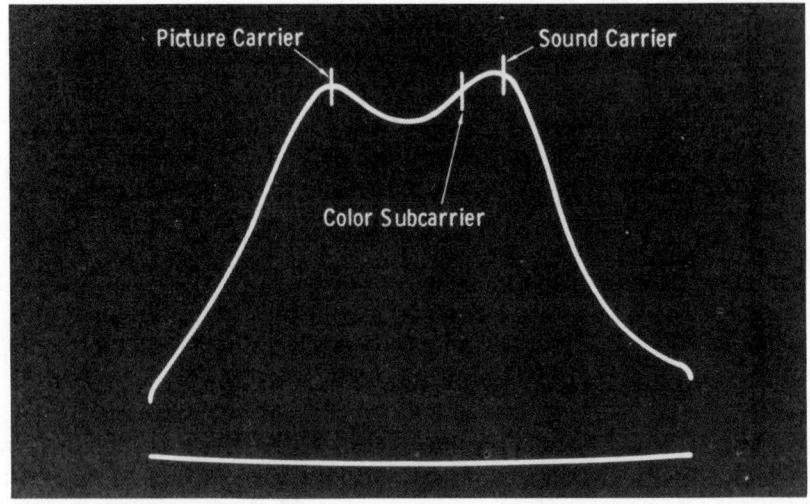

Figure 5-17 Frequency response of a field-strength meter.

5-7 Response of Basic AC Voltmeters to Complex Waveforms

All rectifier-type ac voltmeters are subject to indication error if the applied waveform is nonsinusoidal. This indication error is not arbitrary, but follows established laws of electricity. If the nonsinusoidal waveform has a known waveshape, the voltmeter response will be meaningful. As an illustration, consider the basic complex waveforms depicted in Fig. 5-18. In the case of a square wave without a dc component, a peak-responding ac voltmeter will indicate 0.707 of the peak value of the square wave. Or, if the voltmeter is of the half-wave average responding type, the scale indication will be 1.11 times the peak value of the square wave. Again, if the voltmeter is of the full-wave average responding type, the same scale indication will be obtained. Scale indication for sawtooth waveforms without a dc component, half-rectified sine waves, and full-rectified sine waves are also tabulated in the chart.

Note: "A" is the peak amplitude of the waveform

Waveform	Meter response	Scale reading
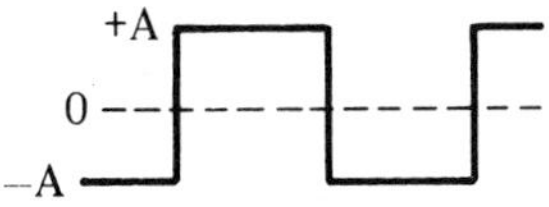 Square wave	Peak	0.707 A
	$\frac{1}{2}$ wave average	1.11 A
	Full-wave average	1.11 A
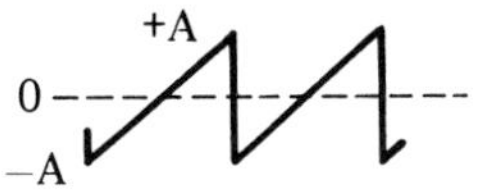 Sawtooth wave	Peak	0.707 A
	$\frac{1}{2}$ wave average	0.555 A
	Full-wave average	0.555 A
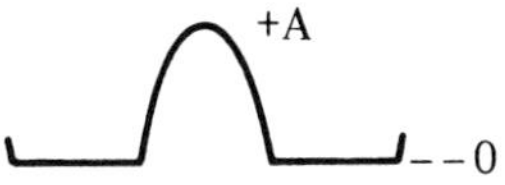 $\frac{1}{2}$ rectified sine wave	+ Peak	0.707 A
	$+\frac{1}{2}$ wave average	0.707 A
	Full-wave average	0.354 A
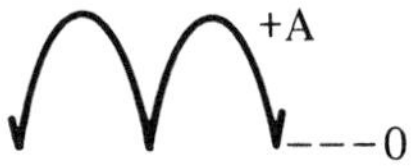 Full rectified sine wave	Peak	0.707 A
	$+\frac{1}{2}$ wave average	1.414 A
	Full-wave average	0.707 A

Figure 5-18 Response of basic ac voltmeters to common complex waveforms.

5-8 RMS Values of Basic Complex Waveforms

Any complex waveform has an rms value. This rms voltage will be indicated directly on a true-rms type of ac voltmeter, such as an electrodynamometer instrument. However, a peak-response, half-wave average response, or full-wave average response ac voltmeter does not usually indicate the rms value of a complex waveform. For example, with reference to Fig. 5-19, the rms value of a square wave without a dc component is equal to the peak value of the waveform, whereas no rectifier type of ac voltmeter will indicate this value. There is one possible exception that may be noted: If a peak-responding ac voltmeter happens to be provided with a peak-indicating scale, it will indicate the rms value of the square wave. However, only an exceptional ac voltmeter has these features.

Next, consider the rms value of a full-rectified sine wave. This is 0.707 of its peak value. In the case of a full-wave average responding ac voltmeter, the scale indication will be equal to this rms value. On the other hand, if a half-wave average-responding ac voltmeter is utilized, the scale indication will not be equal to the rms value of the complex waveform. But if a peak-responding ac voltmeter is employed, its rms scale will correctly indicate the rms value of the full-rectified sine wave. With reference to the sawtooth waveform depicted in Fig. 5-19, its rms value is 0.577 of peak. Although no rectifier type of ac voltmeter will indicate this value correctly, the error is comparatively small in the case of a half-wave average responding ac voltmeter, or of a full-wave average responding meter.

5-9 AC Power Measurements

AC power measurements are often made with ac voltmeters across resistive or reactive loads. There are 12 basic relations among voltage, current, resistance, and power, as shown in Fig. 5-20. There are three different power units, as follows.

1. *Apparent power.* The power value expressed in volt-amperes, obtained by multiplying the rms voltage by the rms amperage, without regard to the phase angle that occurs between voltage and current.
2. *Real power.* The component of apparent power that represents true work in an ac circuit or load. It is expressed in watts and is equal to the apparent power multiplied by the power factor of the circuit or load. $P_R = EI \cos \theta$.

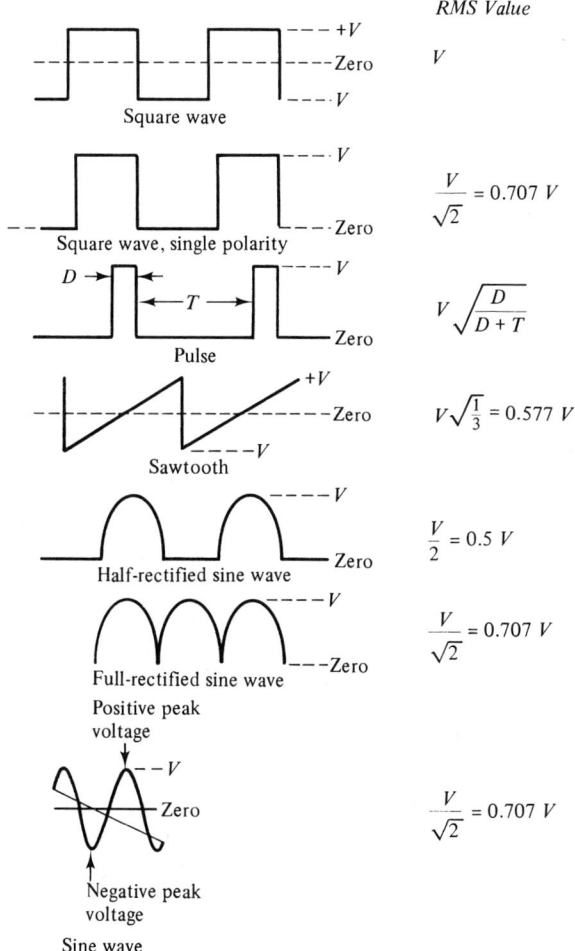

Figure 5-19 RMS values of some basic complex waveforms.

3. *Reactive power.* Also called wattless power, imaginary power, or volt-amperes reactive. It is expressed in VARS, and is equal to the rms reactive voltage multiplied by the rms current, or to the rms voltage multiplied by the rms reactive current.

Practical examples of the foregoing kinds of power are

1. A power transformer operating with its secondary open draws a small amount of current from the ac line. No power is being

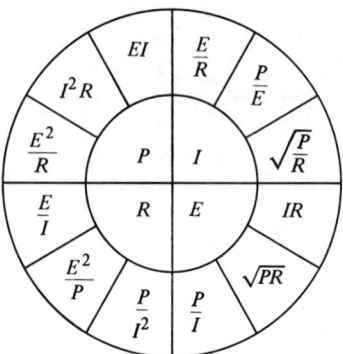

Figure 5-20 There are 12 expressions for I, E, R, and P.

supplied by the secondary, although the primary draws a small rms current at a certain rms voltage. The product of primary current and primary voltage is a resultant value of apparent power in volt-amperes. Note that this apparent power consists of some reactive power and some real power.

2. A power resistor connected across an ac line draws a certain number of rms amperes at the given rms voltage. The product of these voltage and current values is the real power in watts consumed by the resistor, because all of the electrical energy in a resistive load is converted into heat energy.
3. A capacitor connected across an ac line draws a certain number of rms amperes at the given rms voltage. The product of these voltage and current values is the reactive power in VARS that is present in the capacitor, because none of the electrical energy in a capacitor is converted into any other form of energy.

5-10 Test for "Above-ground" Soldering Gun

When connections are being soldered in solid-state receivers or other electronic equipment, semiconductor devices can be damaged if there is appreciable leakage voltage in the soldering gun. To check for an "above-ground" ("hot") soldering gun, the test shown in Fig. 5-21 may be made. If there is more than a fraction of a volt leakage from the soldering gun, it should be discarded, or it should be operated with a flexible grounding wire run from the soldering tip to a good earth ground such as a cold-water pipe.

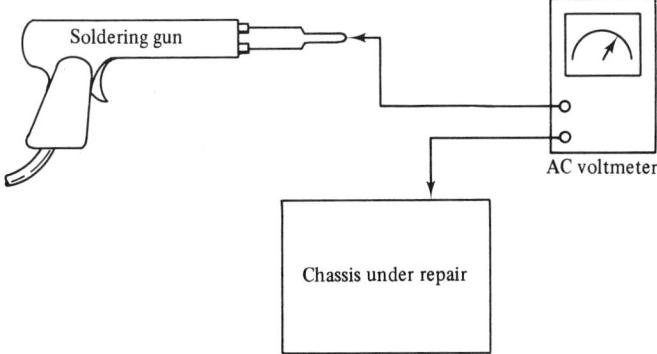

Figure 5-21 Test for an "above-ground" soldering gun.

5-11 AC Converters in Service-type Voltmeters

Several forms of ac converters are encountered in service-type voltmeters. In a conventional multimeter, the ac converter usually consists of a pair of instrument rectifiers (semiconductor diodes). A TVM generally utilizes the same form of converter. On the other hand, a VTVM usually employs a pair of vacuum diodes in a peak-to-peak responding configuration. A DVM may have either of two forms of ac converter. The first arrangement is the same as in a conventional multimeter; the second arrangement uses a single semiconductor diode and capacitor in a peak-responding configuration. All forms of ac converters provide indication on rms scales, and peak-to-peak scales are occasionally provided also.

When a pair of instrument rectifiers is utilized in a conventional multimeter, either half-wave or full-wave rectification may be used. In either case, the meter movement responds to the average value of the rectified waveform, and the ac scales indicate rms values of a sine waveform. Peak indication or peak-to-peak indication cannot be provided, unless it is specified that the input voltage has a sine waveform. Most TVM's have these same limitations in ac voltage measurement. On the other hand, most VTVM's employ an ac converter arrangement that develops the peak-to-peak value of the voltage waveform. In turn, both rms and peak-to-peak scales are customarily provided. The peak-to-peak scale indicates correctly, regardless of the input voltage waveform. However, the rms scale indication is correct only for sinusoidal input waveforms.

Most service-type DVM's provide the same ac converter arrangement as in a conventional multimeter. However, an occasional DVM utilizes a peak-output rectifier arrangement. In this case, peak-indicating scales could be provided in addition to rms-indicating scales. However, it is usual practice to provide rms scale indication only. Peak response results in an rms scale calibration that is accurate only for sine-wave voltages. Note in passing that accessory RF probes for DVM's, TVM's, and VTVM's almost always utilize a peak-output rectifier arrangement. In turn, a series dropping resistor is provided to obtain correct rms ac voltage indication on the dc-voltage function of the instrument. This feature results in the necessity for using an accessory RF probe only with the instrument for which it was designed; otherwise, the scale indication may be in substantial error.

6 Hi-fi Stereo Troubleshooting

6-1 General Considerations

Audio amplifiers are the central section in a hi-fi stereo system. Trouble symptoms caused by amplifier malfunctions include "dead" amplifier, weak output, excessive distortion at rated power output, excessive distortion at all output levels, excessive distortion at low output levels, noise and/or hum interference, and poor balance. Defects in the stereo decoder section can cause poor separation, "dead" channel, or weak channel operation. Any section may become intermittent. Most troubleshooting procedures involve meter tests of various kinds. For example, an ac voltmeter may be used to signal-trace an amplifier, to measure stage gains, and to measure power output. A dc voltmeter may be utilized to check transistors and integrated circuits. An audio TVM is used to check signal levels in preamp circuitry. Harmonic distortion meters are used to measure the percentage of distortion of an amplifier, or system.

6-2 Amplifier Frequency Response

One of the most important characteristics of a hi-fi amplifier is its frequency response. A uniform response from 20 Hz to 20 kHz has been standardized. This response is measured in dB with respect to the output level at 1 kHz. Thus, the output level at 1 kHz is termed 0 dB. In turn, a high-fidelity frequency response is defined as less than ±1 dB variation from 20 Hz to 20 kHz. A test setup as shown in Fig. 6-1 is used to measure the frequency response of an amplifier. Observe that

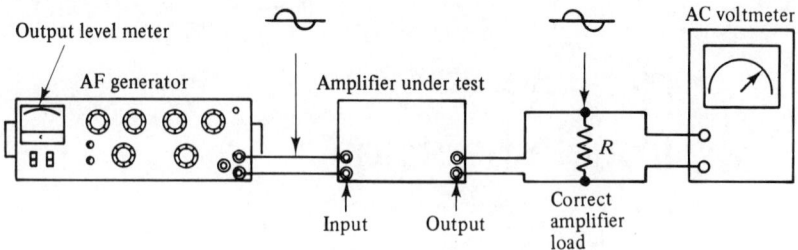

Figure 6-1 Amplifier frequency response test setup.

the audio-frequency generator is provided with an output level meter. This meter is required to ensure that the amplifier is driven with a constant amplitude over its entire frequency range. Note that if an AF generator does not contain a built-in level meter, an external ac voltmeter may be utilized in its stead.

Typical amplifier frequency response curves are shown in Fig. 6-2. Observe that the tone controls, the loudness control, the scratch filter, and the rumble filter will modify the basic flat response of the amplifier markedly. It is good practice to check out these responses, in addition to the frequency characteristic of the amplifier alone. Frequency response is plotted on semi-log graph paper in Fig. 6-2. In turn, the high-frequency ends of the curves are compressed. This representation has the advantage of showing a wide frequency range in comparatively compact form. Observe that the amplifier is terminated with a power resistor with a value equal to the rated load impedance for the amplifier, as shown in Fig. 6-1. The basic frequency response curve is obtained at maximum rated power output from the amplifier. Power output is equal to the square of the rms output voltage divided by the resistance of the load.

To plot the curves exemplified in Fig. 6-2, the dB scale of the ac voltmeter is read as the generator frequency is advanced by suitable increments. The power output level is first adjusted to rated value at 1 kHz. Then, the dB readings may be recorded at frequencies of 10, 100, 10 kHz, and 100 kHz. Additional readings are generally required within intervals of rapid output variation. Many hi-fi amplifiers are provided with a phono input jack. The higher recorded frequencies are pre-emphasized. In turn, the frequency response of an amplifier when driven from its phono jack normally shows a complementary deemphasis characteristic. If a magnetic phono cartridge is used, the amplifier frequency response normally plots as depicted in Fig. 6-3.

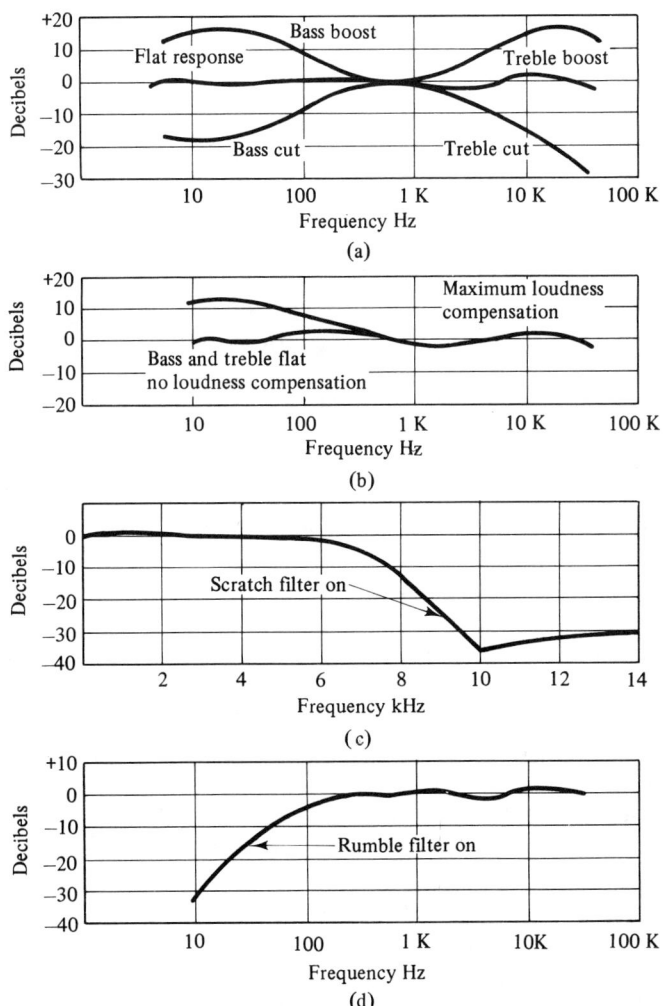

Figure 6-2 Typical amplifier frequency response curves: **(a)** flat response with effects of bass and treble controls; **(b)** flat response with effect of loudness control; **(c)** frequency response of a scratch filter; **(d)** frequency response of a rumble filter.

6-3 Harmonic and Intermodulation Distortion

After the frequency response of an amplifier has been verified (or corrected), another of the most important reproduction characteristics is

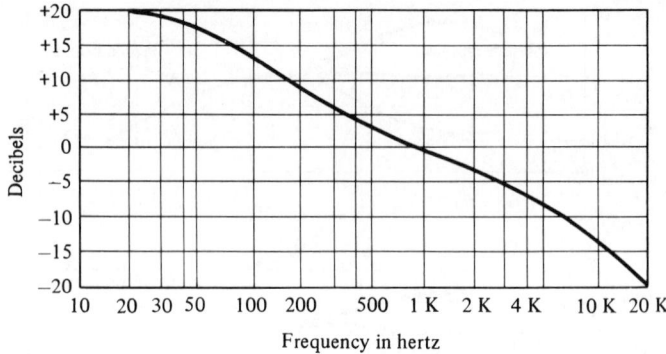

Figure 6-3 Standard equalization frequency response for phono input with magnetic cartridge.

checked, viz., percentage of distortion. Harmonic distortion is always measured, and intermodulation distortion may also be determined. Harmonic distortion is measured with a test setup such as that shown in Fig. 6-4. The measurement is usually made at a frequency of 1 kHz, and at maximum rated power output from the amplifier. Harmonic distortion increases as the high-frequency cutoff point of an amplifier is approached. If the amplifier is ac-coupled, harmonic distortion will increase as the low-frequency cutoff point is approached. However, in the case of a direct-coupled amplifier, harmonic distortion tends to remain constant down to the lowest test frequency. The power bandwidth of an amplifier is defined as the span between the frequency limits at which the percentage of distortion starts to exceed its reference value at 1 kHz, when the power output is reduced 3 dB.

All high-fidelity amplifiers employ negative feedback to minimize distortion. Examples of amplifier operation with and without negative feedback are shown in Fig. 6-5. Out-of-phase feedback is utilized from output to input in order to predistort the input waveform in such a manner that the output waveform will have reduced distortion. It is sometimes supposed that any form of amplifier distortion can be reduced by means of negative feedback. However, this is not so. As an illustration, Fig. 6-6 shows why clipping distortion is unaffected by negative feedback; since the amplifier has no output past the clipping level, it is impossible to force any output past this level. As a practical note, observe that when stage-by-stage distortion measurements are made, negative feedback can produce misleading test results, as shown in Fig. 6-7. Thus, the technician must take negative-feedback action into consideration when tracking down a distorting stage in an amplifier.

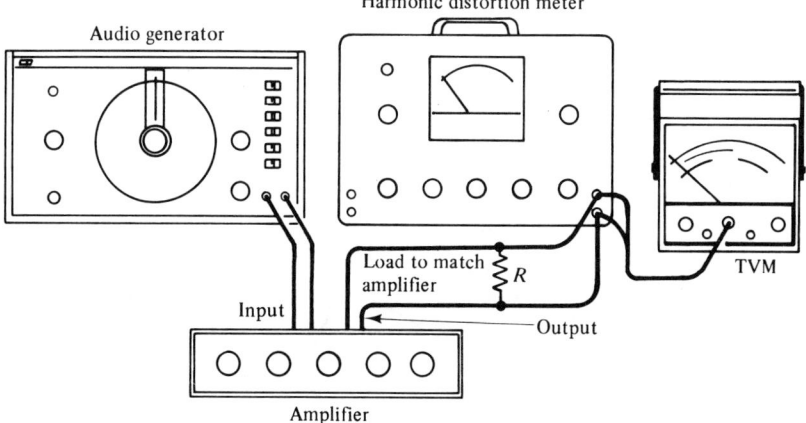

Figure 6-4 Harmonic distortion test setup.

Harmonic distortion also results from crossover irregularities, as exemplified in Fig. 6-8. A quick check for crossover distortion is to reduce the drive to the amplifier under test. In turn, the percentage distortion indication will increase, if the dominant component is crossover distortion. The reason for this response is that crossover distortion increases in percentage at low volume levels, whereas compression or clipping distortion decreases in percentage at low volume levels. Note that amplifier hum and noise are also indicated as harmonic distortion. A quick check for hum is to set the audio generator to a frequency of 59 or 61 Hz, and then to a frequency of 119 or 121 Hz. If hum voltage is present in the output, the pointer will "wiggle" on the scale of the harmonic distortion meter. Observe also that hum and noise often affect percentage distortion indications in the same way as crossover distortion. That is, hum and noise often become a greater percentage of the total output as the volume level is reduced.

Preamplifiers normally operate in class A. A quick check for nonlinear operation is shown in Fig. 6-9. The amplifier is driven by a generator, and a dc voltmeter is connected at the output of the amplifier. In turn, the level of the dc voltage is observed with no input signal applied, and then with an input signal that drives the amplifier to maximum rated output. If the amplifier is operating in class A, there will be no change in dc-voltage indication. On the other hand, if the dc-voltage level shifts, it is known that compression or clipping distortion is occurring. This is not a completely conclusive test, because equal and opposite peak compression or clipping will not change the

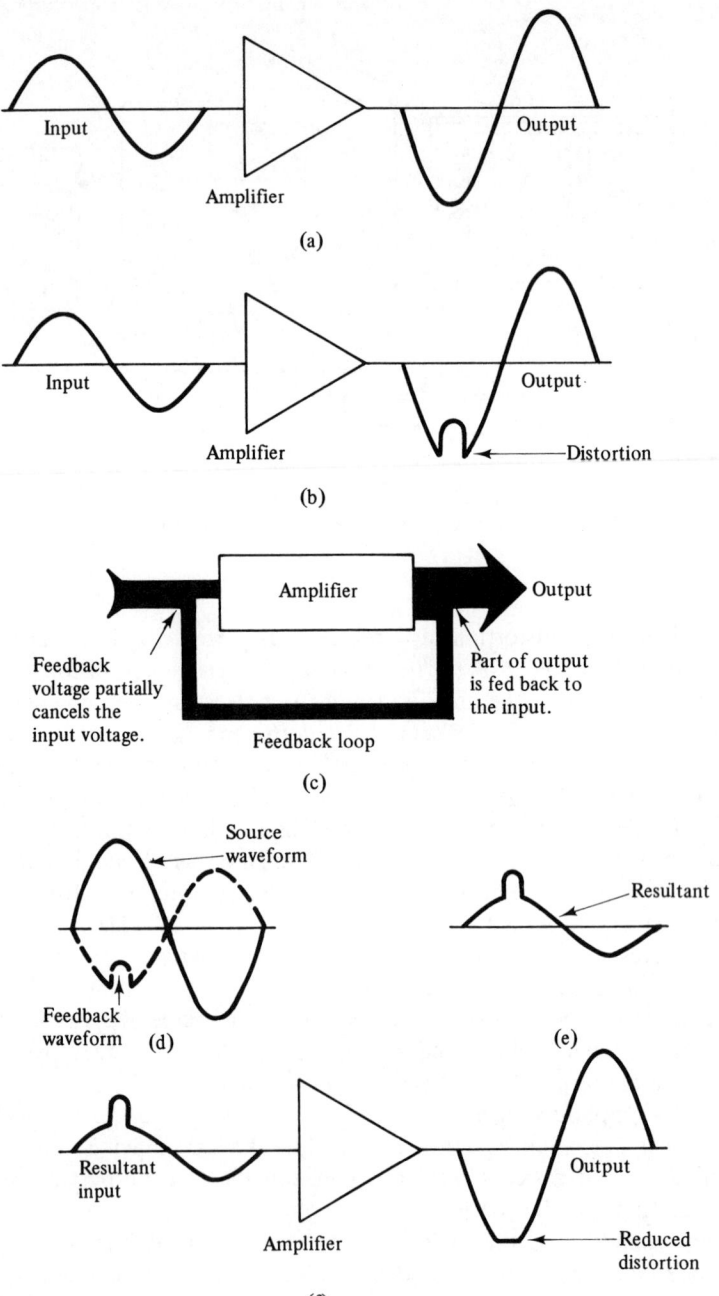

Figure 6-5 Examples of amplifier operation, with and without negative feedback: **(a)** distortionless output; **(b)** distorted output; **(c)** negative-feedback arrangement; **(d)** feedback waveform mixed with source waveform; **(e)** resultant input waveform; **(f)** output waveform has reduced distortion.

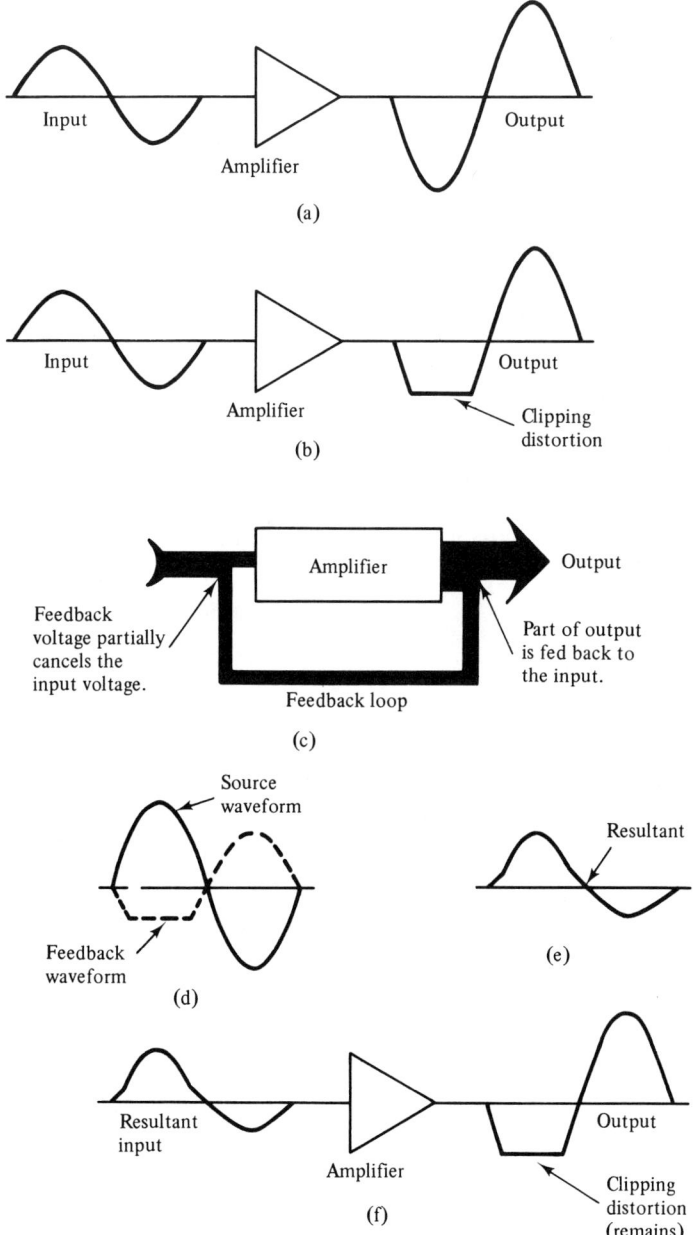

Figure 6-6 Clipping distortion cannot be corrected by negative feedback: **(a)** distortionless output; **(b)** clipped output; **(c)** negative-feedback arrangement; **(d)** source waveform mixed with feedback waveform; **(e)** resultant waveform; **(f)** resultant input waveform produces clipped output waveform.

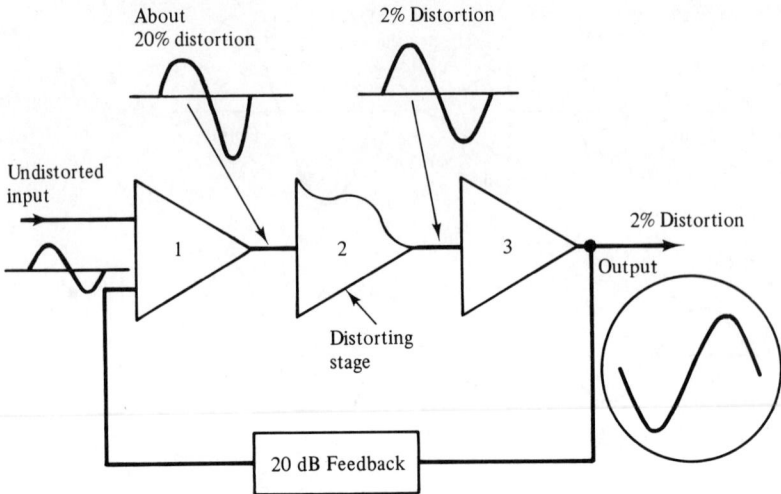

Figure 6-7 Feedback loop makes the first stage "look like" the distorting stage.

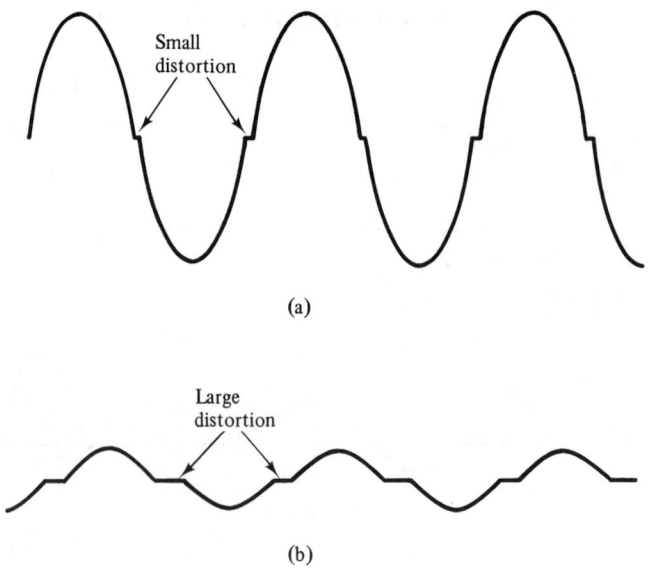

Figure 6-8 Examples of crossover distortion: **(a)** output at higher amplitude; **(b)** output at lower amplitude.

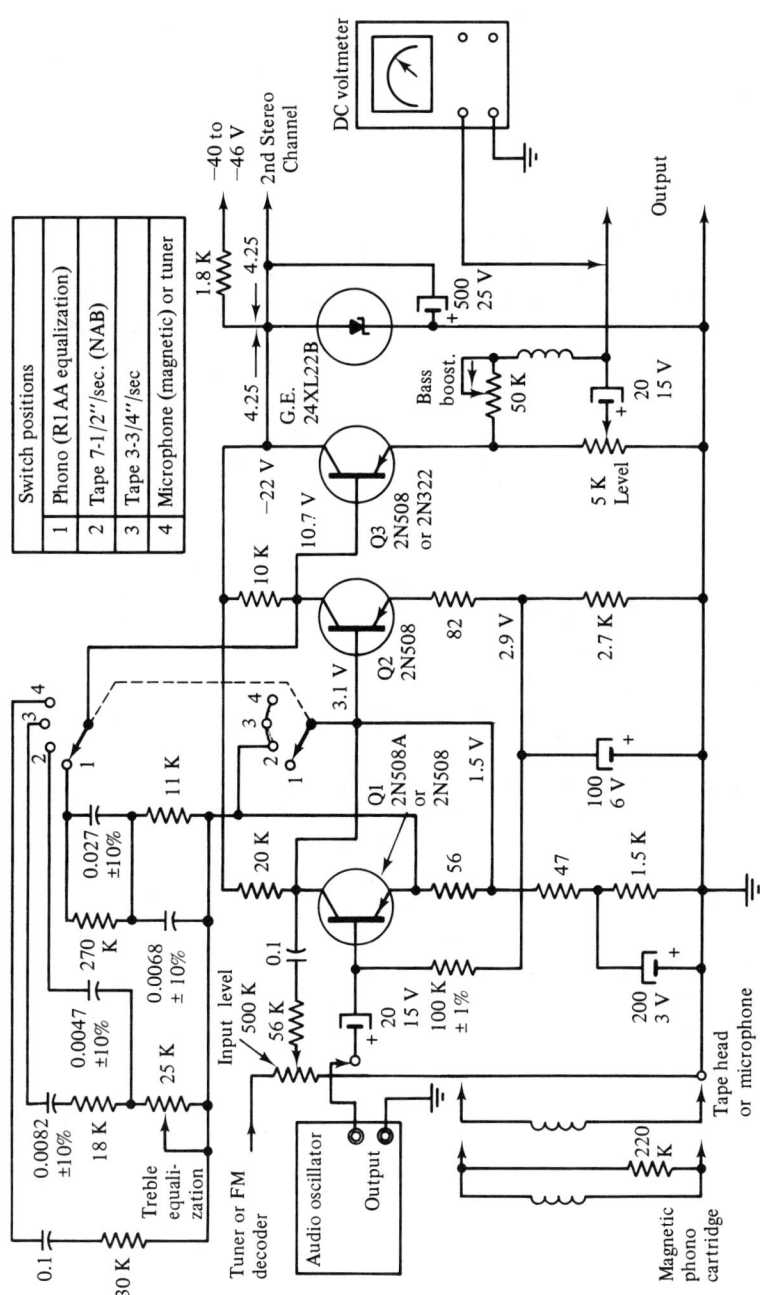

Figure 6-9 Quick check for class A operation. (*Courtesy of General Electric Co.*)

dc-voltage level. However, symmetrical peak distortion is the exception, and not the rule. Observe in Fig. 6-9 that the transistors are direct-coupled. In turn, peak distortion in the first stage or in the second stage will be transferred as a dc-voltage shift in the third stage. Accordingly, this quick check can only show that distortion is occurring, but does not identify the malfunctioning stage.

6-4 Classes of Amplifier Operation

An amplifier may operate in one of the four basic classes: In class A, an amplifier transistor develops collector-current flow over the complete cycle of the input sine waveform; in class B, a transistor develops collector-current flow over one-half cycle of the input sine waveform; in class C, a transistor develops collector-current flow over less than one-half cycle of the input sine waveform. In class D operation, the amplifier employs high-frequency pulses (in excess of 200 kHz) that are first modulated by the audio signal to be amplified, and then are decoded by an integrating circuit that restores the audio envelope (waveshape). (See Fig. 6-10.) Inasmuch as the duty cycle of each high-frequency pulse is relatively short, conduction of the amplifier output transistors is such that their heat dissipation is only a fraction of that encountered with class B circuitry. Over-all efficiency (particularly at maximum rated power output) is also high. A class D amplifier is also called a switching amplifier.

An amplifier may also operate in class AB. In this mode of operation, a small amount of forward bias is applied to what would otherwise be a class B configuration. This forward bias serves to reduce or eliminate crossover distortion. In the strict sense, a bipolar transistor operating in class A is technically a class A_1 amplifier, because base current is drawn in response to the input signal. Similarly, a bipolar transistor operating in class B is technically a class B_1 amplifier. On the other hand, a unipolar transistor operating in class A or class B cannot draw gate current at any time. Accordingly, these arrangements are true class A and class B configurations.

6-5 Amplifier Voltage Gain and Sensitivity

Amplifier voltage gain is defined as the ratio of output rms voltage to input rms voltage. An example is shown in Chart 6-1. Amplifier sensitivity is defined as the value of rms input voltage that provides maximum rated power output from the amplifier. The test setup shown in

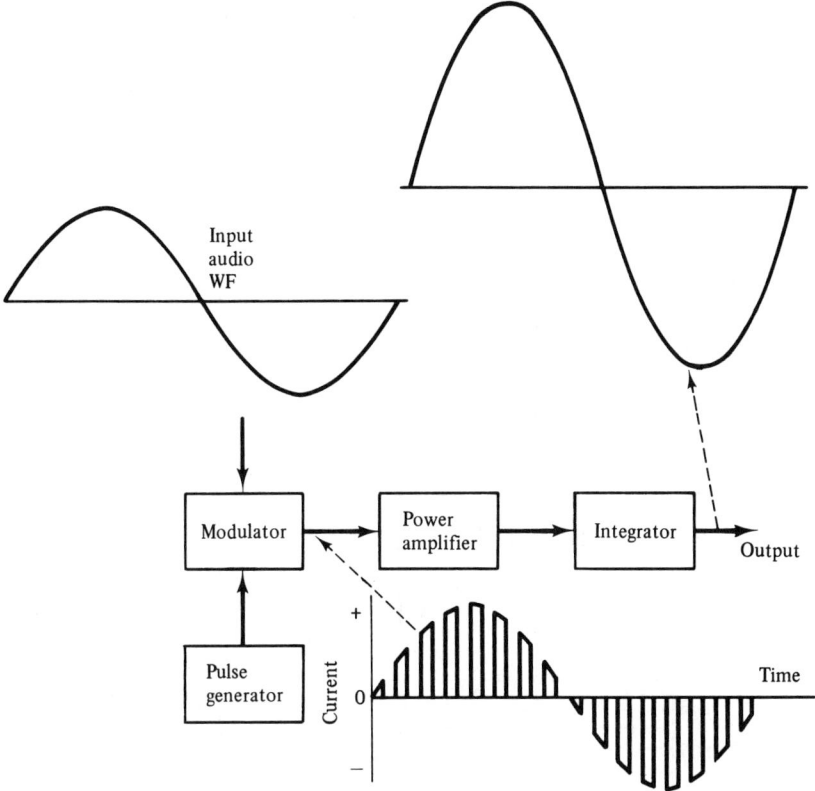

Figure 6-10 Principle of a class D amplifier.

Fig. 6-1 is employed. A frequency of 1 kHz is suitable, with an input signal level that drives the amplifier to approximately half of its maximum rated power output. In turn, the output/input signal-voltage ratio is equal to the amplifier voltage gain. Next, the input signal level is advanced to obtain maximum rated power output. Amplifier sensitivity is then equal to the rms reading of the output level meter on the audio generator. Note that if the amplifier is operating normally, its voltage gain at maximum rated power output will be practically the same as at half of maximum rated output.

Amplifier voltage gain should not be confused with dB voltage gain. Only a few amplifiers are designed with equal input and output resistance values. In turn, the dB gain of most amplifiers cannot be measured directly by utilizing the dB scale on an ac voltmeter. The dB gain of an amplifier is equal to $10 \log_{10} (P_o/P_i)$, where P_o is the audio output power, P_i is the audio input power, and $\log_{10}$ is the loga-

Chart 6-1 Signal-voltage levels in a basic high-fidelity system.

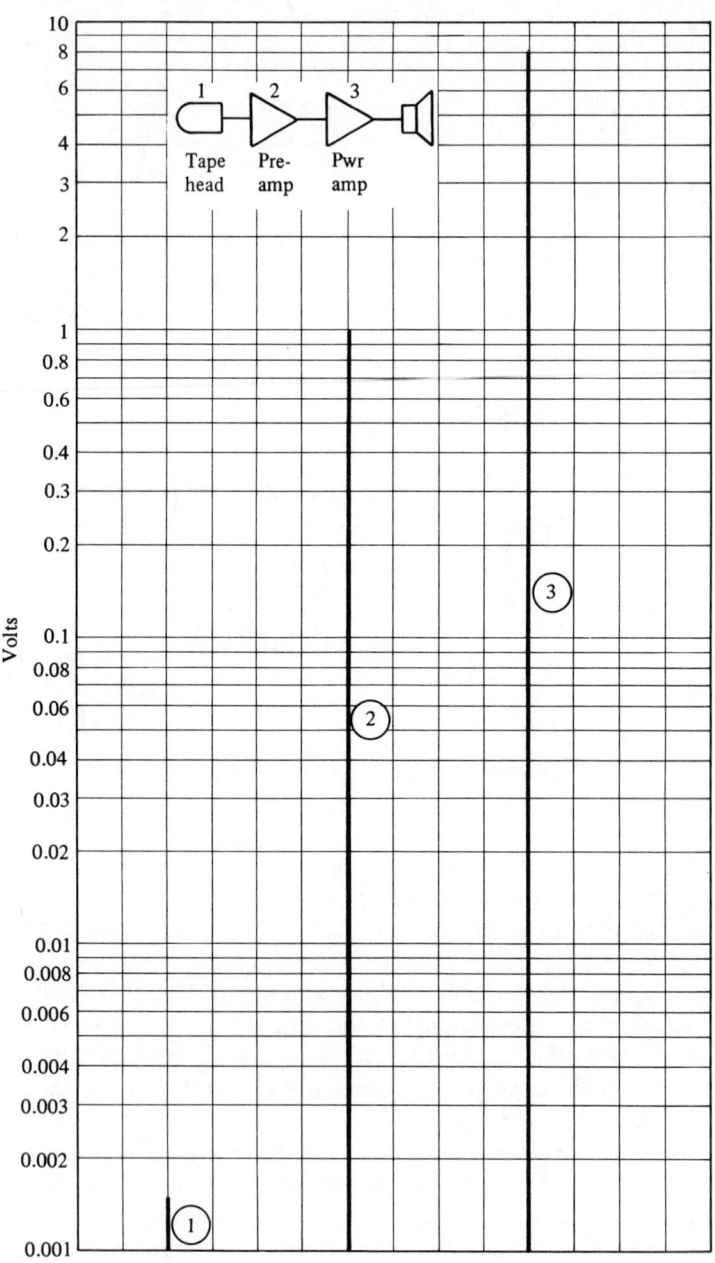

rithm to the base 10 of this power ratio. Audio power across a resistance can be calculated in terms of E^2_{rms}/R, or as the product of E_{rms} and I_{rms}. However, since the measurement of small AF currents is difficult with conventional service-type equipment, it is generally most expedient to calculate input audio power in terms of rms voltage and resistance values. When the input resistance of an amplifier is unknown, it can be measured as shown in Fig. 6-11. First, with R equal to zero, the audio generator is set to produce maximum rated power output from the amplifier. Second, the value of R is adjusted to provide half-voltage output across R_L. Then the input resistance of the amplifier is equal to the value of R plus the output resistance of the audio generator (typically 600 ohms). A comparatively low test frequency, such as 100 Hz, ensures that R will not introduce spurious reactive effects. Comparative over-all signal-power gain values for typical amplifier, radio, and TV systems are given in Chart 6-2.

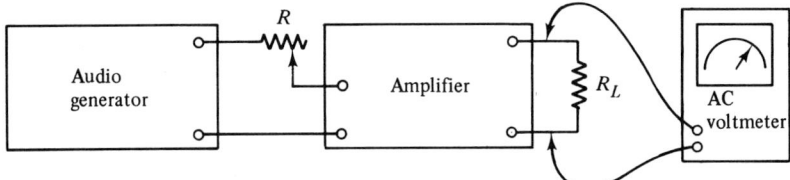

Figure 6-11 Measurement of amplifier input resistance.

6-6 Intermodulation Distortion Measurement

An intermodulation distortion test employs a two-tone signal, whereas a harmonic-distortion check utilizes a single-frequency signal. A two-tone test signal consists of a 60-Hz and a 6-kHz sine-wave mixture, for example. These signal sources are generally built into the intermodulation distortion analyzer. An intermodulation distortion test setup is shown in Fig. 6-12a. As in the case of a harmonic-distortion test, the amplifier should be checked at its maximum rated power output. It is also good practice to check intermodulation distortion at lower power levels, inasmuch as the percentage of distortion occasionally tends to increase as the output level decreases. However, as a general rule, IM distortion is greatest at maximum rated power output. In most cases, an amplifier develops comparable values of intermodulation and harmonic distortion, as exemplified in Fig. 6-12b. Both forms of distortion

Chart 6-2 Relative over-all signal-power gains for a high-fidelity amplifier system, AM radio receiver, FM radio receiver, and television receiver.

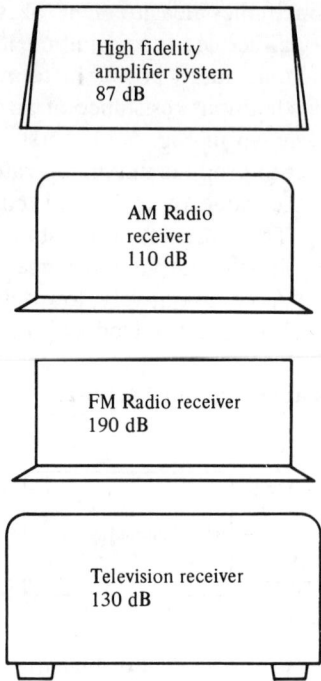

are observed to increase rapidly as the high-frequency cutoff point of the amplifier is approached.

6-7 Pinpointing Defective Components and Devices

Defective components and devices are most often pinpointed by means of dc-voltage measurements. Normal dc operating voltages are specified in amplifier servicing data, as exemplified in Fig. 6-13. When a component or device becomes defective, the fault often becomes apparent as a change in dc voltage(s). For example, if C13 becomes leaky, the base voltage on Q11 will increase. In turn, the transistor is likely to be driven into saturation by the audio signal, resulting in objectionable distortion. It is good practice to check the supply voltage of an amplifier before checking the dc-voltage distribution. In other words, if the supply voltage is subnormal, voltages throughout the circuit will also be subnormal.

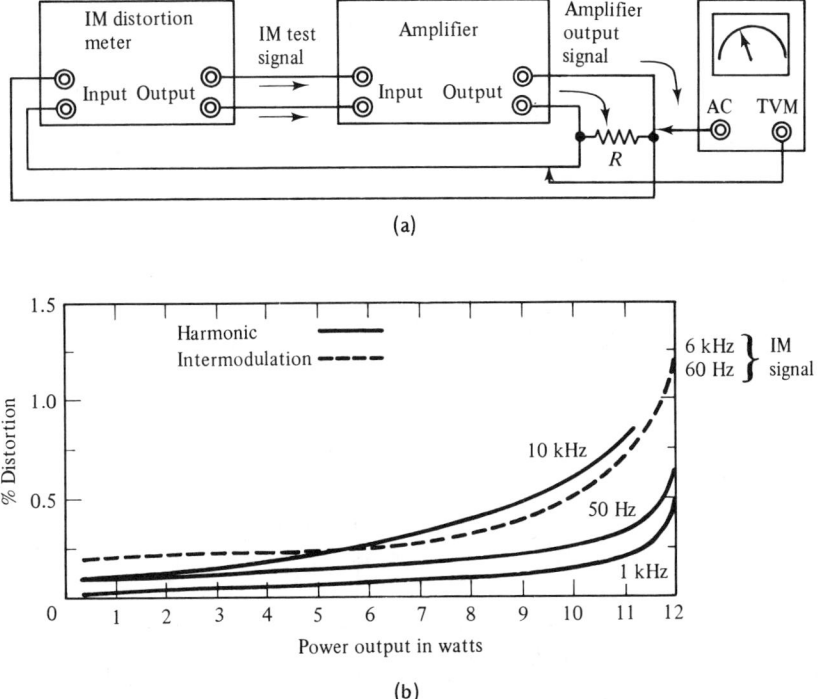

Figure 6-12 (a) An intermodulation distortion test setup. (b) A typical comparison of IM and HD percentages. (*Courtesy of General Electric Co.*)

All of the transistors in Fig. 6-13 are silicon types, and each transistor has a forward bias of 0.6 volt, with the exception of Q12, which operates with a forward bias of 1.2 volts. Base-emitter bias voltages are comparatively critical, and should measure closely to specified values. Note, for example, that Q10 has a base-emitter bias voltage of 0.6 volt, whereas its base voltage is 1.6 volts, and its emitter voltage is 1 volt. Base, emitter, and collector voltages are subject to reasonable tolerances, whereas base-emitter bias voltages have comparatively tight tolerances. Bias stabilization is provided by various means, such as emitter resistors and varistors. A varistor is a diode device that has a voltage-dependent nonlinear resistance which decreases rapidly as the applied voltage increases.

Sectionalization of a malfunction to a particular stage is usually possible by signal-tracing procedures. As an illustration, an audio voltmeter (or often a conventional ac voltmeter) can be used to check the progress of an audio signal from points 1 through 7 in Fig. 6-13. This

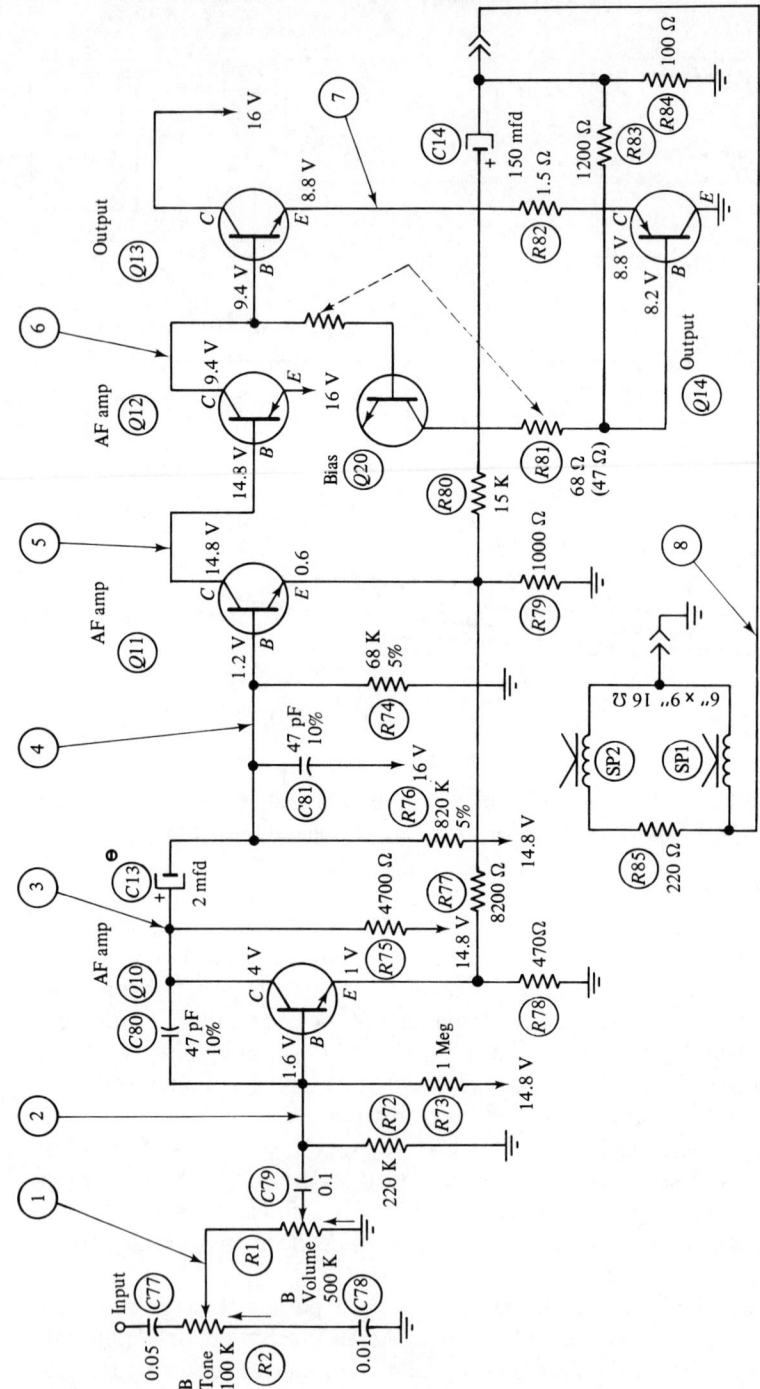

Figure 6-13 Example of dc voltage distribution in an AF amplifier.

technique shows up faults that cannot be found by dc-voltage or resistance measurements, such as open coupling capacitors. To confirm the suspicion of an open coupling capacitor, another capacitor can be temporarily "bridged" across the suspect. Then, if normal operation is resumed, the suspicion is confirmed. Signal-tracing procedures also locate "weak" stages in many cases. Since normal stage gains are seldom specified in amplifier service data, the technician must fall back upon his experience and knowledge of circuit action in various situations. That is, if the output from a stage is less than the input level, it is obvious that the stage is "weak." On the other hand, a level increase of 10, 20, or 30 times may be difficult to evaluate. In turn, the technician is thrown upon his own resources.

Integrated-circuit audio amplifiers are serviced in basically the same manner as transistor amplifiers. With reference to Fig. 6-14, one-third of an IC package is utilized as an audio preamplifier. Integrated circuit terminal voltages are key check points, both for dc and for signal-level measurements. However, normal signal levels are not specified. Note that class A amplification is provided throughout, in this example. In other words, the dc voltages normally do not change with signal present or absent. Transistors are utilized in the higher-level audio stages. Q10 is a direct-coupled amplifier with feedback across its emitter resistors R49 and R50, in series with the NPN-PNP output transistors. Q4 is a PNP audio driver to the base of Q5 and through the crossover compensating diode D4 to the base of Q11. Note that Q5 conducts on positive half-cycles of the signal, whereas Q11 conducts on negative half-cycles. Increased stability of operation is provided by feedback from the audio output lead to the base of Q11.

Note that integrated circuits (IC's) are fabricated in transistor-type (TO-5) packages, and in flat-pack or in-line packages, as shown in Fig. 6-15. An IC may plug into a socket, or it may be soldered into a printed-circuit board. Signal input and output terminals are arranged as depicted in functional diagrams, exemplified in Fig. 6-16.

6-8 Quasi-complementary Amplifier

Quasi-complementary amplifiers are in very wide use. With reference to Fig. 6-17, malfunctions in this type of circuit are usually caused by a defective output or driver transistor. When troubleshooting this configuration, check the resistance between ground and point J, and between $+V_{cc}$ and point J before applying power. For the first measurement, connect the negative lead to ground and the positive lead to point J, and note the reading. Next, reverse the leads; the resistance

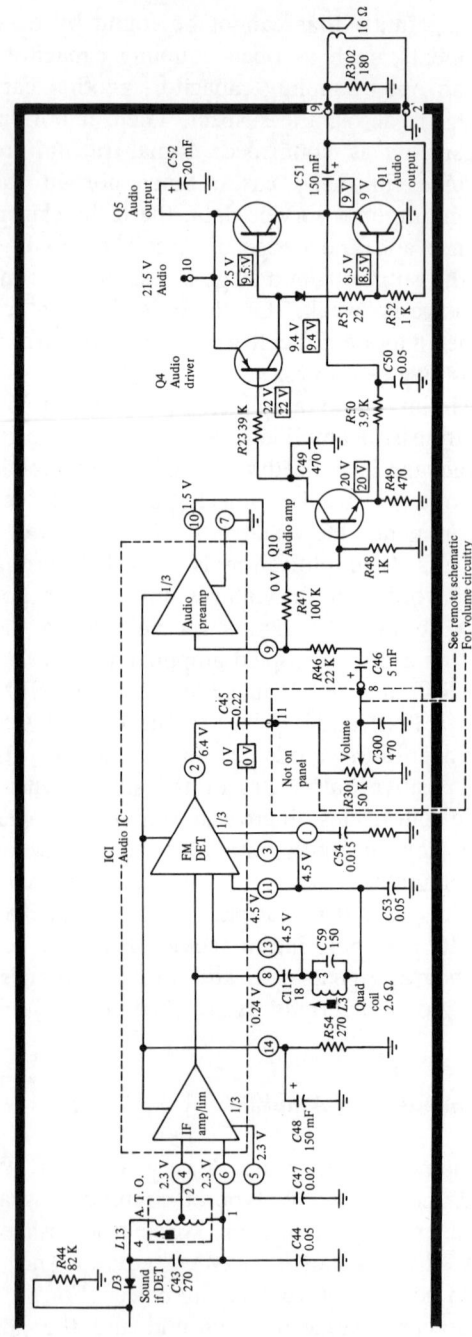

Figure 6-14 An audio amplifier with integrated circuits. (Courtesy of Motorola Semiconductor Products Inc.)

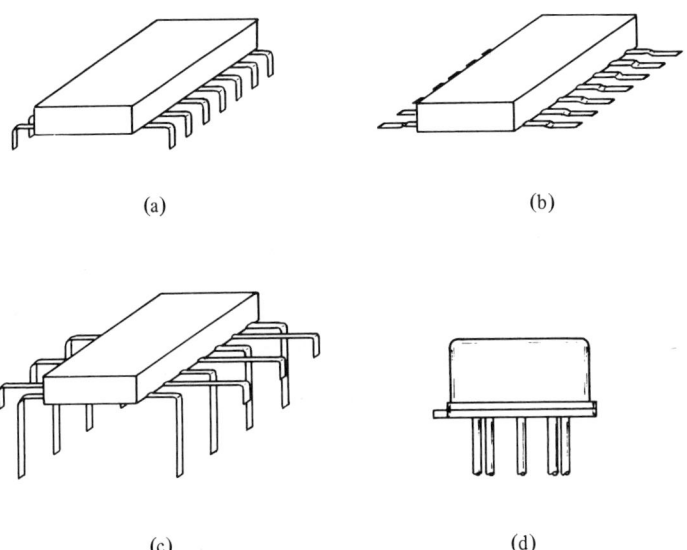

Figure 6-15 Typical flat-pack and TO-5 style IC packages: **(a)** in-line through-board mounting type; **(b)** in-line surface mounting type; **(c)** quad-formed-lead mounting type; **(d)** transistor-TO-5-style package.

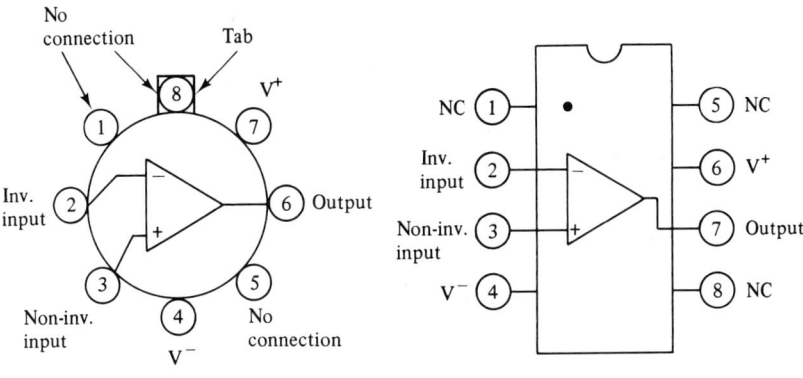

Note: Pin 4 is connected to case top view
Top view

(a)   (b)

Figure 6-16 Functional diagrams for IC packages: **(a)** TO-5 case; **(b)** flat pack.

157

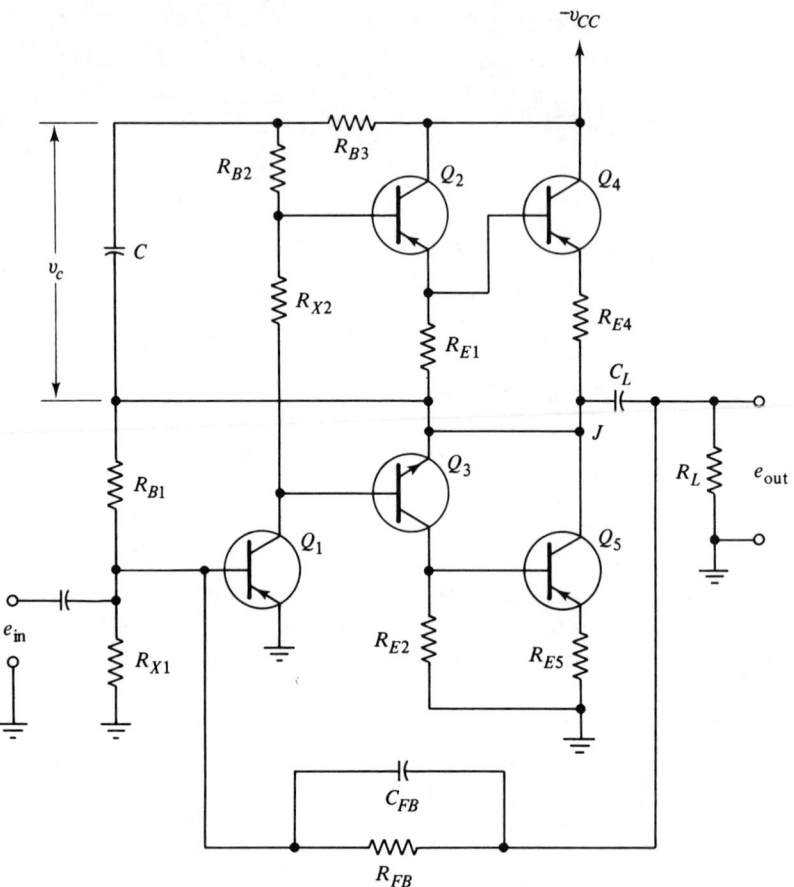

Figure 6-17 Basic quasi-complementary amplifier circuit.

reading should be much less. A "dead short" points to a short-circuited Q_3 and/or Q_5 transistor. Repeat the foregoing procedure to check the resistance between $+V_{cc}$ and point J. A lower resistance should be indicated when the positive lead is applied at V_{cc}. A "dead short" points to a defective Q_2 or Q_4 transistor.

Next, R_{X2} should be short-circuited. Turn the power on to the amplifier, and measure the voltage at point J. This reading should be one-half of $+V_{cc}$. If an error of less than 25 percent is observed, check the values of R_{B1} and R_{X1}. If these resistors have correct values, the voltage at J should be within 10 percent of $+V_{cc}/2$. However, other malfunctions can be responsible for an incorrect voltage at point J. Check the voltage across Q_3. If this value is equal to V_{cc}, it is indicated that Q_4 is short-circuited. Similarly, if the voltage across Q_2 is $+V_{cc}$, Q_5 is short-circuited. A somewhat more elaborate quasi-complementary amplifier configuration is shown in Fig. 6-18. Here, a differential amplifier is used to drive the quasi-complementary section. The same general tests may be made in this configuration for preliminary analysis of trouble symptoms.

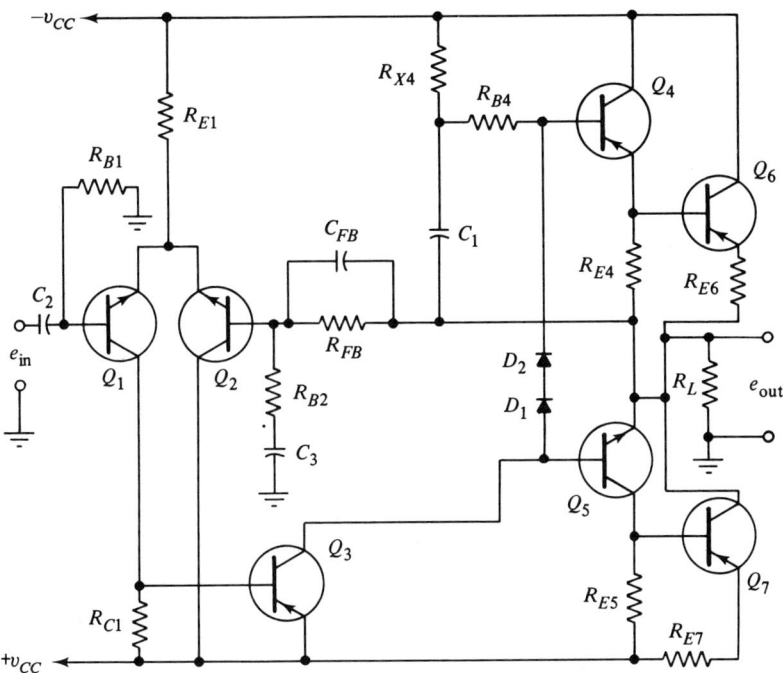

Figure 6-18 Quasi-complementary amplifier driven by a differential amplifier.

6-9 Stereo-multiplex Separation Test

A configuration for an electronic-switching stereo decoder is shown in Fig. 6-19. The most basic test of a decoder concerns the number of dB separation that it provides between left (L) and right (R) signals from a stereo signal generator. Figure 6-20 shows an arrangement for a stereo decoder separation test. L and R signals are applied in turn from a stereo generator. When an L signal is applied in the separation test, there is theoretically zero output from the R channel, and maximum output from the L channel. Similarly, when an R signal is applied,

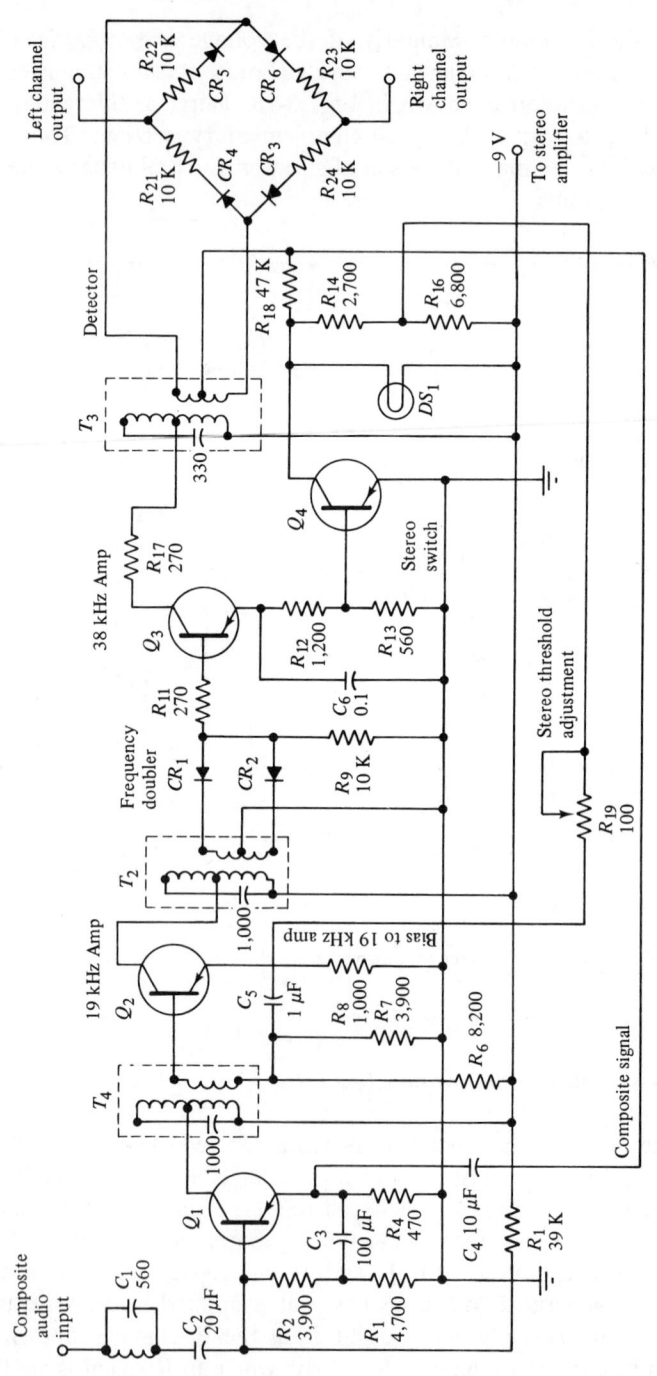

Figure 6-19 Configuration of an electronic-switching stereo decoder.

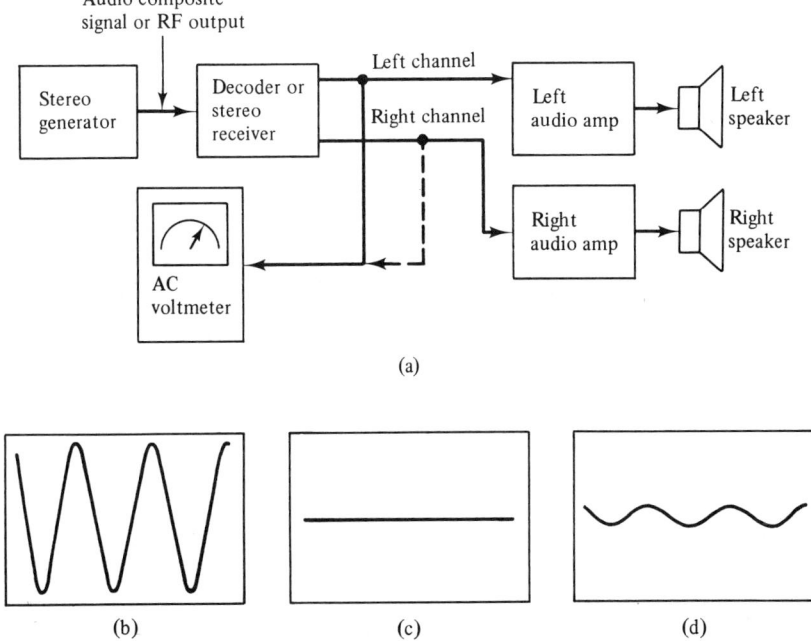

Figure 6-20 Stereo separation test: **(a)** equipment connections; **(b)** R-channel output; **(c)** ideal L-channel output; **(d)** L-channel output with incomplete separation.

there is theoretically zero output from the L channel, and maximum output from the R channel. In practice, however, about 30 dB difference will be observed between the two ac voltmeter readings for a normally operating decoder. A separation of 10 dB is considered to be on the borderline of a trouble symptom. Basic decoder arrangements are depicted in Fig. 6-21.

Inadequate separation is usually caused by a component defect in the decoder unit, such as an "open" or "shorted" capacitor. Misalignment can also cause poor separation. This is a particularly important consideration in a matrix-type decoder that operates with a frequency response curve such as that shown in Fig. 6-22. Decoder alignment procedure will be detailed in the service data for the stereo unit. Other common causes of poor separation are unmatched diodes, diodes with subnormal front-to-back ratios in a switching-bridge configuration, leaky capacitors, off-value resistor in a matrix circuit, or collector leakage in a transistor.

Distorted output from a stereo decoder can be caused by a defec-

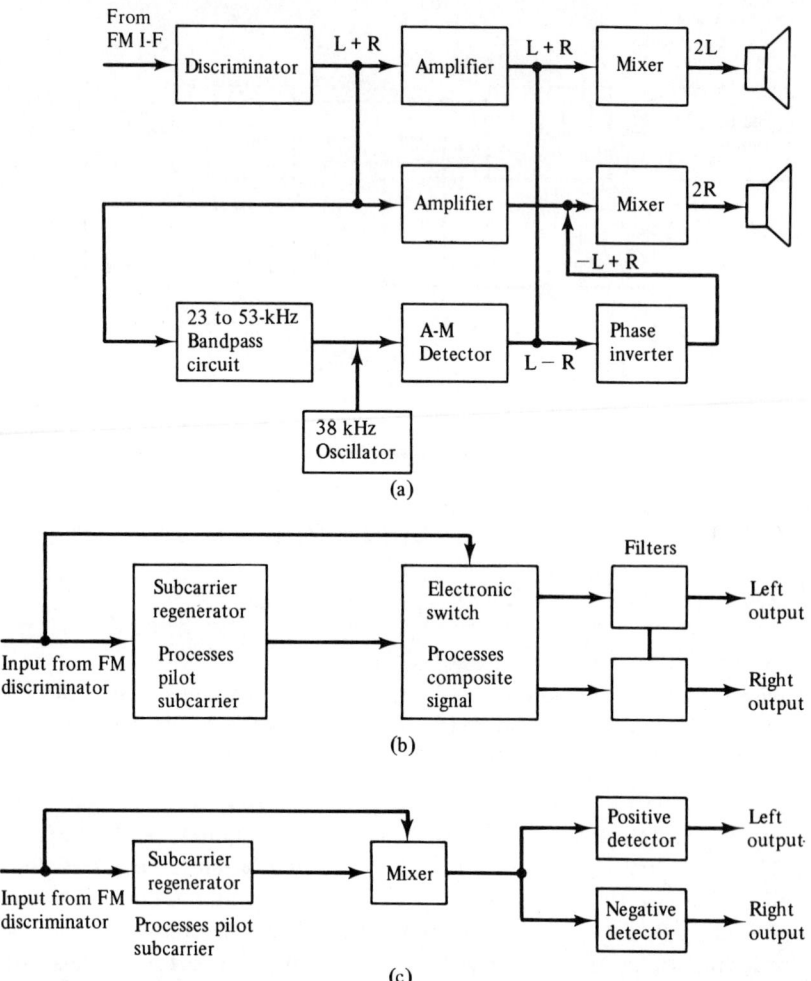

Figure 6-21 Basic types of stereo-multiplex decoders: **(a)** bandpass-and-matrix arrangement; **(b)** electronic switching method; **(c)** envelope detection method.

tive capacitor in a switching bridge, faulty electrolytic capacitor in the subcarrier regenerator section, a defective diode in a switching bridge or an envelope detector configuration, or a failing transistor in an amplifier section. Note also that stereo-indicator failure is a comparatively common trouble symptom in a stereo decoder. For example, the stereo-indicator lamp in Fig. 6-19 may fail to glow when a stereo signal is present. If adjustment of the stereo threshold control is ineffective, the

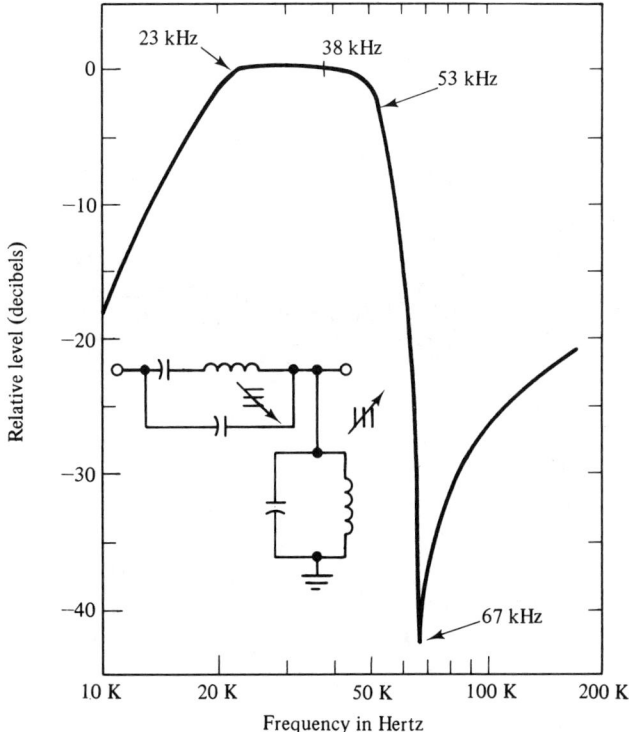

Figure 6-22 Standard frequency response curve for the bandpass filter circuit in a stereo decoder. (*Courtesy of Heath Co.*)

indicator lamp may have burned out. Otherwise, transistor Q_4 should be checked. Malfunction will result if C_6 is seriously leaky, or if a resistor in the indicator circuit changes value substantially. DC-voltage and resistance measurements provide the most useful clues in pinpointing the defective component or device.

6-10 Applications of a Stereo Analyzer

Troubleshooting stereo systems can be facilitated by the use of specialized analyzers such as that illustrated in Fig. 6-23. A check of stereo separation is made as shown in Fig. 6-24. Many factors can influence a receiver's stereo separation. Antenna signal, IF alignment, stereo decoder alignment, and even the power supply can, in some cases, be responsible for a complaint of poor stereo separation. An on-the-air

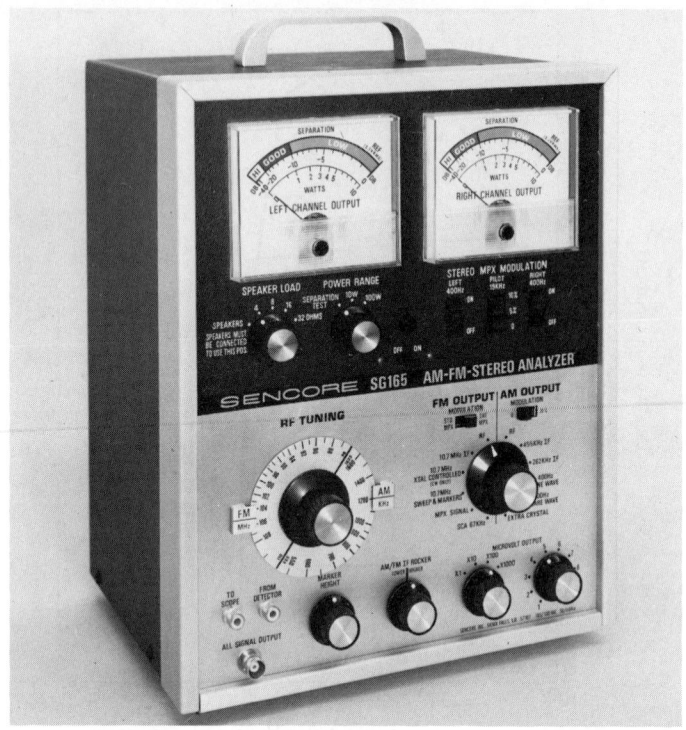

Figure 6-23 An AM-FM-stereo analyzer. (*Courtesy of Sencore, Inc.*)

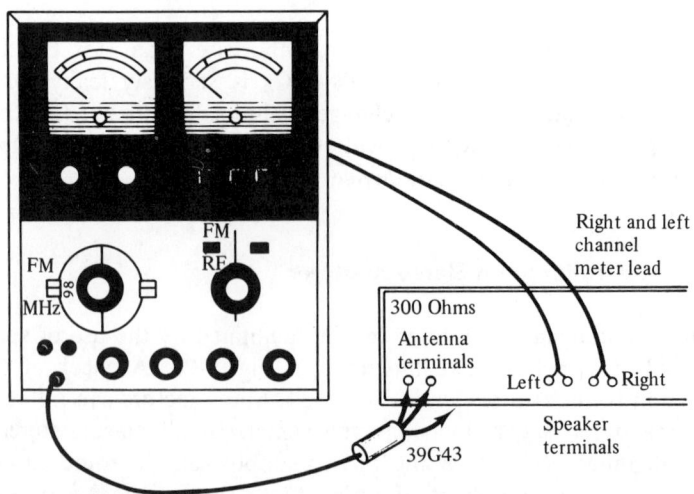

Figure 6-24 Analyzer connections for stereo separation test.

164

signal is a poor vehicle for determining a receiver's performance, whereas a check with a stereo analyzer will quickly indicate the extent of the problem. In most cases, it is not even necessary to remove the receiver from its cabinet to check stereo separation with a stereo analyzer. Procedure is as follows:

1. Remove the antenna from the receiver, but leave the speakers connected. Turn the receiver volume or loudness control up so that even a weak signal is audible, and with the receiver's AFC off, and all other controls set for a flat audio response, tune the receiver to a quiet spot on the dial near 98 MHz.
2. Turn the receiver power off, remove the speakers, and connect the analyzer's left and right meter leads to the receiver speaker terminals, as depicted in Fig. 6-24. Set the analyzer's speaker-load switch to the load required by the speaker. If the load requirements are not given, the technician may use the 16-ohm position. Then set the meter-watts-range switch to its separation-test position.
3. Connect the analyzer's all-signals-output cable (with matching pad) to the antenna terminals of the receiver. In the case of an auto radio, the associated dummy antenna should be used, instead of the matching pad. Then set the analyzer's output-selector control to FM-RF, and set the modulation control to STD MPX. This provides a standard test signal for checking stereo separation.
4. Adjust the analyzer's microvolt-output controls for a 500-μV signal level.
5. Switch the L and R 400-Hz switches "on," and set the pilot 19-kHz switch to its 10 percent position.
6. Turn the receiver power on, and adjust the analyzer's RF-tuning control for a maximum indication on the L and R channel-output meters. Reduce the receiver volume or loudness control as necessary to keep the output meters indicating below full scale.
7. Fine-tune the receiver as follows: (a) For receivers with a zero-center tuning indicator, adjust the receiver's tuning control for zero-center indication. (b) For receivers with a peak tuning meter, adjust the receiver's tuning control for maximum indication on the tuning meter. Note that if the receiver has both a zero-center and a peak tuning meter, the foregoing indications should coincide. If they do not coincide, misalignment of the receiver's IF amplifiers or detector is indicated. (c) For receivers with no visible tuning indicator, adjust the receiver's tuning control for maximum indication on the analyzer's output meters.

8. Adjust the receiver's volume or loudness control and balance control for a full-scale (0-dB reference) indication of the L and R channel output meters. (See Fig. 6-25.) Note that if it is necessary to adjust the receiver's balance control more than 20 percent from center, a defect is indicated either in one of the audio channels, or in the multiplex decoder circuit. Note also that, if a receiver or tuner is not capable of a full-scale indication on the output meters, separation must then be measured by comparison of dB values.

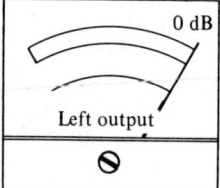 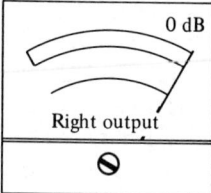

Figure 6-25 Reference adjustment for both stereo channels in receiver.

9. Alternately turn on and off the L and R 400-Hz switches, and observe the L and R channel output meters. Read the separation directly in dB for the channel with the 400-Hz signal "off." With reference to Fig. 6-26, the readings for both channels should be at least 20 dB, with a maximum difference between readings of 10 dB. Otherwise, receiver malfunction is indicated.

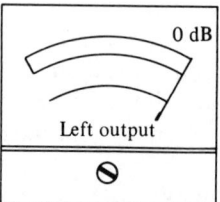 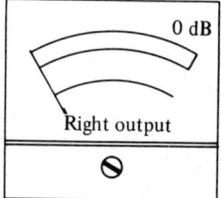

Figure 6-26 Separation measurement, with R channel "off."

10. Check the effect of the receiver tuning control on the separation reading. If better separation is obtained at some tuning point other than that observed in the foregoing step 7, misalignment of the receiver's IF stages or detector is indicated.
11. Check the "lock-in" range of the receiver's multiplex system by switching the 19-kHz pilot to 5 percent level, and observ-

ing the effect on receiver separation. A properly operating receiver should produce nearly the same separation with the 19-kHz pilot at the 5 percent level as at the 10 percent level. In case of poor lock-in, the alignment of the multiplex decoder's 19-kHz and 38-kHz circuits should be checked.
12. For a more thorough check, the foregoing steps may be repeated at RF frequencies of 90 MHz and 106 MHz.

Next, consider measurement of the sensitivity of an FM receiver. This is equal to the number of microvolts signal input required to produce a 30-dB signal-plus-noise to noise ratio. The diagram in Fig. 6-27 shows the effect of the noise and signal outputs of a receiver with respect to the input signal. Observe that the noise decreases and the signal increases as the input signal increases. At some point, while the output signal is increasing and the noise is decreasing, the ratio between them will be 31 to 1 (30 dB). This is the point at which the sensitivity measurement is made.

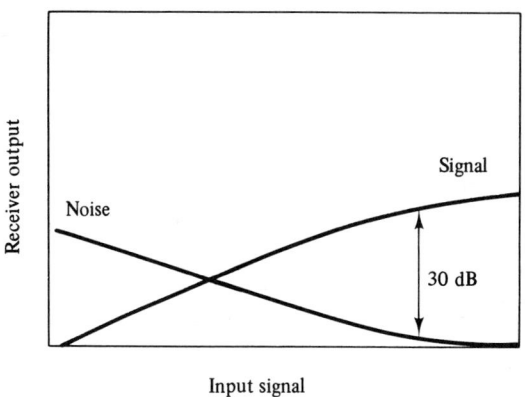

Figure 6-27 Relative receiver output versus noise and signal inputs.

Proceed as follows:

1. Set up the receiver and analyzer by making the connections as detailed in the foregoing first three steps.
2. Switch the L and R 400-Hz signal "on," and set the 19-kHz pilot level to 10 percent. Set the FM-modulation control to its IHF MPX position.
3. With the microvolt-output controls set for a low signal level of 100 μV or less, fine-tune the receiver as in steps 6 and 7 of the foregoing procedure. Note that if it is not possible to

eliminate the third-harmonic distortion from the signal by careful fine tuning, insufficient FM IF bandwidth is indicated. Realignment is in order in this situation.

4. Turn the analyzer's power switch "off." Adjust the receiver's volume control for an indication of -30 dB on the output meters of the analyzer.
5. Turn the analyzer's power switch "on." Without changing the output settings of the receiver, adjust the analyzer's microvolt-output controls to produce a full-scale indication of 0 dB on the output meters.
6. Note the setting of the microvolt-output controls, and multiply the setting of the coarse control by the corrected output of the fine control as listed in Fig. 6-28. When the pad or the dummy antenna is used, the foregoing result is multiplied by 0.5 to determine the input signal level in microvolts.

Control setting	Corrected output
10	10
9	8.1
8	6.4
7	5.1
6	4.0
5	3.0
4	2.4
3	1.8
2	1.4
1	1.2

Figure 6-28 Example of corrected microvolts output table.

6-11 Applications of an Impedance Bridge

An impedance bridge is useful for checking R, C, and L values in audio work. It is the only type of instrument that can measure inductance values directly. A service-type impedance bridge generally includes four separate bridge circuits for measuring resistance, capacitance, low inductance, and high inductance values. Null indication is typically provided by a 100-μA galvanometer. Resistance values from 0.1 ohm to 10 megohms, capacitance values from 100 pF to 100 μF, and inductance values from 0.1 mH to 100 H can be measured. Readout of Q values from 0.1 to 1000, and of dissipation factors from 0.002 to 1 is also provided. A built-in 1-kHz sine-wave generator is provided. Alternatively, a service-type bridge can be externally driven by a 60-Hz sine-wave source, for example. An impedance bridge is particularly valuable for selection of inductive replacement components, when exact replacements are unavailable.

Basic bridge circuitry is shown in Fig. 6-29, and the appearance of a typical impedance bridge is seen in Fig. 6-30. Resistance measurements are made with a Wheatstone bridge, pictured in basic form in Fig. 6-29(a). A dc-voltage source is utilized, with balance indication by a galvanometer. Figure 6-29(b) shows the arrangement of an elementary capacitance bridge. Balance is indicated by an ac galvanometer that usually employs semiconductor rectifiers. An elementary inductance-bridge arrangement is depicted in Fig. 6-29(c). This arrangement is chiefly of theoretical interest. A somewhat more useful arrangement is shown in Fig. 6-29(d). This is an inductive impedance-bridge arrangement in which the resistance R1 associated with inductor L1 can be balanced by resistance R2 in combination with inductor L2.

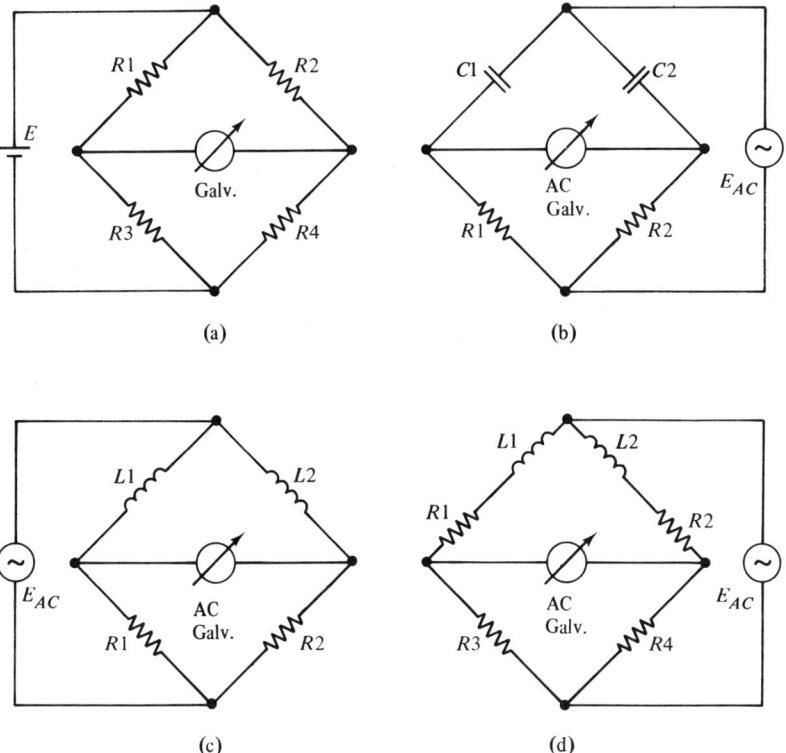

Figure 6-29 Basic impedance-bridge circuits: **(a)** Wheatstone bridge; **(b)** elementary capacitance bridge; **(c)** elementary inductance bridge; **(d)** skeleton inductive impedance bridge.

170   Hi-Fi Stereo Troubleshooting

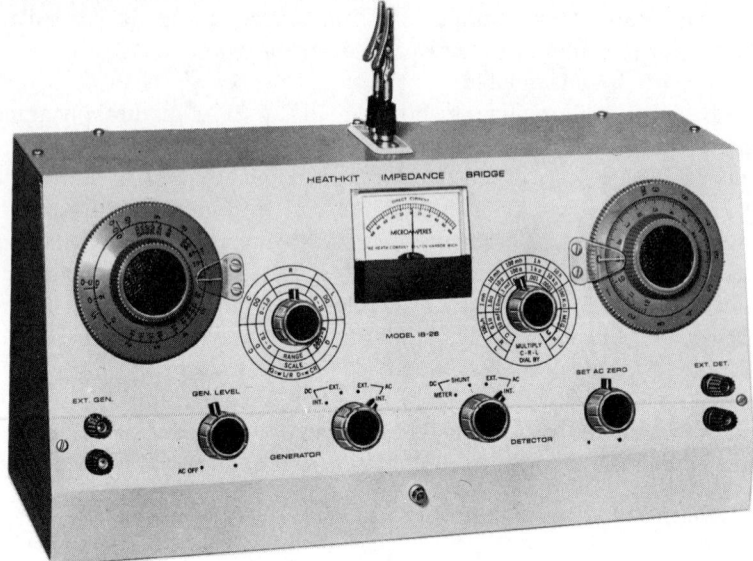

Figure 6-30   A service-type impedance bridge. *(Courtesy of Heath Co.)*

An inductive impedance bridge commonly used in service-type bridges is the basic Hay bridge diagrammed in Fig. 6-31. This arrangement is more practical than the former configuration, because a standard capacitor is utilized instead of a standard inductor for obtaining bridge balance. A standard capacitor is not only much more economical than a standard inductor, but it is also relatively immune to stray-field pickup that could upset the null indication. A Hay bridge is most useful for measuring high Q values, although it can also be used to measure low Q values with a tradeoff in measurement accuracy. When coils with low Q values are being checked, it is advisable to switch the bridge

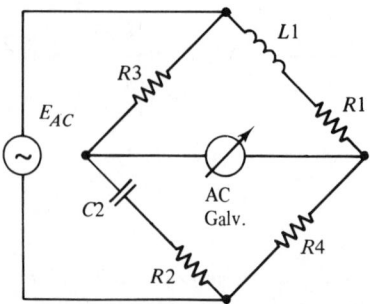

Figure 6-31   Plan of the Hay inductive impedance bridge.

configuration to the basic Maxwell-bridge configuration, shown in Fig. 6-32. Although a Maxwell bridge can be used to check coils with high Q values, a tradeoff in indication accuracy is encountered.

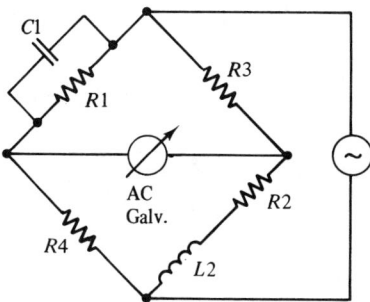

Figure 6-32 Plan of the Maxwell inductive impedance bridge.

6-12 Varistor Response to AC Voltage

Varistors are used in some types of ac circuitry. There are two basic types of varistors, as shown in Fig. 6-33. An unsymmetrical varistor is exemplified by a conventional germanium diode. Or, a symmetrical varistor is exemplified by two germanium diodes connected back to back. If a sine-wave voltage is applied to an unsymmetrical varistor, rectification occurs, and a dc component of current flows. On the other hand, if a sine-wave voltage is applied to a symmetrical varistor, no rectification occurs, and a dc component of current is developed. Again, if a pulse-type ac voltage is applied to a symmetrical varistor, partial rectification occurs, and no dc component of current flows. In this situation, partial rectification takes place owing to the unsymmetrical waveform of the applied voltage, as shown in Fig. 6-34. Typical applications for symmetrical varistors are found in television sweep circuitry. Typical applications for unsymmetrical varistors are found in temperature-compensating bias networks for transistors.

6-13 Notes on Tools and Accessories

Troubleshooting of solid-state circuitry on printed-circuit boards requires a well-tinned, pencil-type soldering iron. Care should be taken not to overheat transistors, diodes, and integrated circuits. Printed-circuit boards can also be damaged by overheating. Solder with 60 percent tin and 40 percent lead (60/40) solder is preferred. Heat sinks,

172   Hi-Fi Stereo Troubleshooting

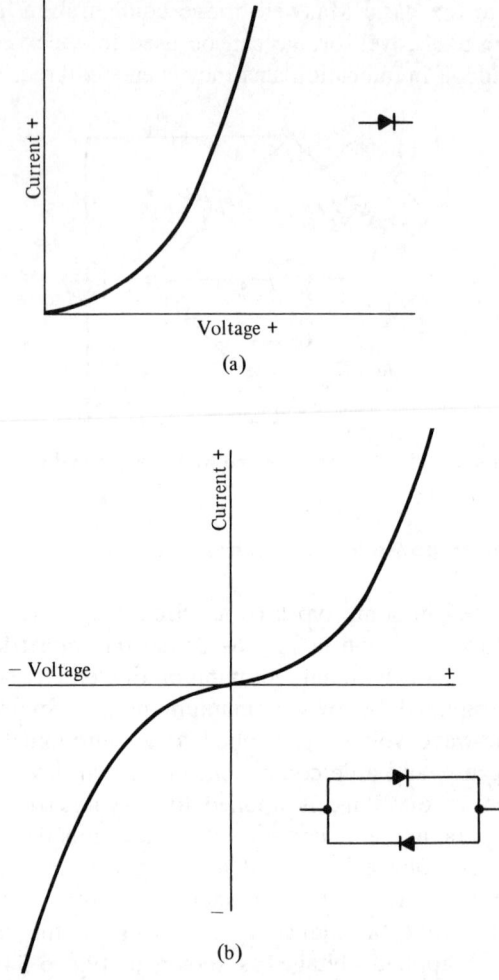

Figure 6-33 Basic varistor characteristics: **(a)** unsymmetrical varistor; **(b)** symmetrical varistor.

or their equivalent, should always be attached to leads of solid-state devices during soldering or desoldering operations. Electrical leakage between the heating element and the tip in a soldering iron can cause the tip to be above ground potential. This spurious voltage can cause transistor or IC damage if the chassis is returned to ground in any way. It is good practice to connect a flexible grounding strap from the metallic neck of the soldering iron to a low-resistance ground, such as a water pipe.

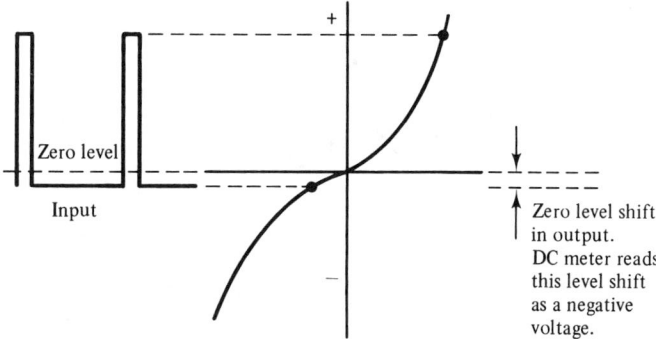

Figure 6-34 Dissymmetry of input voltage waveform produces a dc component in the varistor output.

Meter test probes with sharp tips are preferred, both to avoid the possibility of accidental short circuits to adjacent conductors, and to pierce resin, varnish, or surface corrosion on PC conductors. If poor test connections are made, the false readings will be attributed to nonexistent malfunctions in the equipment under test. Solid-state equipment is very compact. In turn, miniature tools and connectors are a practical necessity. Small needle-nosed pliers, tweezers, a magnifying glass, and thin-nosed diagonal cutters will be found to be very useful. A vacuum-type desoldering tool should be used to remove components and devices from a PC board. Note that molten solder collected by a desoldering tool should be discarded, and not reused. In other words, the remelted solder becomes objectionably oxidized, and cold-solder joints are likely to result from its reuse.

7
Radio Receiver Troubleshooting

7-1 General Considerations

Radio receivers can be classified fundamentally into AM and FM types. Although many of the same troubleshooting procedures are used with both types of receivers, there are also some basic differences. It is instructive to start with consideration of AM troubleshooting techniques, and then to show which of these techniques are inapplicable to FM receivers, with explanation of specialized FM troubleshooting techniques. Various kinds of trouble symptoms are encountered in AM radio reception, such as "dead" receiver, weak reception, distorted output, interference (poor selectivity), oscillator dropout, erratic operation, and intermittent reception. Localization tests are made in a preliminary checkout of receiver response. These tests include visual inspection, battery voltage measurement, or power-supply voltage measurement, total current drain measurement, speaker-circuit click test, local-oscillator test, and current variational analysis.

A visual inspection will often reveal physical damage, such as broken fine wires from the loopstick antenna, breaks at the tuning-capacitor terminals, or breaks at circuit-board terminals. The loopstick should be inspected for cracks, and the tuning capacitor should be turned through its range to ensure that it does not bind and that plates do not touch. Leads to the volume control, speaker, earphone jack, and battery should be inspected for damage. Circuit boards should be carefully inspected for cracks. A circuit board may be flexed at several points by applying light pressure with the eraser end of a pencil. If momentary return of sound occurs, it is likely that there is a crack in a printed-circuit conductor or a poor connection to a component lead.

It is sometimes possible to pinpoint such defects without making any electrical tests. It is good practice to check the battery polarity in battery-type receivers, because a replacement battery is sometimes accidentally reversed. If this accident occurs, it may be necessary to replace various transistors and electrolytic capacitors that have been damaged by the polarity reversal. Note that if a battery weakens to the point that it supplies less than two-thirds of its rated voltage, the receiver is likely to become inoperative. As noted previously, battery voltage must be measured under normal load.

7-2 Current Drain and Operating Voltages

Each transistor in a receiver imposes a certain current drain. Each transistor terminal operates at a specified dc voltage. All tuned circuits and transistors operate at typical signal-voltage levels. A block diagram for a representative six-transistor AM radio receiver, with general signal-voltage levels is shown in Fig. 7-1. It is usually impractical to measure these signal voltages directly with high-frequency ac voltmeters. Instead, the technician ordinarily determines the signal-voltage levels indirectly. For example, if a 100-μV signal is injected at the input of the first IF amplifier, an ac voltmeter connected across the audio output circuit normally indicates 0.7 volt (if a 10-ohm load is assumed). In any case, the audio output stage will normally supply 50 mW of audio signal to the load.

As would be anticipated, many component and device defects result in a change of the dc operating voltages in one or more stages. If the operating point of a single stage is greatly changed, the total current drain of the receiver usually changes also. This measurement is easily made, as shown in Fig. 7-2. Rated current drain is often specified in the receiver service data. Otherwise, the technician may be able to make a comparison test on a similar receiver that is in good operating condition. An abnormal current drain generally indicates a short-circuited or leaky filter or decoupling capacitor. A subnormal current drain is likely to be caused by an open-circuited transistor or a break in the dc path of a stage. If the normal current drain is not known for a receiver, it can be estimated by the following rule of thumb: With reference to Fig. 7-3, a converter stage will draw from 0.5 mA to 1 mA; each IF stage will draw about 1.5 mA, as will an audio driver stage; a class B push-pull stage will draw approximately 4 mA. If a single class A audio-output stage is employed, it will draw approximately 4 mA.

If the current drain is zero, the supply circuit is probably open-cir-

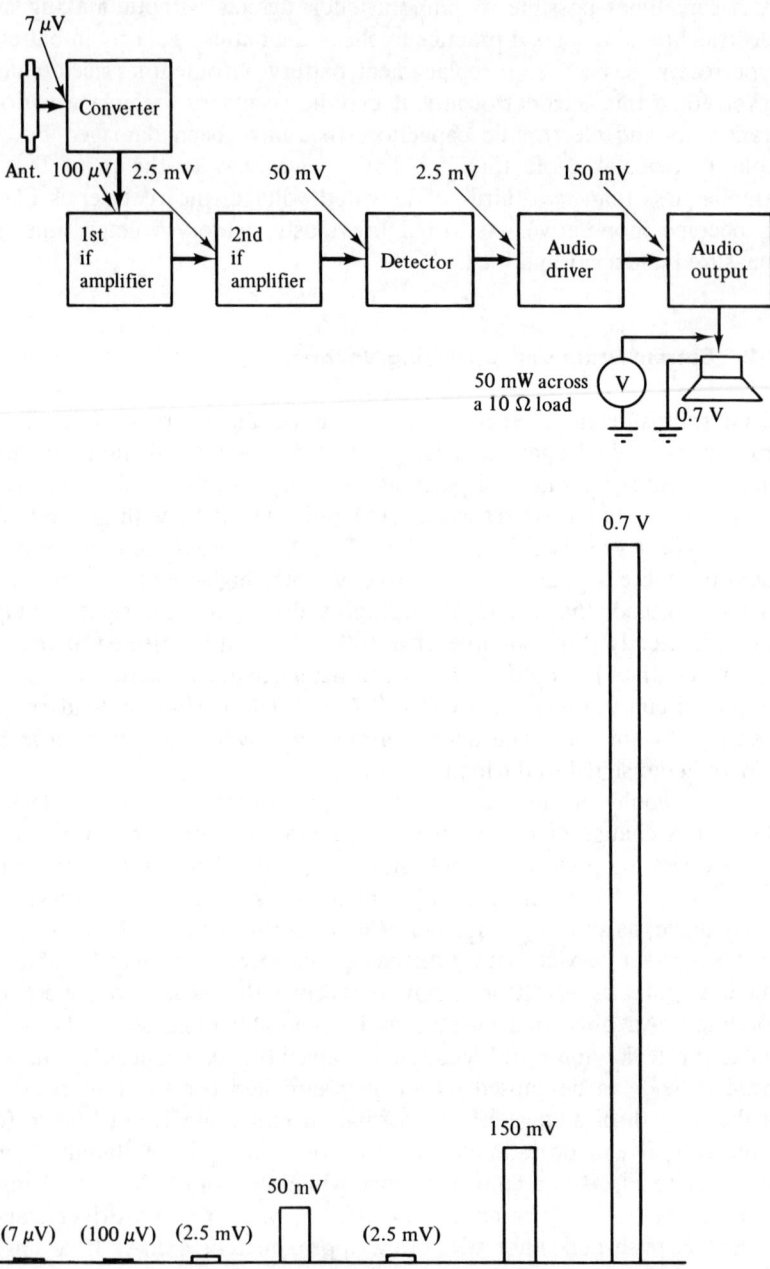

Figure 7-1 Signal-voltage progression for a typical six-transistor AM radio receiver.

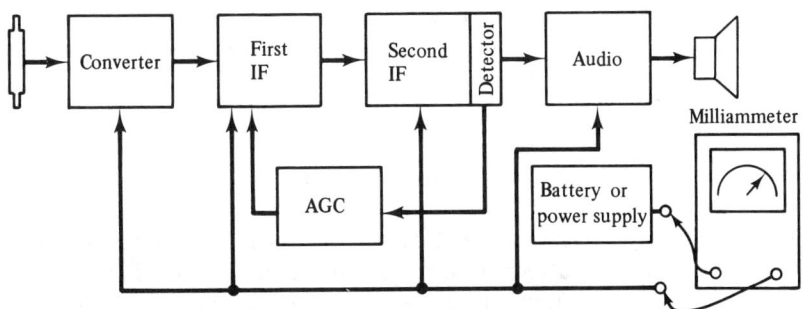

Figure 7-2 Current-drain check for AM radio receiver.

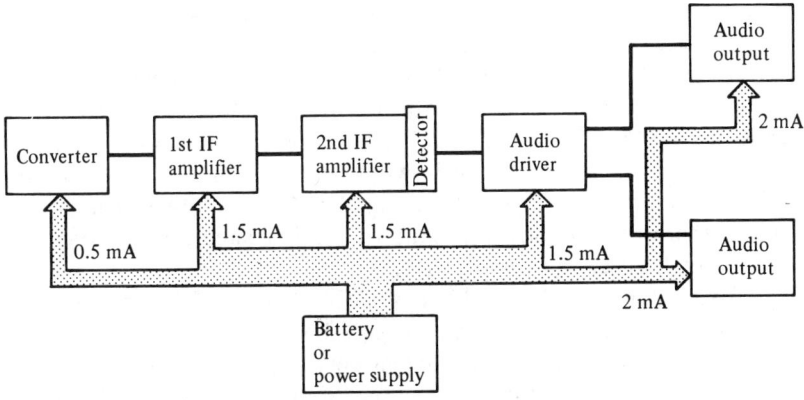

Figure 7-3 Current distribution for a typical six-transistor receiver under no-signal conditions.

cuited. For example, off-on switches may become worn and defective. If the no-signal current drain is as much as ±50 percent incorrect, there is probably a short-circuited decoupling capacitor or a very large change in the operating conditions of a stage. DC current measurements often suffice to pinpoint a faulty component or device. On the other hand, if the current drain is approximately normal, additional test methods must be utilized to track down the malfunction. One of these methods is called the speaker-circuit click test. It can localize a defect in the collector circuit of the output stage. The technician holds the speaker close to his ear, and listens for a click as the receiver is turned off and on. If no click is heard from the speaker, it is concluded that an open circuit will be found in the speaker, output transformer, or associated connections. For example, there may be a defective normally closed contact in the earphone jack, or possibly a broken connection to the jack.

7-3 Systematic Troubleshooting Procedures

If preliminary analysis does not reveal the cause of receiver malfunction, systematic troubleshooting procedures are required. A "dead" receiver can result from failure of the local oscillator. As shown in Fig. 7-4, normal dc operating voltages are specified in receiver service data. If the local oscillator fails, its base and emitter voltages will change. Note also that a useful quick check of oscillator operation can be made with another receiver that is in normal operating condition. Tune the other receiver to some broadcast station in the range between 1 MHz and 1.5 MHz. Then place the two receivers near each other and turn the tuning dial of the "dead" receiver through the frequency range from 545 kHz to 1.45 MHz. If the oscillator in the "dead" receiver is operating, a loud whistle will be heard from the good receiver, owing to interference from the oscillator. On the other hand, if no whistle (zero-beat action) is produced, it is indicated that the oscillator in the "dead" receiver is inoperative.

Next, consider the analysis of AGC trouble. With reference to Fig. 7-4, dc-voltage measurements may be made and compared with specified values. However, a current-demand test is informative and easier to use for preliminary analysis. Malfunction in the AGC section(s) can cause weak reception, overloading, or "dead receiver" trouble symptoms. In normal operation, the collector current of AGC-controlled IF stages varies from one milliampere to a few tenths of a milliampere, depending on the amplitude of the incoming signal. Note that this change in IF collector-current demand can be checked by monitoring the total current drain of the receiver while tuning through the band. With reference to Fig. 7-5, if the signal proceeds through the receiver to the speaker, and a class B output stage is used, the current demand increases appreciably as the receiver is tuned through a station frequency, with the volume control advanced. On the other hand, if the volume control is turned down, the current demand decreases noticeably as the receiver is tuned through a station frequency, provided that the signal progresses through the IF system.

If the receiver utilizes a class-A audio-output stage, the setting of the volume control will normally have very little effect on the total current demand. One method of varying the gain of a transistor is to vary

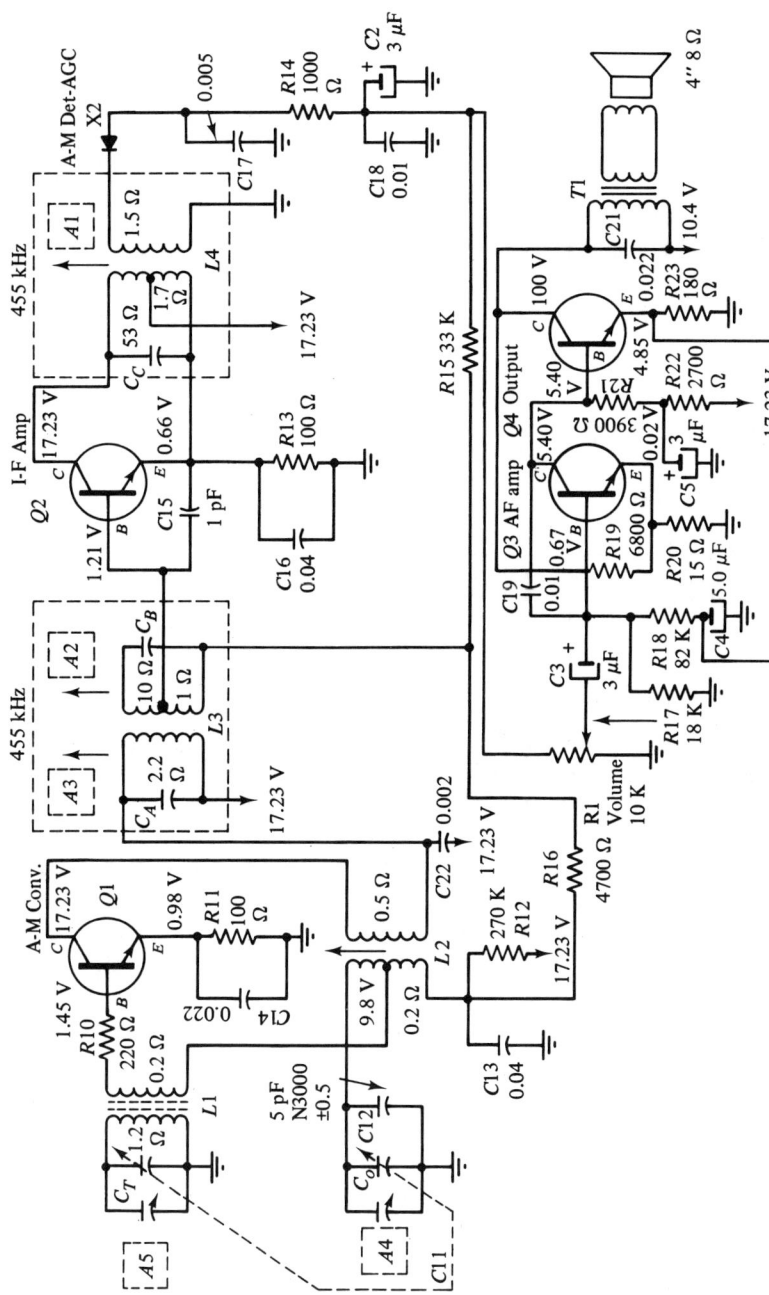

Figure 7-4 Normal dc operating voltages for a typical AM radio receiver.

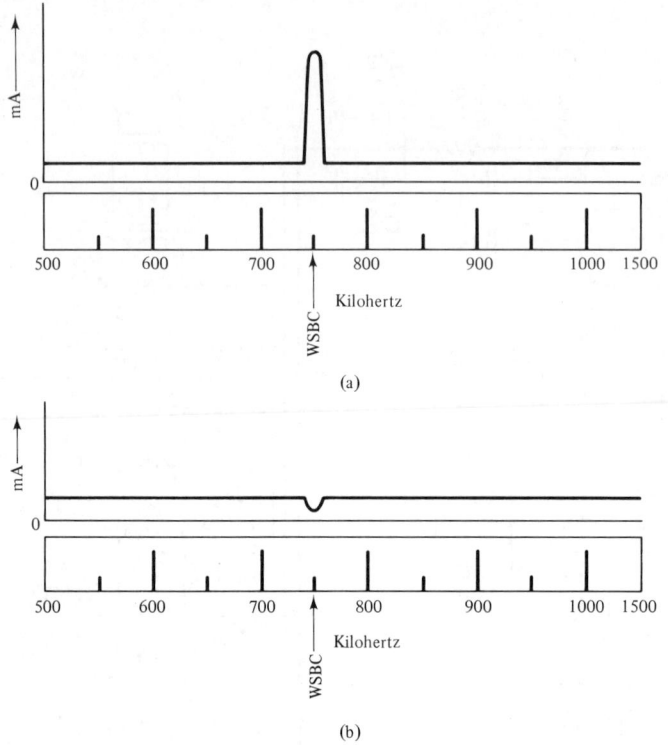

Figure 7-5 Current variation versus signal condition in normal operation: (a) with volume control advanced; (b) with volume control turned down.

the dc emitter current, as shown in Fig. 7-6(a). Another method of varying the gain of a transistor is to vary the dc collector voltage, as shown in Fig. 7-6(b). These curves are for 455-kHz IF amplifiers operating with a normal emitter current of 1 mA. In the configuration of Fig. 7-7(a), the AGC voltage varies the emitter current of the transistor. The AGC voltage is positive, and reverse-biases the transistor to reduce its emitter current and lower its gain. Next, in the configuration of Fig. 7-7(b), the AGC voltage is negative, and forward-biases the transistor to increase its emitter current and reduce its collector voltage, thereby lowering its gain. Note that R5 has a value of 10 kΩ, or more. In turn, the increased collector current produces a large voltage drop across R5, which reduces the collector voltage on the transistor. In turn, the stage gain decreases. Note that if forward AGC is used, the current demand does not decrease in a variational test [Fig. 7-5(b)], but instead the current demand will increase appreciably.

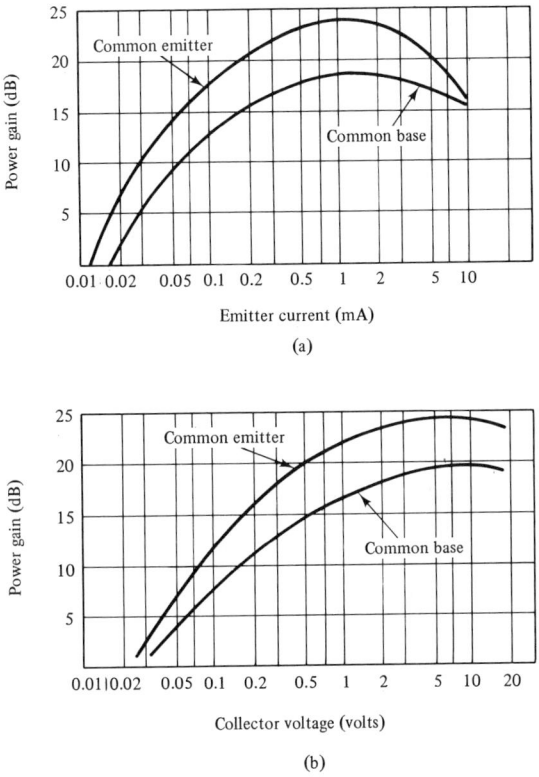

Figure 7-6 Reverse-AGC and forward-AGC characteristics: **(a)** transistor power gain with reverse-AGC variation; **(b)** transistor power gain with forward-AGC variation.

7-4 Signal-injection Tests

Signal-injection tests are made by connecting an ac voltmeter at the output of the audio amplifier, and applying signal voltages at successive stages from a signal generator, as shown in Fig. 7-8. The voltmeter indicates whether the signal is progressing through the receiver section under test, and it also indicates the voltage gain of each stage. In other words, as the generator-signal injection point is moved from the collector to the base of a transistor, the change in ac-voltage output is a measure of the stage gain. Audio stages are checked with an AF signal; a frequency of 1 kHz is typical. IF stages are checked with a modulated-IF signal; 30 percent modulation is standard. Similarly, RF stages are checked with a modulated RF signal; again, 30 percent modulation is

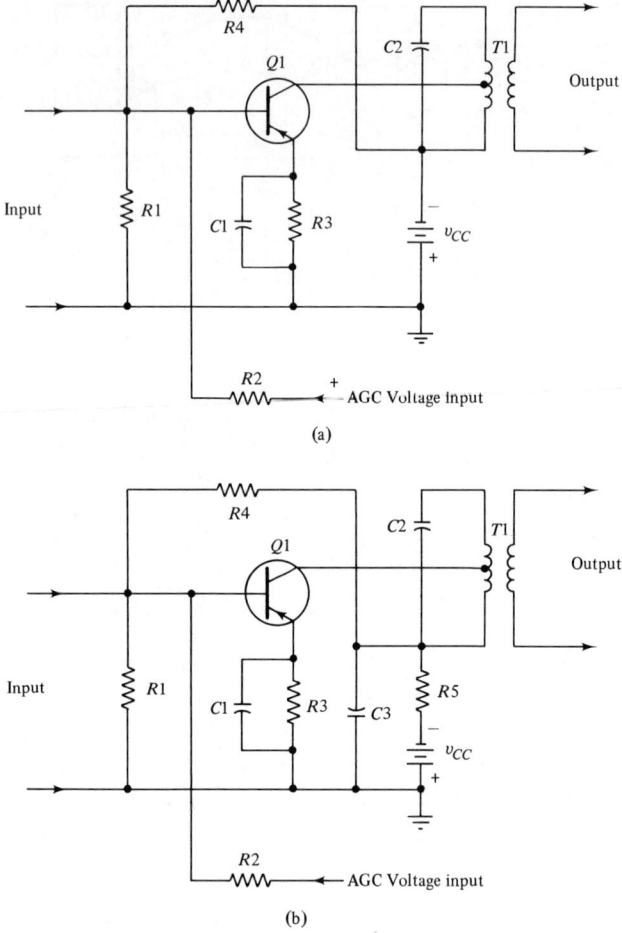

Figure 7-7 Reverse-AGC and forward-AGC circuitry: **(a)** reverse-AGC configuration; **(b)** forward-AGC configuration.

standard. Note that a blocking capacitor is connected in series with the "hot" output lead from the generator. This capacitor prevents bias-voltage or supply-voltage drainoff from the point of signal injection.

Most service-type signal generators do not have calibrated output facilities. Since the technician occasionally has need to know the approximate level of an injected signal, a field-strength meter that has a suitable frequency range is a helpful instrument. As shown in Fig. 7-9, the RF output from the signal generator is applied to the field-strength

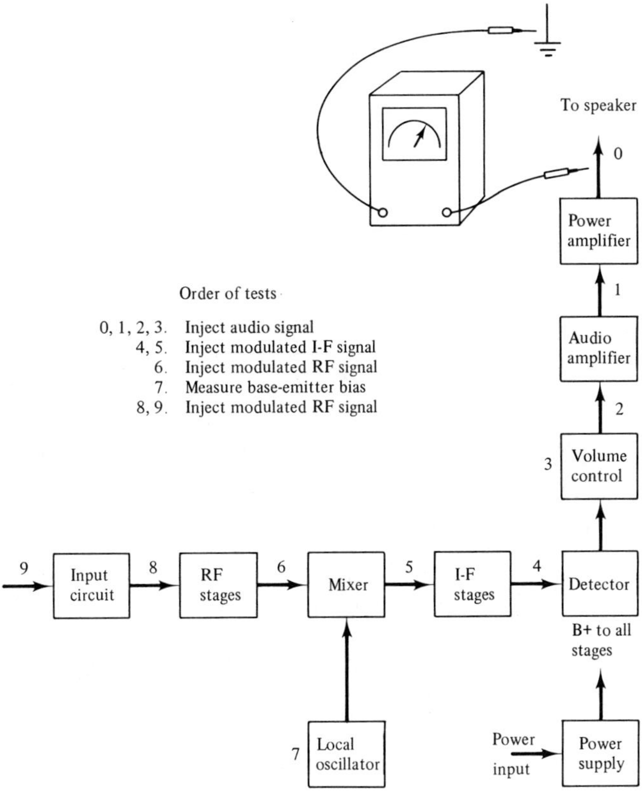

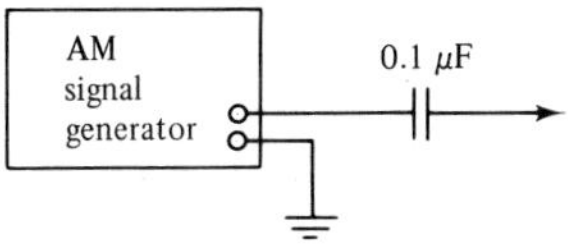

Order of tests

0, 1, 2, 3. Inject audio signal
4, 5. Inject modulated I-F signal
6. Inject modulated RF signal
7. Measure base-emitter bias
8, 9. Inject modulated RF signal

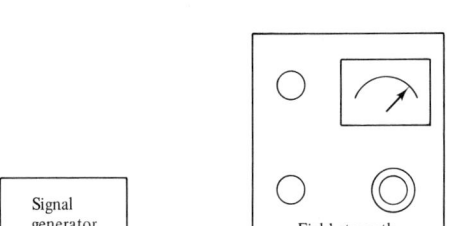

Figure 7-8 Basic signal-injection procedures.

Figure 7-9 Measuring the output voltage from a signal generator.

meter. In turn, the FS meter indicates the signal amplitude in microvolts. Note also that accessory calibrated attenuator units are available, such as that shown in Fig. 7-10. When connected in series with a signal generator and the receiver under test, an accessory attenuator provides means for adjusting the injected signal to a known level. An RF voltmeter is connected at the input of the attenuator to measure the signal-voltage level. Then the attenuator is adjusted to reduce this level by a factor of 10, 100, 1000, 10,000, and so on.

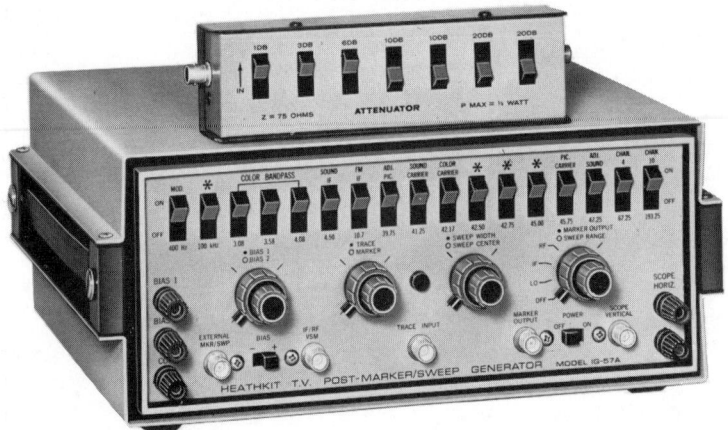

Figure 7-10 Step attenuator for use with a signal generator. (Courtesy of Heath Co.)

An RF voltmeter, a signal generator, and a calibrated attenuator are arranged as shown in Fig. 7-11. Connection is made to the voltmeter by means of a T connector in the output cable from the generator to the attenuator. The generator may be adjusted for a reading of 0.1 volt rms on the voltmeter, for example. Then, if the attenuator is adjusted for a factor of 100 times, the RF output voltage will be 1,000 microvolts rms. Or, if the attenuator is adjusted for a factor of 10,000 times, the output voltage will be 10 microvolts rms. Note that the output cable from an attenuator is terminated in a specified resistance, such as 50 ohms. If the cable termination is omitted, the attenuator calibration indication will be inaccurate.

7-5 AM Alignment Procedures

Symptoms of misalignment include poor sensitivity, broad tuning (poor selectivity), and possible interference. As a general rule, alignment

7-5 AM Alignment Procedures   185

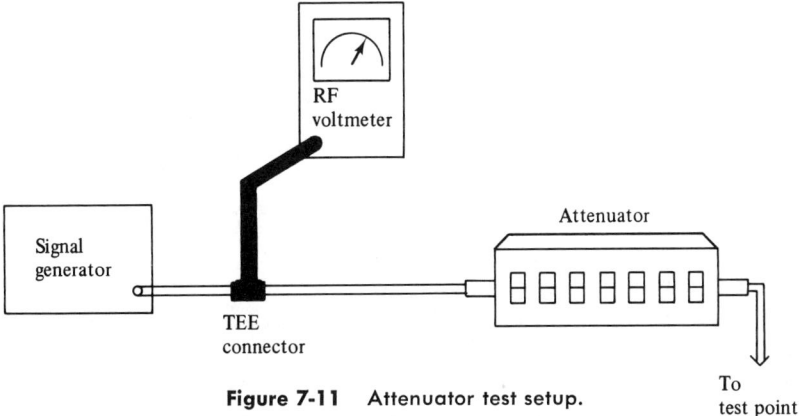

Figure 7-11   Attenuator test setup.

adjustments are made after all troubleshooting procedures have been completed. An exception to this rule is observed when it is known or suspected that alignment adjustments have been tampered with. Although receiver alignment tends to drift slightly after extended periods of time, substantial misalignment results from component defects. As an illustration, if C1 or C2 becomes open-circuited in the IF configuration depicted in Fig. 7-12, the stage will become misaligned. After a tuned-circuit component or its associated transistor is replaced, realignment is often required because of component tolerances and the resulting effective change in circuit capacitances.

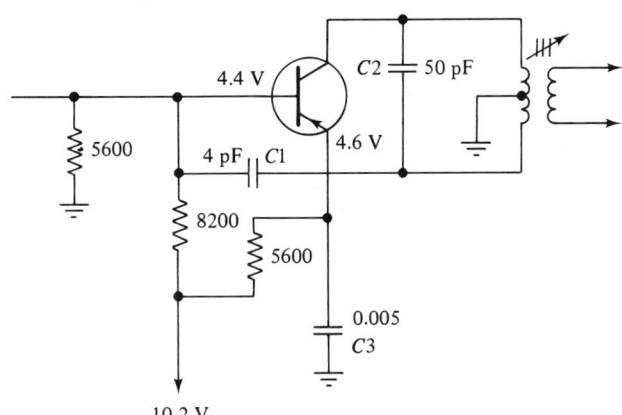

Figure 7-12   Stage misalignment can be caused by open capacitors.

AM radio receivers are usually aligned by the peak-response method. A typical alignment setup is depicted in Fig. 7-13. Alignment adjustments are made with respect to individual specified frequencies

186 Radio Receiver Troubleshooting

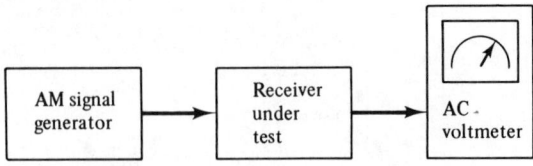

Figure 7-13 Basic alignment setup.

in the receiver service data. A weak generator signal is customarily employed, so that the AGC section of the receiver remains essentially inoperative. In other words, if a strong alignment signal were used, the AGC section would automatically reduce the receiver output as the tuned circuits were brought into peak alignment, and the maximum response point would tend to be masked. Unless the receiver service data specify otherwise, alignment procedure for an AM radio follows these consecutive steps:

1. Each tuned-IF circuit is peaked at the specified frequency (such as 455 kHz); a 30 percent modulated signal is used. (See Fig. 7-14.)
2. The oscillator trimmer capacitor (or slug) is adjusted for maximum receiver response at 1600 kHz on the tuning dial.
3. The oscillator padding capacitor (if provided) is adjusted for maximum response at 600 kHz on the tuning dial.
4. Each tuned-RF circuit is adjusted for peak response at the frequency (or frequencies) specified in the receiver service data.

7-6 Troubleshooting Distortion and/or Regeneration

When distortion occurs in an AM receiver, such as mushy or raspy sound output, sometimes accompanied by squeals or whistles, the trouble is most likely to be found in the audio section. Sometimes malfunction is caused by overdrive, or a shift may have occurred in the dc operating point. Some detector-circuit faults cause distorted sound output. The IF amplifier can also distort the sound if regeneration occurs. It is often helpful to listen to the quality of the sound output with the control turned high, and then turned low. If the distortion remains unchanged at various volume levels, it is probable that the defect is located ahead of the volume control. It is advisable to check out the detector circuit in this event. On the other hand, if the distortion increases as the volume control is advanced, the trouble is most likely to

Alignment Instructions

Maintain line voltage at 117 volts. Use only enough generator output to obtain a suitable indication. Allow a 15 minute warmup for receiver and equipment. CAUTION: Use isolation transformer, if available. If not, observe polarity when connecting test equipment. Suggested Alignment Tools: A1 thru A13..........GENERAL CEMENT #8868, 8987, 9089....WALSCO #2531-X, 2541, 2587	

AM alignment

Fashion loop of several turns of wire and connect generator across loop. Set volume control at maximum.					
	Generator frequency	Dial setting	Indicator	Adjust	Remarks
1.	455 kHz (400~ Mod.)	Tuning gang fully open.	Output meter across voice coil.	A1, A2, A3	Adjust for maximum. Repeat until no further improvement can be made.
2.	1600 kHz	Tuning gang fully open.	Output meter across voice coil.	A4	Adjust for maximum
3.	1400 kHz	Tune to signal.	Output meter across voice coil.	A5	Adjust for maximum.
4.	600 kHz	Tune to signal.	Output meter across voice coil.	A6	Rock tuning gang and adjust for maximum. Repeat steps 2 thru 4 until no further improvement can be made.

Figure 7-14 Typical alignment instructions for an AM receiver. (*Courtesy of Howard W. Sams & Co., Inc.*)

be in the audio section, and overdrive is indicated, owing to a dc level shift. However, if the distortion increases at low volume levels, suspect crossover distortion.

If distortion is the same at various volume levels, but increases for weak-station reception, it is likely that the detector does not have its needed small forward bias. It is also possible that a replacement detector diode may have been installed in reverse polarity. In such a case, strong signals will also be distorted owing to associated AGC malfunction. A detector diode normally operates with approximately 0.1 volt of forward bias. Distortion accompanied by critical tuning, often with a hissing sound or whistles, is most likely to result from IF or RF regeneration. If squealing persists without a change in frequency, but only changes in volume as the volume control is turned, it may be assumed that regeneration is occurring within a comparatively restricted area, such as the RF, IF, and the detector/AGC sections. On the other hand, if the squeal changes in frequency as the volume control is turned, it is indicated that the audio section is also involved in the regenerative loop. Also, if the squealing starts or stops at some setting of the volume control, it may be assumed that the audio amplifier is involved.

Consider the analysis of distortion that is confined to the audio section. DC voltage measurements should be made at the transistor terminals. A small forward bias of 0.09 to 0.12 volt is customarily employed by the class B output transistors to prevent crossover distortion. When bias voltages are incorrect, off-value resistors may be the cause. Collector-junction leakage can cause reduced gain and shift of the operating point. Note that the ac audio voltage should be the same at the collectors of both class B output transistors. Unequal collector signal voltages point to unmatched transistors. If driver and output transformers are utilized, a transformer defect may be present. Measure the winding resistances on each side of the center tap(s). Unequal winding resistances are evidence of transformer breakdown.

Next, if distorted sound output appears to be caused by regeneration, the technician usually inspects ground connections first. In other words, any common ground connection that has appreciable resistance develops a common impedance across which feedback can occur. Decoupling capacitors are also ready suspects; they may be quick-checked by "bridging." However, always turn the receiver off before the test capacitor is connected across the suspect. Otherwise, current surges may damage transistors in the associated networks. Sometimes corroded battery connectors develop a common impedance that causes regeneration. Other possibilities are "open" neutralizing capacitors, or a replacement transistor that has abnormally high gain (beta value).

7-7 Troubleshooting FM Receivers

From the viewpoint of the troubleshooter, the chief difference between an FM receiver and an AM receiver is that the former utilizes a detection arrangement that is unresponsive to an AM signal. Another practical difference is in the much greater RF frequency range of an FM receiver, from 88 to 108 MHz. Although an FM detector is unresponsive to amplitude modulation, most service-type AM generators can be used to inject a test signal at any point in the signal channel, and a tone output will be obtained from the speaker. The reason for this perhaps unexpected action is that most service-type signal generators have amplitude-modulated oscillators. When an oscillator is amplitude-modulated, incidental frequency modulation also occurs, as depicted in

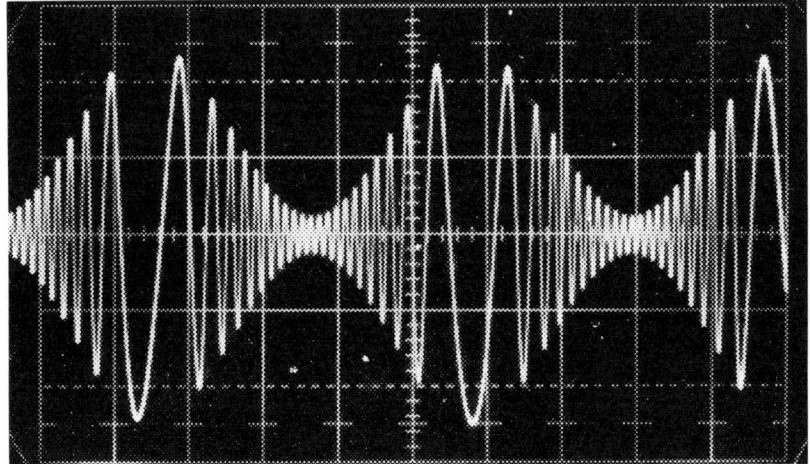

Figure 7-15 Amplitude modulation with incidental frequency modulation.

Fig. 7-15. This incidental FM is passed by the limiter and demodulated in the FM detector of the receiver.

Signal-injection procedure for an AM/FM receiver with common IF stages is shown in Fig. 7-16. If the receiver operates on its AM function, but not on its FM function, the IF section can be eliminated from preliminary suspicion. It is probable that the trouble will be found in the FM detector, limiter, converter, oscillator, or RF stage(s). Localization can almost always be made on the basis of signal-injection tests. An IF signal is first applied at the input of the second AM and FM IF stage. If a high percentage of modulation is utilized, incidental FM will be available for preliminary checks of the limiter and FM detector stages. The TVM is operated on its ac voltage function, and the signal is checked first at the input to the limiter (A). If the test signal is present, the TVM is next connected at the output of the FM detector. Again, if the test signal is present, the limiter and FM detector are eliminated from suspicion.

To proceed with signal-injection tests in Fig. 7-16, the TVM may be returned to the input of the limiter (A). Then the 10.7-MHz signal from the generator is injected at the input of the first AM and FM IF stage (2). In normal operation, an increase in the meter indication will occur, corresponding to the voltage gain of this IF stage. If the receiver has three or four IF stages, each stage is checked out in this manner. However, it is necessary to reduce the generator output progressively, to avoid overload of the IF section. An overload condition causes a stage to seem to have zero voltage gain. If all of the IF stages are eliminated from suspicion, the converter is checked next. A 10.7-

190   Radio Receiver Troubleshooting

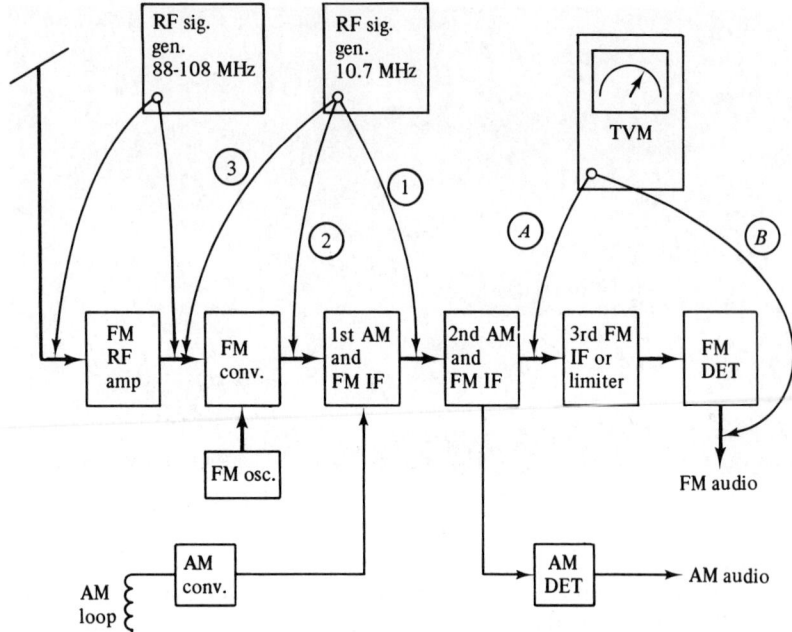

Figure 7-16   Signal injection procedure for an AM/FM receiver with common IF stages.

MHz signal from the generator is injected at the input of the FM converter stage (3). An increase in the meter reading occurs normally, corresponding to the converter stage gain.

If the converter stage is eliminated from suspicion, the RF stages are checked out next with an amplitude-modulated RF signal in the range of 88 MHz to 108 MHz. The receiver and generator are operated at the same frequency, such as 100 MHz. If the meter does not respond, it is concluded that the fault will be found in either the RF amplifier stage or in the FM oscillator circuit. Proceed by applying the 100-MHz signal at the input of the FM converter. In the event that the meter does not respond, it is concluded that the FM oscillator is at fault. Note that if two generators are available, a continuous-wave (CW) output can be employed from the second generator to substitute for the FM oscillator output. This substitute signal is applied at the output terminal of the oscillator circuit. A meter reading confirms the initial conclusion that the oscillator is "dead."

Pinpointing defective components or devices after sectionalization tests involves the same principles that were explained for troubleshooting AM receivers. In other words, dc-voltage measurements are basic. Supplementary resistance measurements are often advisable; a high-low ohmmeter permits many measurements to be made in circuit, without opening conductors or unsoldering components or devices. Suspected "open" capacitors can be quick-checked by bridging tests. Keep in mind that some malfunctions are deceptive. For example, an "open" capacitor in the FM oscillator section can shift the oscillating frequency so high that the receiver is "dead," although the oscillator circuit appears to be operating normally.

7-8 Weak Output in AM and FM Receivers

When a weak-output trouble symptom is encountered in an AM or FM receiver, do not overlook the possibility of a faulty overload diode. With reference to Fig. 7-17, D1 is reverse-biased on weak and medium strength signal levels, but becomes progressively forward-biased at strong signal levels. In other words, the overload diode is effectively out of the circuit until its cathode becomes forward-biased owing to the collector of Q2 rising to a higher negative potential than the collector of Q1. When V2 is more negative than V1, D1 loads down the primary of T1 and causes the gain of the Q1 stage to decrease considerably. The essential considerations are that D1 have a normal front-to-back ratio, and that the diode remain reverse-biased until the limit of the AGC dynamic range is approached.

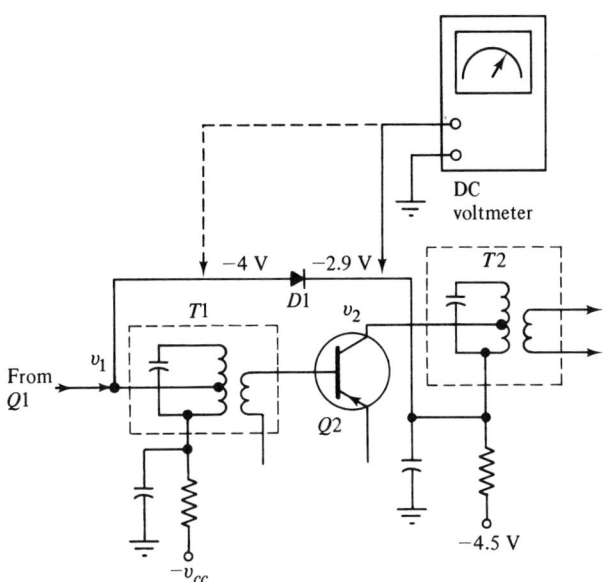

Figure 7-17 Measurement of overload diode bias voltage.

7-9 Troubleshooting CB Transmitters and Receivers

Citizens band (CB) transmitters and receivers are basically similar to broadcast equipment. However, CB transmitters are comparatively low-powered, ranging from 400 mW to 5 W input to the final amplifier. Figure 7-18 shows a configuration for a 5-W CB transmitter. The power rating is defined as the product of dc voltage and current to the collector of the RF power amplifier. In other words, to check this power specification, the technician measures the dc voltage at the collector of Q3, and also measures the collector current. For example, one end of L2 may be temporarily disconnected for a current measurement. In turn, the collector voltage multiplied by the collector current is equal to the dc power input to the final amplifier.

Elaborate CB receivers are essentially the same as short-wave broadcast receivers. Simple CB receivers and walkie-talkie receivers often employ superregenerative detectors, as exemplified in Fig. 7-19. Typical troubleshooting procedures for CB equipment are as follows:

Power Output. If the dc power input to the final amplifier in a CB transmitter is 5 watts, the RF power output is typically 3 watts. When the power output is to be checked, an RF wattmeter is used in place of a transmitting antenna. An RF wattmeter is illustrated in Fig. 7-20. Note that RF wattmeters may or may not provide a 50-ohm load (dummy antenna). In any case, the RF power reading must be made across a 50-ohm resistive load. An RF load resistor (dummy antenna) is pictured in Fig. 7-21. If the RF power output is found to be subnormal, a systematic checkout of the transmitter is required. The power-type transistors are likely suspects, followed by leaky or open capacitors. DC-voltage measurements will usually suffice to close in on a defective component.

Crystal Oscillator. In some cases, subnormal RF power output is caused by a defective quartz crystal. Off-frequency operation is likely to be an accompanying trouble symptom. Occasionally, a defective crystal will operate satisfactorily when power is first turned on, and will then jump to another frequency after a while. Note that a replacement crystal must be ground for the circuit in which it is to be used. If a crystal is suspected of being defective, and the capacitors in the oscillator circuit are all right, it is advisable to check the adjustment(s) in the circuit before the crystal is replaced. For example, the adjustment of T1 in Fig. 7-18 would be checked. The oscillator frequency can be measured accurately with a frequency counter such as that illustrated in

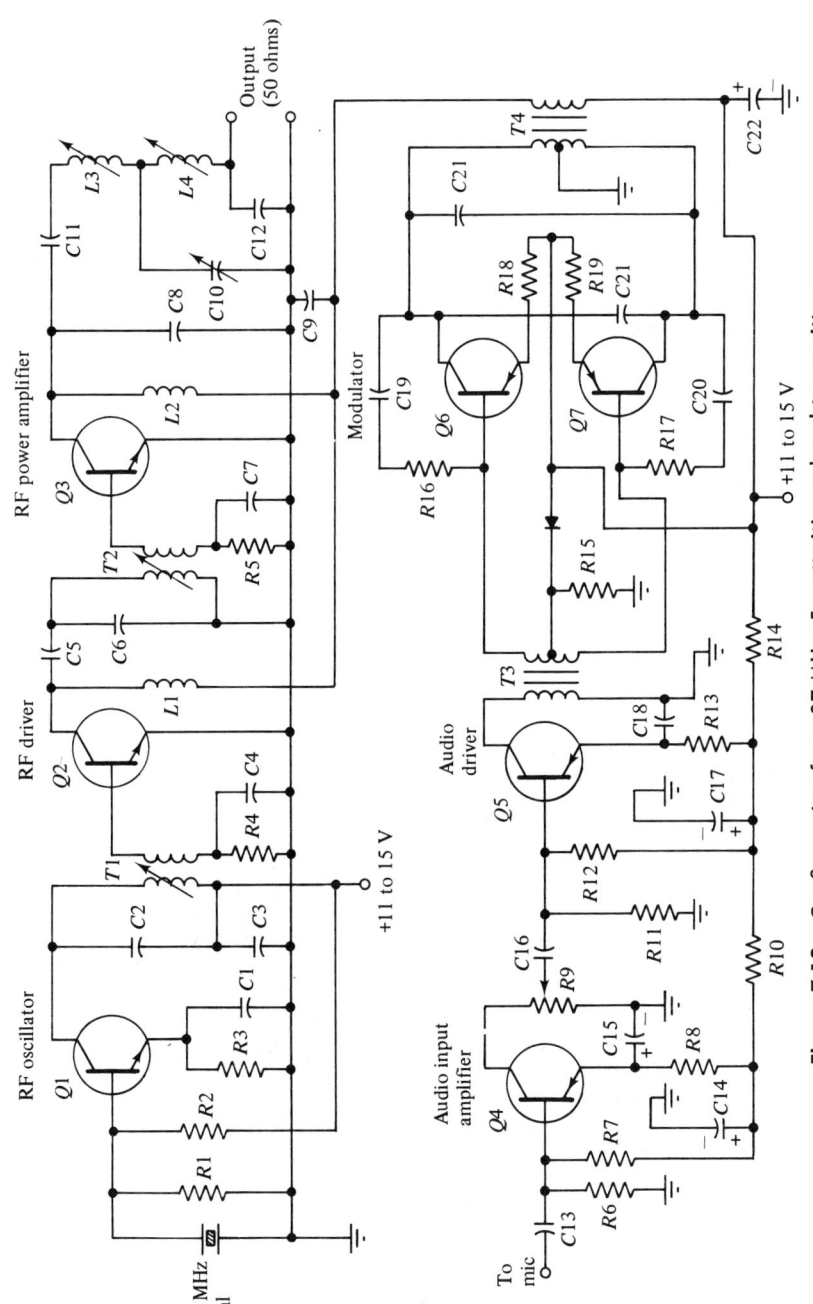

Figure 7-18 Configuration for a 27-MHz, 5-watt citizens band transmitter.

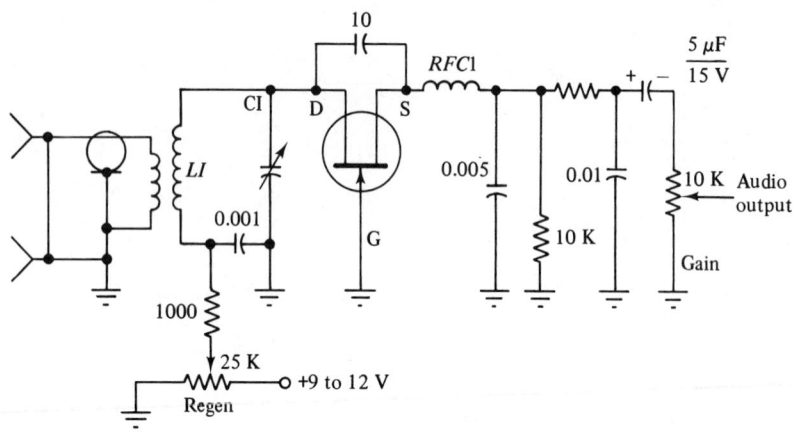

Figure 7-19 Basic superregenerative detector arrangement.

Figure 7-20 An RF load wattmeter. (*Courtesy of Heath Co.*)

Figure 7-21 An RF load resistor. (*Courtesy of Heath Co.*)

Fig. 7-22. A dip meter (Fig. 7-23) is useful for quick checks of tuned circuits.

Modulator. If there is weak or no modulation of the RF carrier, or if there is ample modulation but objectionably distorted output, there is a defect in the modulator section. However, before the modulator circuit is checked, it is advisable to try another microphone of the same type. With reference to Fig. 7-18, the modulator stages comprise Q4 through Q7. If Q6, for example, develops appreciable collector-junction leakage, the output from the modulator will become weak and distorted.

Superregenerative Receiver. A superregenerative receiver (Fig. 7-19) is extremely sensitive, although it has the disadvantages of comparatively poor selectivity and of loud random noise output when an incoming signal is not present. The most common trouble symptom is lack of superregenerative action, with operation as a conventional detector. Since the circuit is quite critical, a systematic checkout of the components and devices is usually required. DC-voltage measurements are less informative than in much other circuitry, and technicians often make substitution tests of suspected components. A useful chart of AM radio troubleshooting procedures is shown in Chart 7-1.

Figure 7-22 A digital frequency counter. (*Courtesy of Heath Co.*)

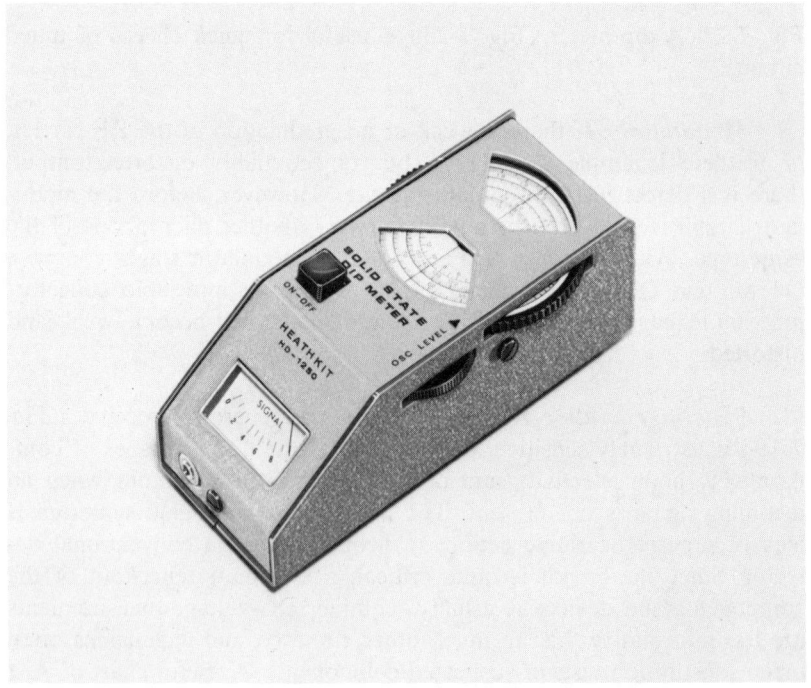

Figure 7-23 A service-type dip meter. (*Courtesy of Heath Co.*)

Chart 7-1 AM radio troubleshooting chart.

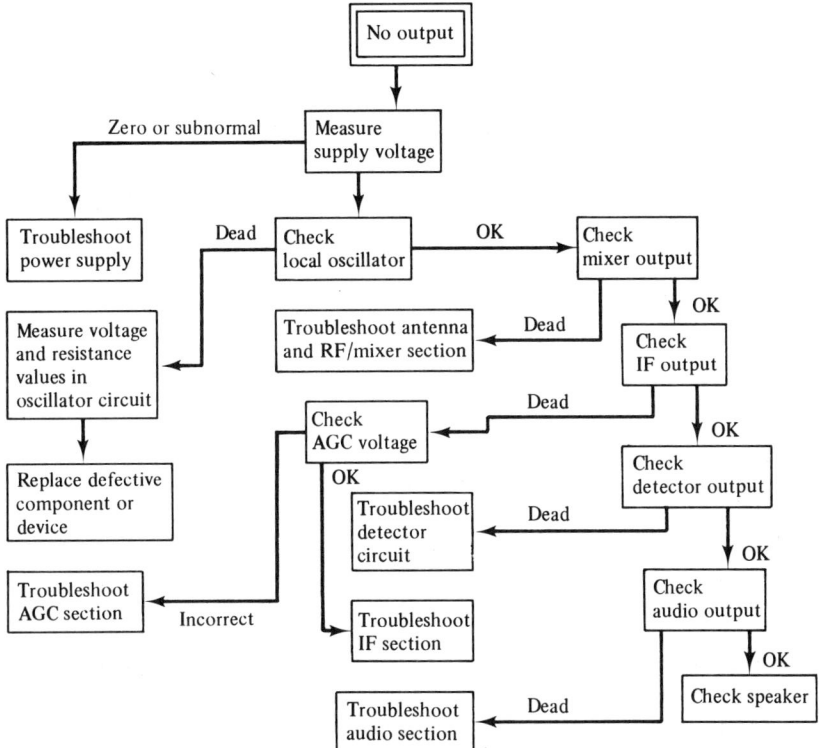

7-10 Optimum Value of Oscillator Injection Voltage

Low gain in the mixer stage does not necessarily indicate that there is a defective component or device in this stage. In other words, the trouble may be found in the oscillator circuit, or in the coupling circuit between the local oscillator and the mixer stages. There is an optimum value of oscillator injection voltage for a mixer stage, as shown in Fig. 7-24. This injection voltage can be measured at the output of the coupling circuit between the oscillator and mixer devices. A TVM with an RF probe is used to measure the oscillator injection voltage. As shown in the diagram, the mixer gain increases 20 dB as the oscillator injection voltage is increased from 0.05 volt to 0.1 volt. Too much injection voltage is as undesirable as too little voltage. Note that as the oscillator injection voltage is increased from 0.1 volt to 0.2 volt, the mixer stage gain decreases approximately 15 dB.

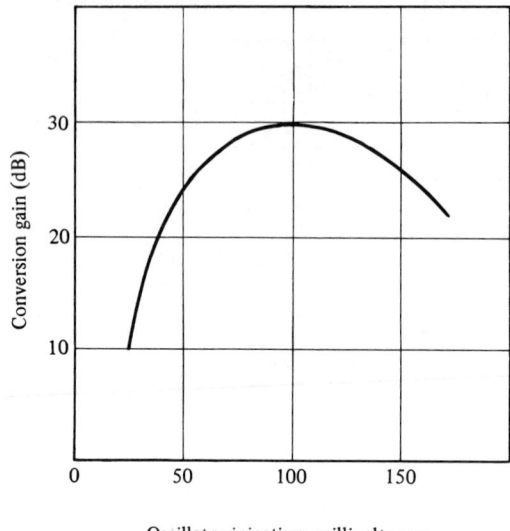

Figure 7-24 Example of conversion gain versus oscillator injection voltage.

This dependence of the mixer stage gain upon suitable amplitude of oscillator injection voltage is the result of nonlinear circuit action characteristics. That is, the heterodyne mixing process is a nonlinear relationship between the incoming signal voltage and the oscillator injection voltage. Maximum heterodyne output occurs on the most rapidly curving portion of the mixer device transfer characteristic. A transfer characteristic is exemplified in Fig. 7-25. When the incoming signal is combined with the oscillator injection CW voltage and applied to the base of a zero-biased mixer transistor, the combined waveform varies over the rapidly curving interval on the transfer characteristic. In turn, the maximum difference-frequency output is obtained. Thus, when a "tough dog" low-gain problem is encountered in the mixer stage, the technician should not forget to check the oscillator injection voltage. This measurement should be made with no signal input from the antenna.

7-11 Diode and Triode Transistor Detectors

An AM detector functions to pass the modulation component and to reject the carrier component of an amplitude-modulated waveform. In many instances, the AM detector also serves as an AGC source. A de-

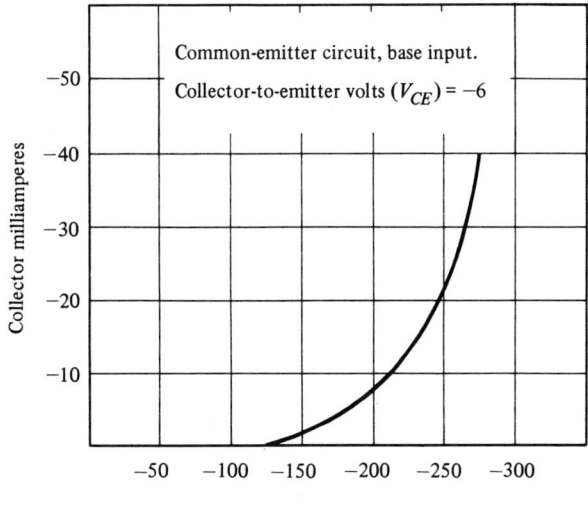

Figure 7-25 Transfer characteristic for a typical bipolar transistor.

tector, or demodulator, comprises a rectifier and a filter with a specified time constant. With reference to Fig. 7-26, a diode device rectifies the modulated-IF signal and capacitor C1 charges to a voltage V1 that is proportional to the prevailing amplitude of the IF signal. This charge potential is almost equal to the peak value of the IF signal at any time. Voltage V1 varies in accordance with the variation of the IF signal—this is a pulsating dc voltage with a polarity that depends on the polarization of the diode in the detector circuit.

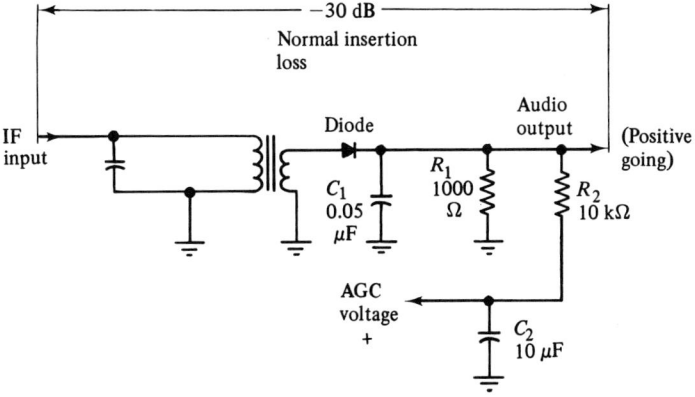

Figure 7-26 Basic diode detector and AGC arrangement.

In the case of an AM radio receiver, voltage V1 varies at an audio-frequency rate. This varying voltage appears across load resistor R1. Note that the input resistance to the following audio-amplifier stage is usually about 1000 ohms. In turn, maximum power transfer is obtained with a 1000-ohm load. For normal demodulator action, C1 must have a value sufficiently large to present a low reactance to the IF signal, but this value must be small enough that V1 can "follow" the highest audio frequency. Otherwise, the high-frequency response of the detector stage will be impaired and the fidelity will be subnormal. A typical value for C1 is 0.05 μF. This provides a time constant of 0.05 msec. The dc component developed across R1 is often utilized for AGC. For this purpose, additional low-pass filtering is provided by R2 and C2, with a time constant of 100 msec. *The power loss of this diode detector arrangement is from 25 to 35 dB.*

Where more AGC power and/or more gain following detection is needed, a receiver may employ a transistor detector arrangement, as shown in Fig. 7-27. Rectification is provided by the base-emitter circuit, and the audio component of the signal is amplified in the collector circuit. Thus, the net power gain of a transistor detector is equal to the loss in the demodulation process as compensated by the gain in the amplification process. *The over-all power gain of this transistor detector arrangement is 5 dB, approximately.* AGC voltage is taken from the emitter branch of the circuit. AGC filter action is provided by R1 and C2. Most trouble symptoms in detector circuits will be tracked down to leaky or "open" capacitors. However, semiconductor junctions may also cause detector malfunction by becoming short-circuited, open-circuited, or leaky.

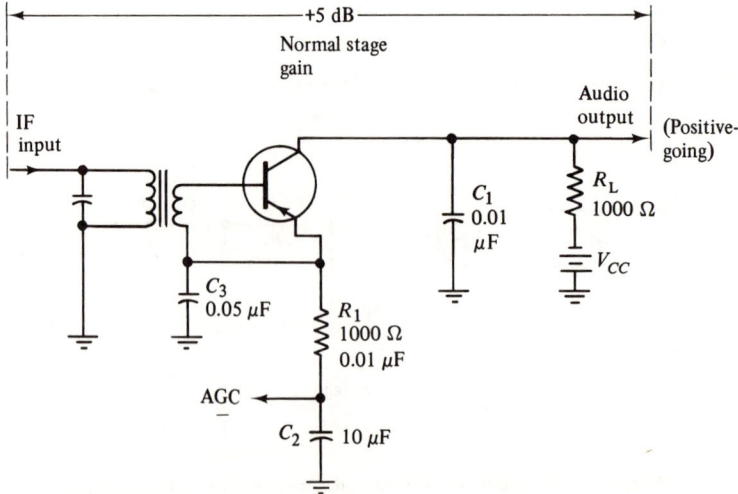

Figure 7-27 Basic transistor detector and AGC arrangement.

7-12 Stage Gain Values for an FM Broadcast Receiver

Normal stage gain values for a broadcast FM IF amplifier are noted in Fig. 7-28. This typical arrangement employs two integrated circuits and a ratio detector. A total signal-voltage gain from input to output of the IF strip is 116,000 times, approximately. The first stage has a gain of 100 times; however, 60 percent of this signal level is lost through the tuned coupling transformer. Since the second IC normally provides a signal-voltage gain of almost 3000 times, a 15-μV input is stepped up over-all to a 1.75-V output. Limiting action in this arrangement is provided by the ratio detector.

Stage gain values for a more elaborate IF strip are noted in Fig. 7-29. Four integrated circuits are utilized. The IF strip provides an overall signal voltage gain of 100,000 times at maximum available gain (MAG). This maximum available gain is realized normally with an input signal voltage of 2 μV. When the input signal level is increased to 50 μV, the first and second IC's start limiting action. The FM front end in this example normally provides a signal voltage gain of 10 times. Each tuned IF coupling transformer imposes a 50% signal loss, approximately. Stage gains can be measured accurately with a lab-type signal generator that has a carrier-level meter and a calibrated output attenuator.

7-13 Automatic Frequency Control

Automatic frequency control (AFC) is used in broadcast-type FM receivers to correct any tendency of the local oscillator to drift off-frequency. In other words, if the local-oscillator frequency happens to drift slightly off its correct value, the AFC circuit reacts accordingly and brings the oscillator back on practically its normal frequency. This control action has definite limits, and if the oscillator drifts substantially in frequency, control action is lost and the receiver will lose the station to which it was tuned. Fig. 7-30 depicts the plan of an AFC system. It consists of a reverse-biased diode connected across the oscillator coil. A diode has junction capacitance, and it is equivalent to a capacitor when it is reverse-biased. If the value of the reverse bias is made variable, the diode becomes the equivalent of a variable capacitor. (See Fig. 7-31.) Thereby, the oscillator frequency can be controlled by variation of reverse-bias voltage on the AFC diode.

As explained previously, the output voltage from a discriminator or a ratio detector will have a positive or a negative polarity, depending on whether the incoming frequency is above or below the center frequency of the tuned transformer. Now, suppose that the local oscillator

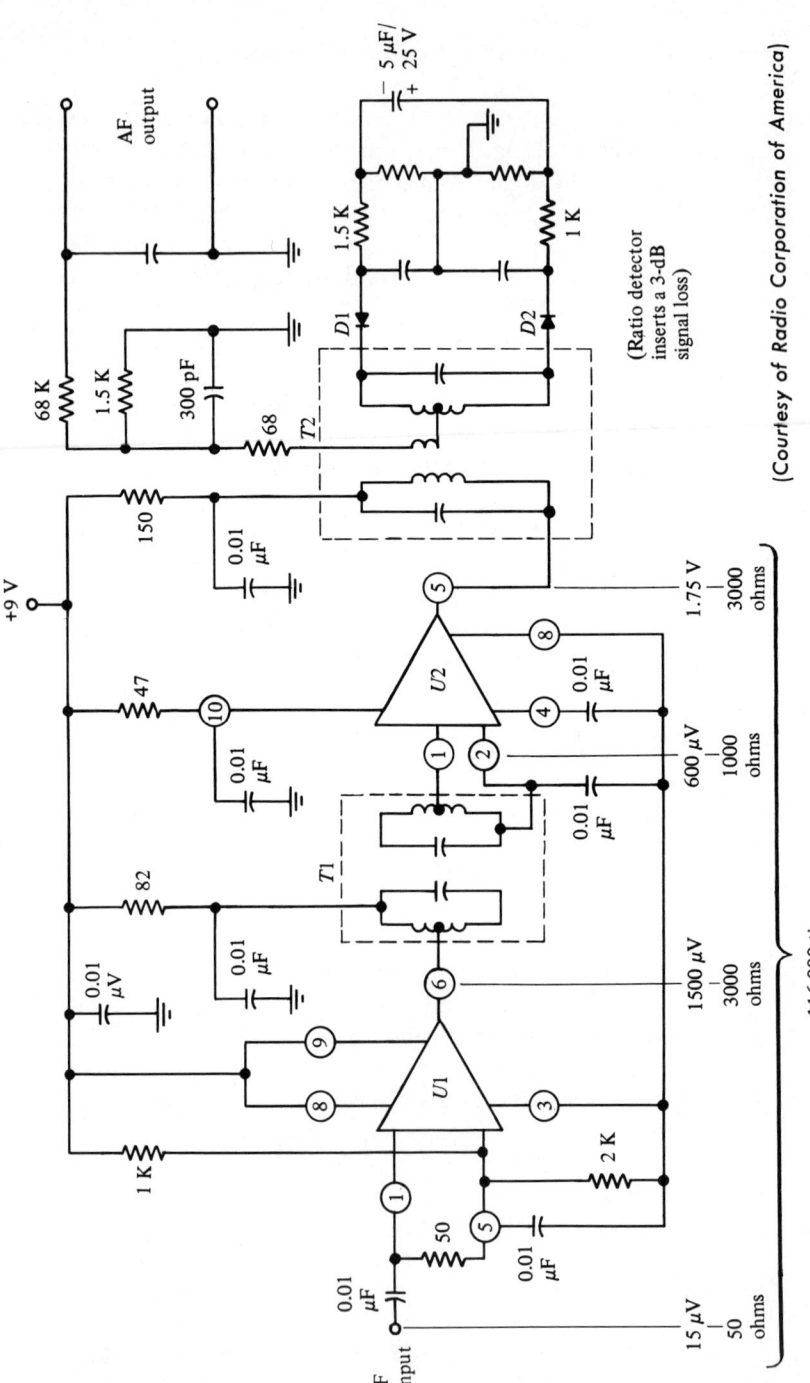

Figure 7-28 Stage gain values for a broadcast FM IF amplifier with two integrated circuits.

(Courtesy of Radio Corporation of America)

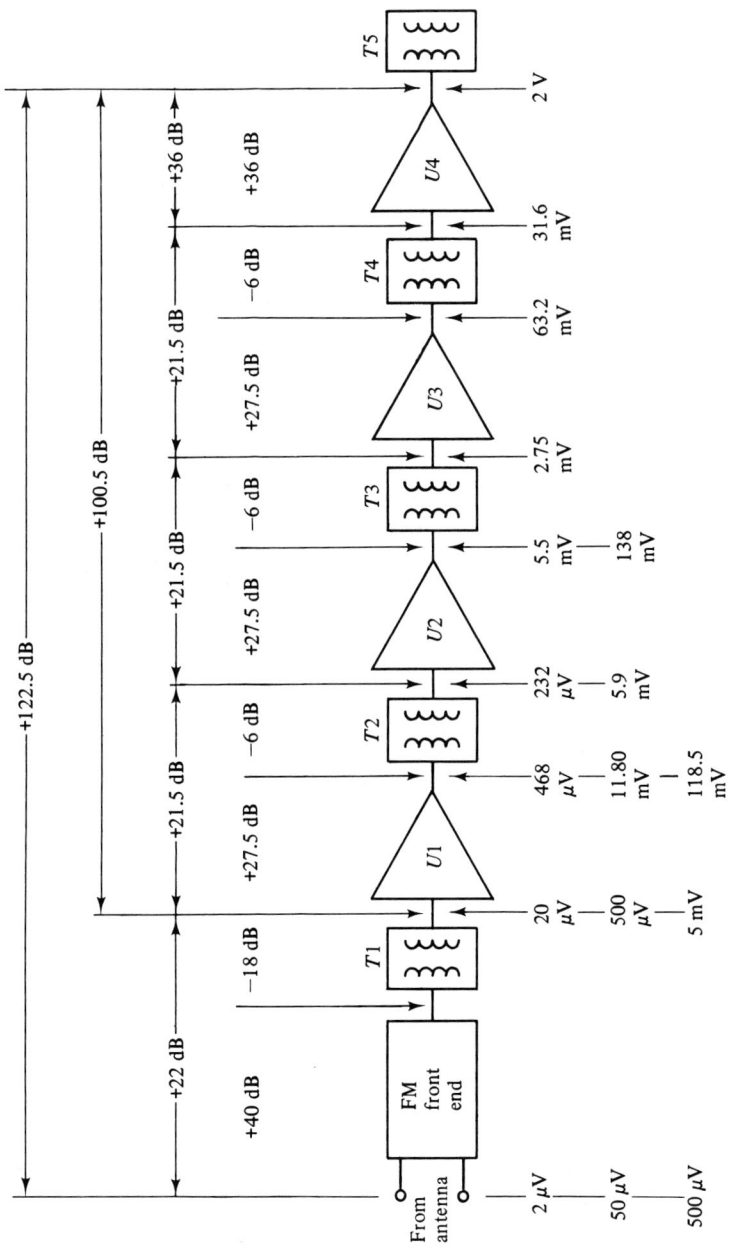

Figure 7-29 Stage gain values for a typical broadcast FM receiver. (Courtesy of Radio Corporation of America)

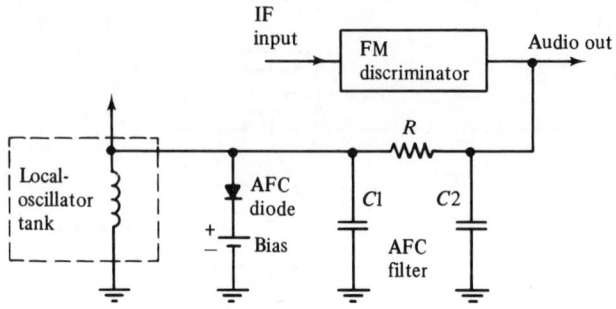

Figure 7-30 Plan of an FM automatic frequency control system.

drifts to a higher frequency. In turn, the IF frequency drifts to a lower frequency. With reference to Fig. 7-30, the discriminator output voltage then drifts in a positive direction, on the average. This average positive dc component is separated from the AF component by the AFC filter in Fig. 7-30, and is applied to the AFC diode. In turn, part of the fixed reverse bias on the diode is cancelled, and an increase in junction capacitance results. This increased junction capacitance lowers the oscillating frequency, thereby returning the IF center frequency to practically its correct value.

Troubleshooting of the AFC system entails dc voltage measurements; the measured values are compared with the specified values in the receiver service data. Resistance measurements are often useful to supplement dc voltage measurements. Since the FM detector is a vital part of the AFC system, the components and devices in this subsection should also be checked out. Poor alignment of the FM detector circuits can result in impaired AFC action. In most situations, AFC malfunction is tracked down to leaky or "open" capacitors. However, the AFC diode may deteriorate eventually and require replacement. Off-value resistors are less likely to cause trouble symptoms, but this possibility should be kept in mind.

In FM reception, the AFC section can be switched in or out. It is helpful to switch out the AFC function while one is tuning a weak FM station, in the event that there is a strong FM station on the adjacent channel. Then, after the weak station is tuned in, the AFC function can be used to advantage. Unless the AFC function is switched out while one is tuning, a weak FM station could be "skipped" because of AFC holding action on an adjacent-channel strong signal. A normally operating FM receiver suppresses weak-station interference on the

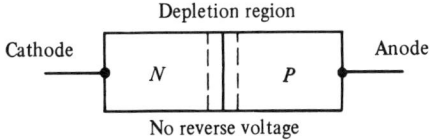

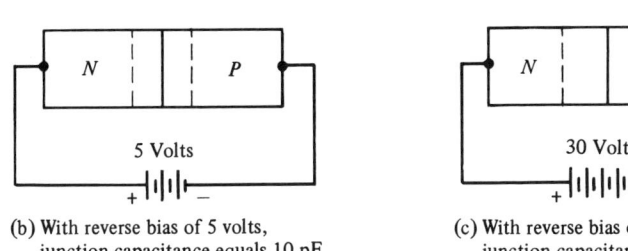

Figure 7-31 Typical varactor characteristics.

same channel with a strong station. In other words, if an FM receiver is tuned to a strong station, the listener will normally be unaware of the presence of a weaker station that is broadcasting on the same carrier frequency. This is called the *capture effect,* and it is an inherent characteristic of the limiting and demodulation processes. Of course, if ratio-detector or discriminator malfunction occurs, co-channel interference can become a problem. Note that co-channel interference cannot be eliminated by a normally operating receiver in the situation where both stations have practically equal signal strength. For the receiver to "capture" one of the stations (and suppress the other station's signal), the captured station must be at least 1.3 dB stronger than the suppressed station signal. In turn, the capture ratio of the normally operating FM detection system is said to be 1.3 dB. In the event of detector malfunction, this capture ratio might deteriorate to 3 dB, 6 dB, or even more.

8
Black-and-White Television Troubleshooting

8-1 General Considerations

A television receiver normally provides a picture-channel gain of over 100 dB, as depicted in Fig. 8-1. For example, an input signal level of 1000 μV will provide a video-output level of 50 V peak-to-peak (p-p). The video amplifier provides a typical signal-voltage gain of 50 times. This gain figure can be measured approximately by driving the receiver with a steady signal, as from an AM generator, and measuring the comparative signal levels at the outputs of the video detector and of the video amplifier. This method is only approximate, because typical ac voltmeters have substantial input capacitance, with the result that video-frequency circuits may be loaded to some extent. However, this is a practical preliminary troubleshooting test. It serves to make a distinction between video-amplifier and picture-tube malfunction when a weak-picture trouble symptom is encountered.

It is not practical to measure signal voltages in the IF-amplifier section, owing to the excessive circuit loading that is imposed by high-frequency voltmeters on 45-MHz circuits. Therefore, IF stage gains are measured as shown in Fig. 8-2. A dc voltmeter is connected at the output of the video detector, and an IF signal is injected in turn at the input and at the output of the stage under test. The difference between the two meter readings corresponds to the signal-voltage gain of the stage. Note that a blocking capacitor should be provided in series with the generator output lead, if the instrument does not have a built-in blocking capacitor. This precaution is required to prevent drainoff of dc bias and supply voltages from the circuit under test.

Note that a CW signal can be utilized in the test shown in Fig. 8-2.

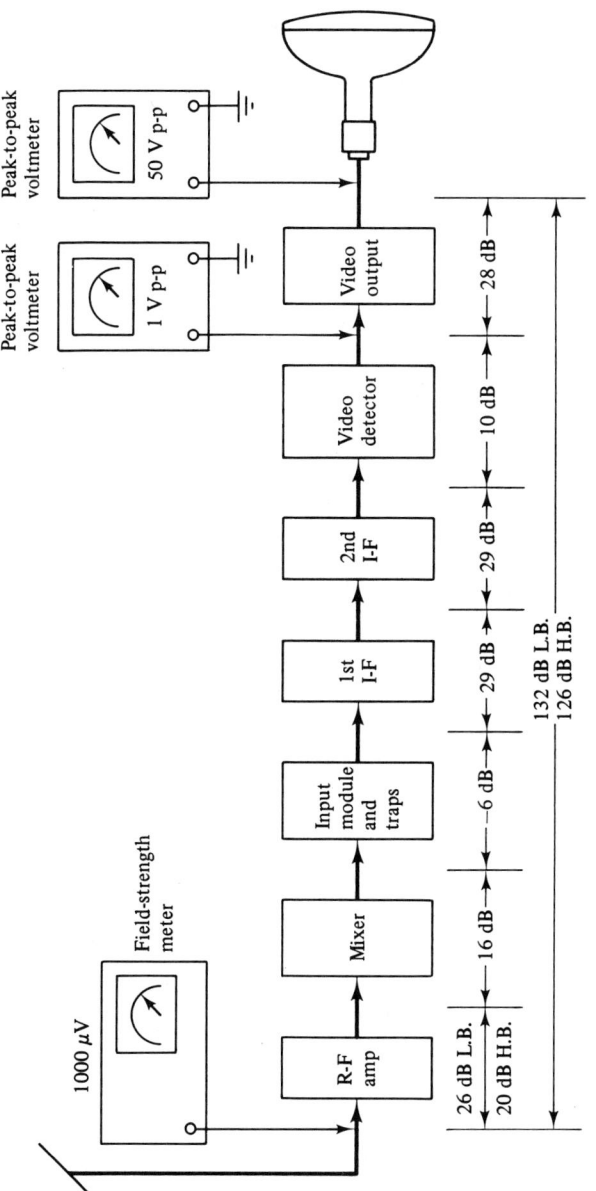

Figure 8-1 Relative video-signal levels in a representative TV receiver.

Figure 8-2 Measurement of signal-voltage gain in an IF stage.

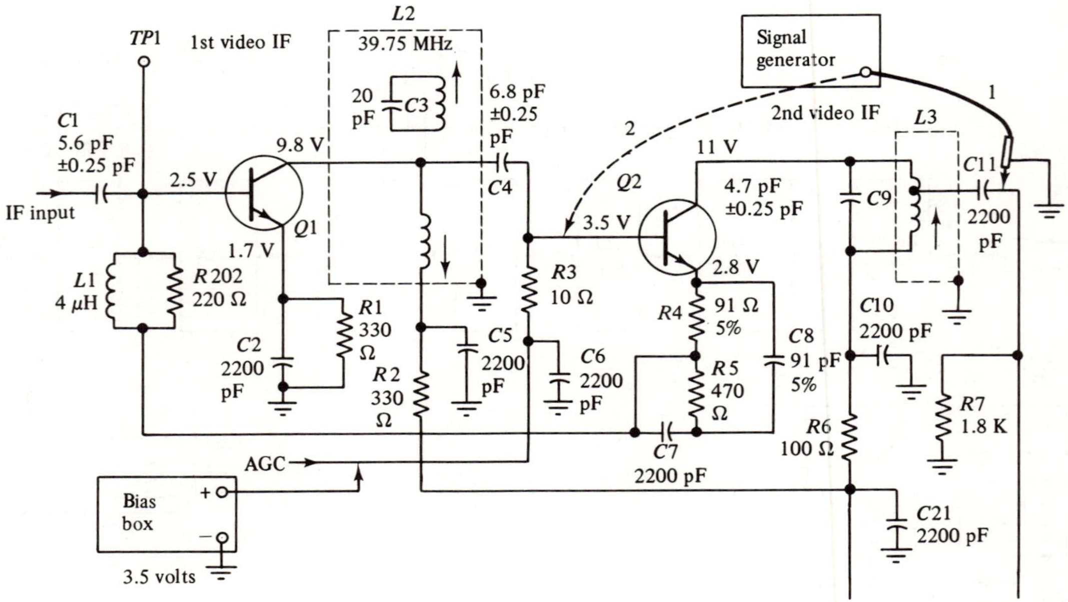

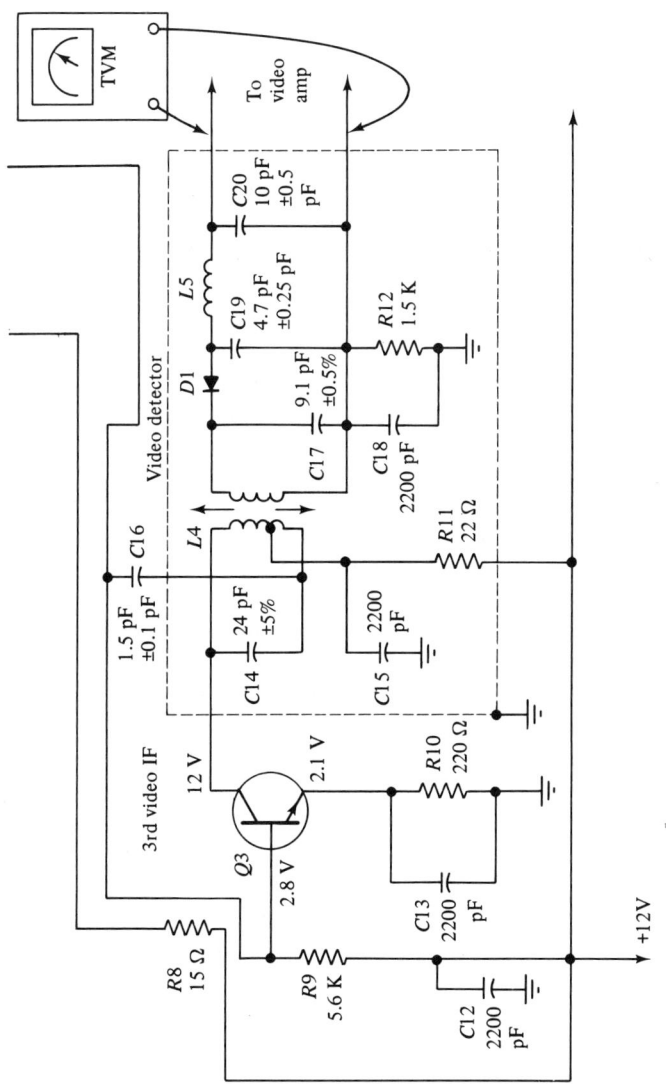

Figure 8-2 Continued

A mid-band frequency of approximately 44 MHz is suitable. Care should be taken to avoid system overload; the test signal should be kept below the level that produces 2 V p-p from the video detector. As shown in the diagram, it is necessary to clamp the IF AGC line when stage-gain measurements are made. Otherwise, AGC action will falsify the test results. Since IF stage gain varies greatly with AGC voltage values, a mid-range value is desirable for preliminary analysis. The receiver service data usually specify a typical AGC operating-voltage value. In normal operation, the signal-voltage gain of an IF stage will vary from nearly unity to about 30 times, depending upon the prevailing signal strength. Note, however, that the last IF stage is not AGC-controlled, and operates at maximum gain at all times.

If normal output is obtained when a signal is injected at the input of the first IF stage, and if the IF system has normal gain, it follows that a weak-picture or no-picture trouble symptom throws suspicion on the RF tuner. A typical tuner configuration with specified dc voltages is shown in Fig. 8-3. Although it might seem that the local oscillator Q3 is reverse-biased, this is not so from a dynamic standpoint. In other words, the oscillator transistor operates in class C, and base current flows in brief pulses. In turn, collector current also flows in brief pulses, and the transistor is cut off only on an average basis. Both the RF amplifier and the mixer transistors operate with forward bias at all times. DC-voltage measurements are basic in tuner troubleshooting procedures. Most trouble symptoms are caused by poor contacts in the turret mechanism.

Local oscillators in TV tuners operate on the "high side." In other words, the oscillator frequency is above the incoming signal frequency. This mode of operation is utilized to minimize the possibility of interference with other services. If the local oscillator "drops out," the receiver will be "dead." When the oscillator stops operating, a change in dc voltages occurs at the transistor terminals. Some circuit defects can cause the oscillator to operate far off normal frequency. In turn, the transistor terminal voltages are not greatly affected, although the receiver is "dead" as if oscillator operation had stopped. When oscillator malfunction is suspected, a quick check can be made as shown in Fig. 8-4. The signal generator is tuned to the normal operating frequency of the oscillator. In turn, if receiver operation resumes, the suspicion of oscillator trouble is confirmed.

8-2 Evaluation of Trouble Symptoms

When there is no sound output and the picture-tube screen is dark, it is advisable to measure the supply voltage. As depicted in Fig. 8-5, a

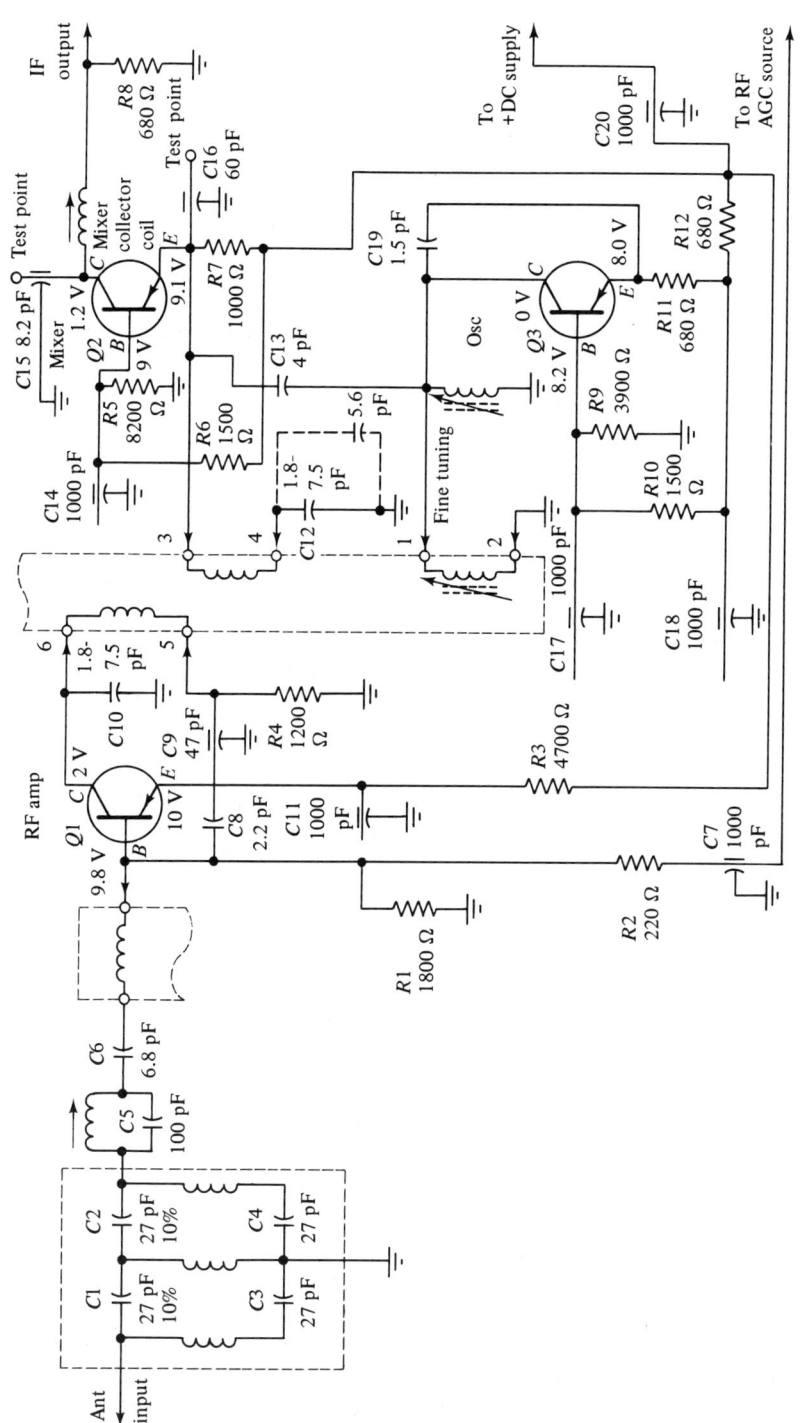

Figure 8-3 Typical TV tuner configuration, with specified dc voltages.

212 Black-and-White Television Troubleshooting

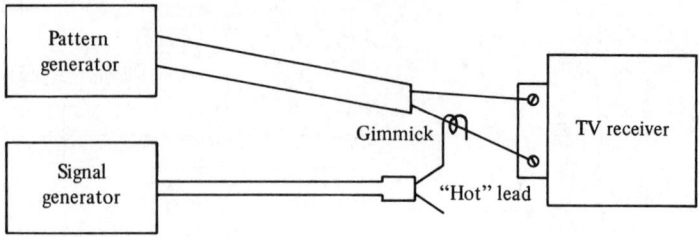

Figure 8-4 Quick check for suspected oscillator malfunction.

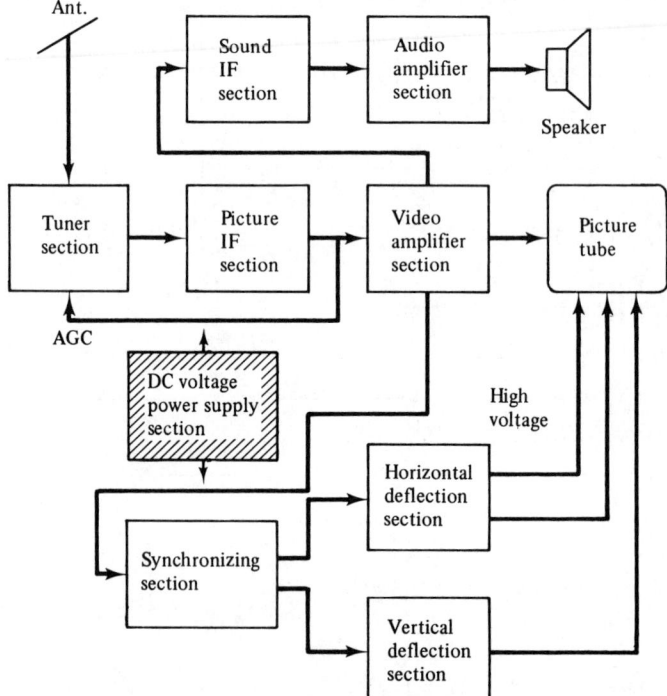

Figure 8-5 Power-supply malfunction affects both picture and sound.

power-supply malfunction affects both picture and sound reproduction. Again, consider a trouble symptom in which there is no sound and no image, although a raster is displayed on the picture-tube screen. In turn, the technician recognizes that the high-voltage supply is operative, that the horizontal- and vertical-sweep sections are normal, and that the dc voltage power supply is working. Since both the sound and image are absent, it is probable that the fault will be found in a section that is

8-2 Evaluation of Trouble Symptoms 213

common to both the video and the sound signals. Thus, he turns his attention to the RF tuner, video-IF, and video-amplifier sections. To check for the possibility of antenna trouble, a signal generator may be used to apply a modulated VHF signal to the tuner.

Next, if picture reproduction is normal, but sound is absent, there is an immediate suspicion that the fine-tuning control is out of adjustment. Otherwise, the trouble is most likely to be found in the sound-IF section or in the audio section. There is also a possibility of speaker failure. If an ac voltmeter indicates the presence of audio signal across the speaker input terminals, the suspicion of speaker trouble is confirmed. In many cases, the voice coil will be "open," with a broken pigtail lead from the coil to the associated speaker terminal. Again, consider a trouble symptom in which sound output is normal and a raster is displayed, but with no image. It is likely that there is a defect in the RF tuner, IF amplifier, or video-amplifier section, even though sound reproduction is normal. In other words, various circuit faults can "kill" the picture signal, but pass enough sound signal that the audio output appears to be practically normal.

When sound reproduction is normal, and the picture-tube screen is dark, the video-signal input to the picture tube should be checked with an ac voltmeter. If there is 50-V p-p drive present, suspicion falls upon the picture tube. However, there is also the possibility that a "shorted" capacitor or other component defect has changed the dc-voltage distribution to the picture tube. Therefore, it is advisable to check the dc operating voltages at the picture-tube socket, as depicted in Fig. 8-6. The high-voltage value should also be measured with the aid of a high-voltage dc probe. If the dc operating voltages are found to be within normal tolerance of ±20 percent, it is concluded that the picture tube is defective. Note that sometimes a "short" occurs in a picture tube, causing the operating voltages to appear to be incorrect. Therefore, it is good practice to double-check, with the socket removed from the picture-tube base.

Consider next a trouble symptom in which the sound output is normal, and an out-of-sync picture is displayed. If sync action has been "lost," the picture can be free-wheeled into frame by careful adjustment of the horizontal and vertical hold controls. On the other hand, if the picture cannot be free-wheeled into frame, it is indicated that a deflection oscillator malfunction is present. Note that if both horizontal and vertical sync are "lost," the trouble is most likely to be found in the sync-separator, sync-amplifier, or sync-limiter sections. However, this trouble symptom may also be caused by sync clipping owing to overload, usually in the video amplifier. It is also possible for a regenerative condition in the IF section to cause sync clipping.

Another common trouble symptom is no vertical sync lock, al-

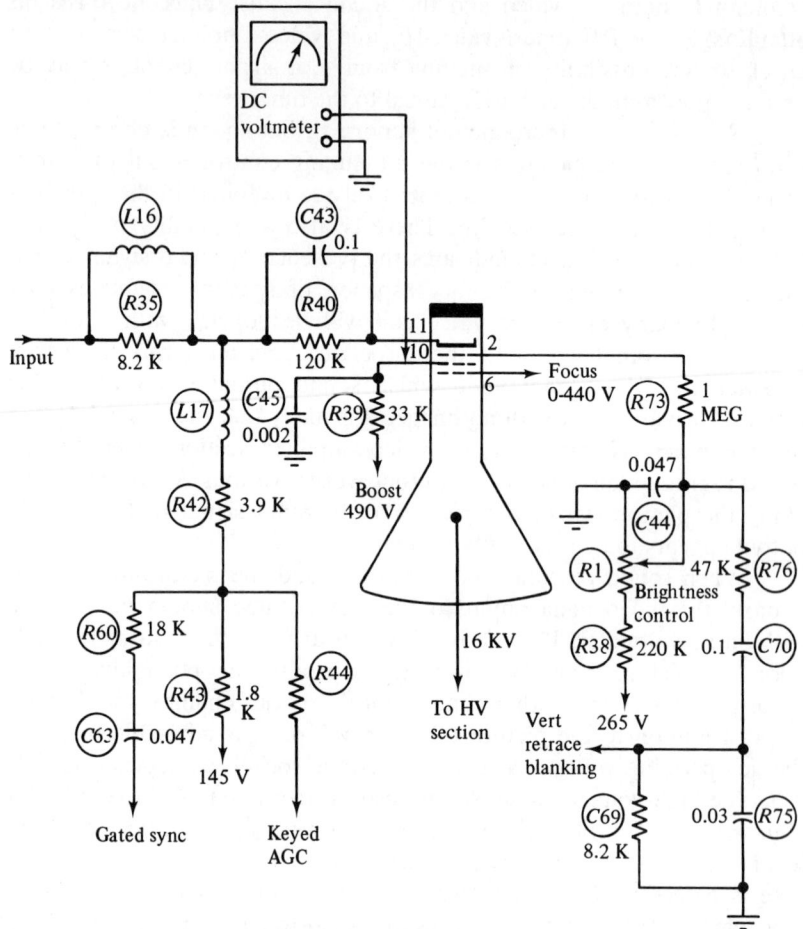

Figure 8-6 Off-value picture-tube voltages can cause a dark screen.

though horizontal locking action is normal. In this situation, suspicion falls on the vertical oscillator. Thus, the defect is localized to the coupling from the output of the sync amplifier (integrator) and the base of the vertical-oscillator transistor, or to an oscillator malfunction that causes off-frequency operation. When the oscillator is off-frequency, dc-voltage measurements are not fully informative. Instead, resistance checks are advisable, a high-low ohmmeter being used. Suspected "open" capacitors can be quick-checked by bridging with a known good capacitor. A leaky collector junction can cause off-frequency operation. In many cases it is practical to make a turn-off test, as exemplified in Fig. 8-7. A dc voltmeter is connected from the collector of Q18 to ground;

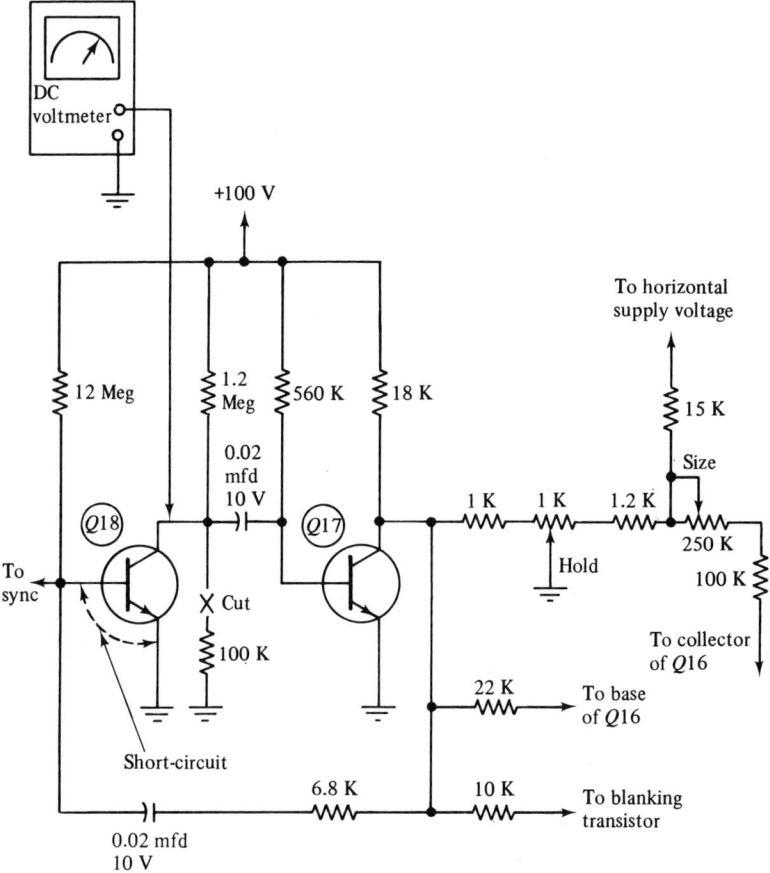

Figure 8-7 Turn-off test of vertical-oscillator transistor.

the meter should have a high input resistance, such as 15 megohms. The PC conductor from the collector to the 100-kΩ resistor is temporarily razor-cut. Then, when a short circuit is applied between the base and emitter terminals of Q18, the voltmeter will indicate virtually 100 volts unless the transistor has a leaky collector junction. Finally, the "open" PC conductor is repaired with a small drop of solder.

Consider next a trouble condition in which vertical sync lock is normal, but horizontal sync lock is marginal or completely lost. It is probable that the trouble will be found either in the horizontal-AFC circuit or in the horizontal-oscillator circuit. In other words, loss of horizontal sync lock is caused by lack of horizontal sync pulses, or by substantial off-frequency oscillation. If the picture can be locked at some unusual setting of the hold controls, or if the image locks in split-picture form, the defect will be found in the oscillator circuit or the AFC circuit. On the other hand, if the picture can be momentarily freewheeled into horizontal frame by adjustment of the hold controls, it can be concluded that the horizontal-sync pulses are not entering the AFC section.

As exemplified in Fig. 8-8, the horizontal AFC, oscillator, and driver transistors may be reverse-biased in normal operation. This is an average dc bias, and the transistors normally conduct in brief pulses. In the case of off-frequency operation, dc-voltage measurements are not fully informative, because development of reverse-bias potentials is frequency-dependent. From a statistical viewpoint, electrolytic capacitors are most likely to become defective. Thus, the technician would check C38, C39, C40, C41, and C42 at the outset. Paper capacitors are the next most likely suspects. Transistors are normally very long-lived, and inductors rarely become defective. However, if a protective diode such as X37 is found defective, Q22 is almost sure to require replacement.

If the foregoing trouble symptom is reversed, horizontal deflection is normal with vertical deflection failed. A single horizontal line is displayed halfway down the picture-tube screen. The technician turns his attention to the vertical-deflection section, and makes tests to determine the subsection that is defective. If a multivibrator-type vertical oscillator is used, as depicted in Fig. 8-9(a), there is no interaction between the oscillator stage and the output stage. In turn, a dc-voltage measurement can be made at the base of the vertical-oscillator transistor in order to determine whether the oscillator is operating. Thereby, the trouble symptom can be sectionalized. On the other hand, if a vertical feedback oscillator is utilized, as shown in Fig. 8-9(b), a fault in the output stage can "kill" the vertical oscillation. Therefore, a 60-Hz signal is customarily injected at the output of the oscillator stage, to determine whether the output amplifier will respond. This test serves to sectionalize the trouble symptom.

8-3 IF Amplifier Troubleshooting Techniques

In Fig. 8-10, a quick check is depicted for a second-IF stage that is suspected of being "dead." A completely "dead" stage blocks the passage of picture, sound, and sync signals. Jumper capacitor C_j has a value of approximately 500 pF, and it is temporarily connected between the base and collector terminals of Q2. Thereby, the incoming signal is bypassed over Q2. Then, if a weak picture is displayed (and sound is

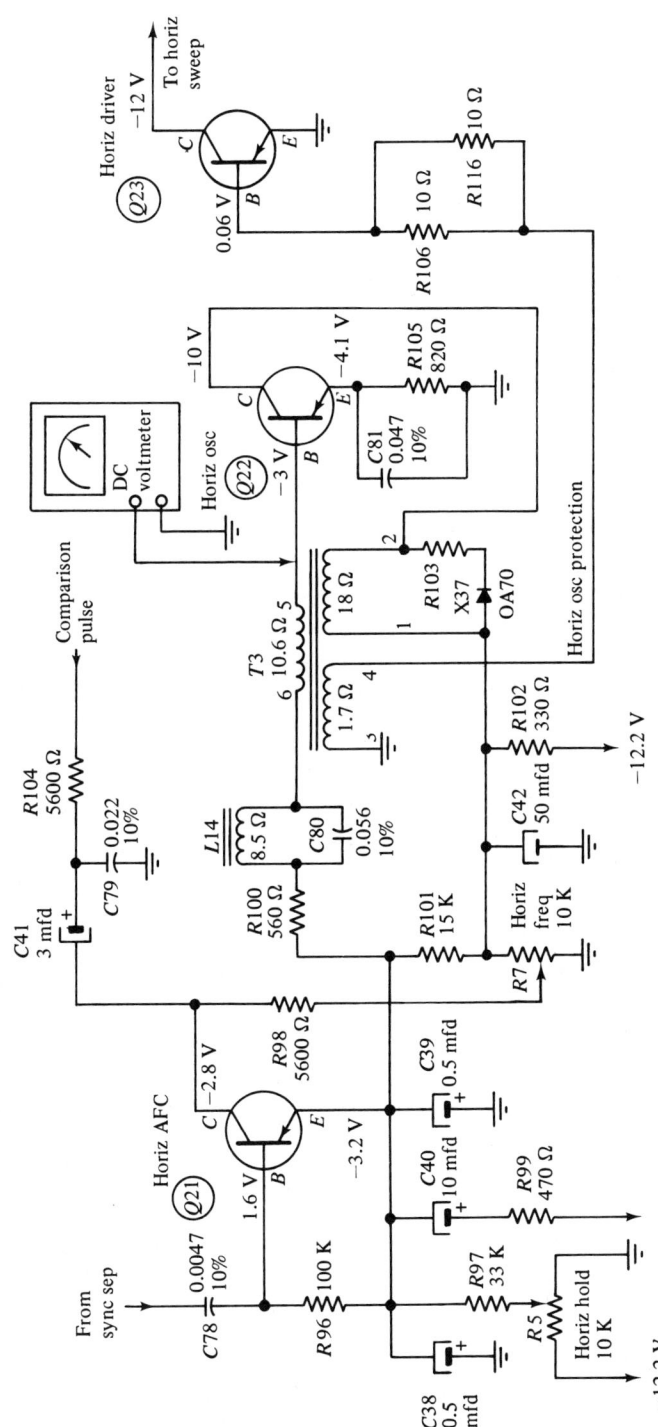

Figure 8-8 AFC, oscillator, and driver transistors are reverse-biased on the average,

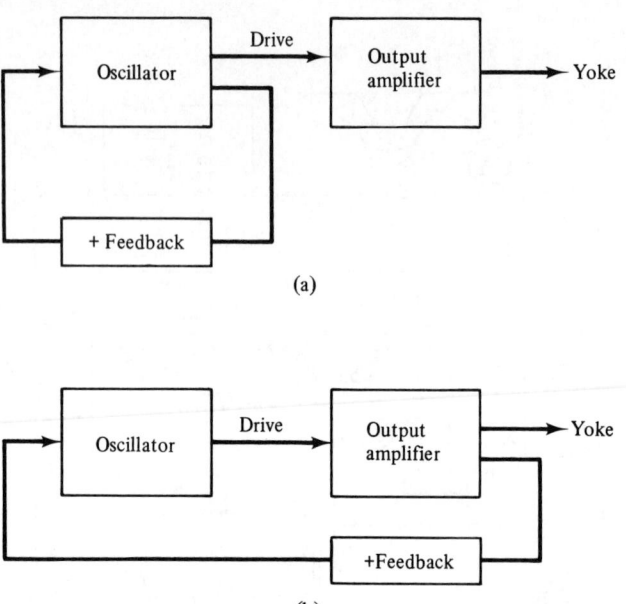

Figure 8-9 Two basic vertical-sweep section arrangements: **(a)** multivibrator oscillator; **(b)** feedback-type oscillator.

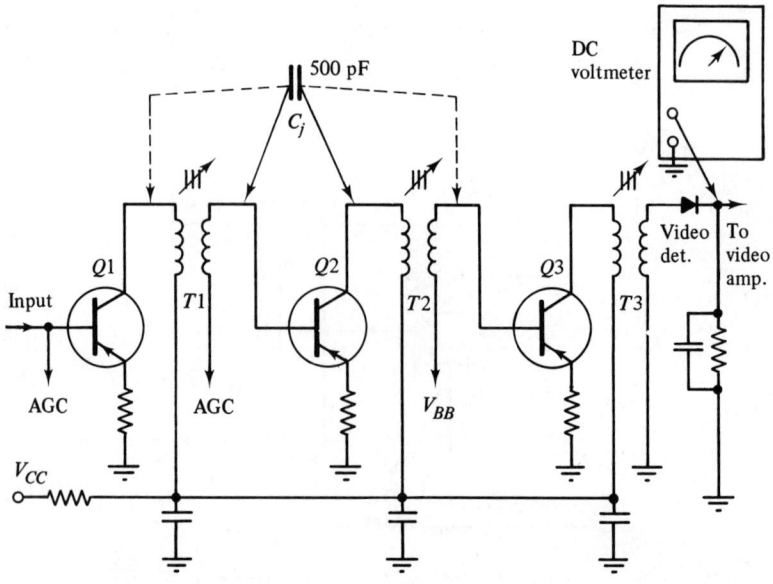

Figure 8-10 Quick check of second-IF stage with jumper capacitor.

reproduced), the suspicion is confirmed that Q2 has zero gain. The transistor may be open-circuited, short-circuited, or biased off; collector voltage may be absent, or the emitter circuit may be open. In any case, the quick check provides a useful preliminary analysis. Note that if the IF amplifier happens to be oscillating, passage of picture, sound, and sync signals will also be blocked. But in this event, the dc voltmeter will indicate several volts output with no incoming signal.

As shown by the dotted lines from C_j in Fig. 8-10, the jumper capacitor may also be connected temporarily from the collector of Q1 to the base of Q2. If a weak picture is then displayed (and sound is reproduced), it is indicated that the secondary of T1 or the primary of T2 is defective. Sometimes a jumper-capacitor test will result in nearly normal sound reproduction, with no visible picture. When this response is obtained, it is the result of very low IF gain. In other words, the sound signal is normally transmitted at a substantially higher level than the picture signal. Another factor in this jumper test is the serious detuning effect of C_j. The essential consideration is whether the sound and/or the picture signal proceeds through the receiver when the jumper-capacitor quick check is made.

After the defective stage is localized, it is advisable to clamp the AGC line to an average operating potential as specified in the receiver service data. This precaution avoids the possibility of confusing AGC trouble with IF-amplifier trouble. For closing in on defective components or devices, dc voltage measurements are often adequate. Transformer windings can be checked with an ohmmeter. Some IF amplifiers employ neutralizing capacitors, as exemplified in Fig. 8-11. An open neutralizing capacitor is very likely to result in stage oscillation. Note

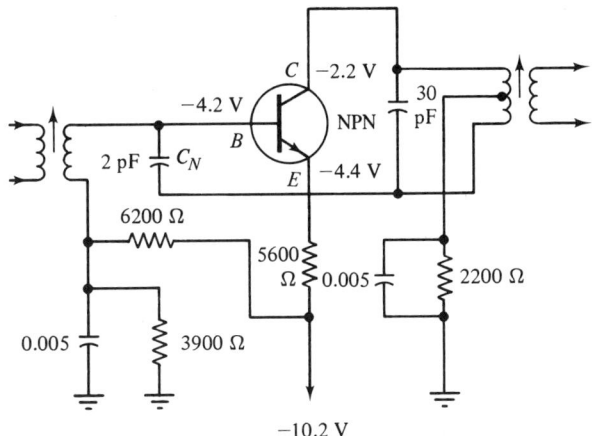

Figure 8-11 Neutralizing capacitor C_N prevents IF oscillation.

that if a "hot" transistor is installed as a replacement, C_N could require a slightly greater value to stabilize the stage. IF oscillation can also be caused by misalignment in some cases. For example, when all of the IF circuits are accidentally peaked to about the same frequency, the system is likely to break into uncontrolled oscillation. As a practical note, realignment is ordinarily done after all troubleshooting has been completed, unless it is known or suspected that the alignment adjustments have been tampered with.

8-4 AGC Section Troubleshooting

Most present-day receivers use keyed-AGC systems. A typical AGC arrangement for a solid-state receiver is shown in Fig. 8-12. It comprises a closed loop with the AGC gate, RF and IF amplifiers, video detector, and first video amplifier. Most AGC troubleshooting procedures involve dc-voltage measurements under signal and no-signal conditions. In other words, the AGC voltage responds in a specified way or ways to various incoming signal levels. With reference to Fig. 8-12, the AGC section maintains a relatively constant output of 1.2 V at the emitter of the first video amplifier transistor. This output voltage will normally vary only a small amount over a wide range of signal-input voltage. However, the small variation that occurs provides a wide range of control action. This is an example of a keyed or gated AGC system in which the AGC control voltage is developed during the horizontal-sync interval, and is sustained by the AGC circuitry for the duration of the forward-scanning interval.

Observe in Fig. 8-12 that as the input signal to the RF amplifier increases, the output from the first video amplifier tends to increase accordingly. However, any increase in video signal level is applied as an input to the AGC gate. A 30-volt negative pulse is applied to Q4, which opens the AGC gate during flyback time, and the sync-pulse component of the video signal is amplified by the transistor. In turn, a positive AGC voltage is developed. To prevent the collector-base junction of Q4 from becoming forward-biased by the AGC voltage, diode D1 is connected in series with the collector circuit. Thus, the AGC voltage is applied as forward bias to the RF-amplifier transistor, which in turn shifts its operating point toward saturation. As the transistor approaches saturation, its collector voltage decreases and its gain decreases. In addition to amplifying the incoming RF signal, Q1 also functions as a dc amplifier for the AGC voltage. Accordingly, the AGC voltage is applied in opposite polarity to the base of Q2. Reverse AGC control of Q2 occurs, which tends to minimize frequency shift from changing collector junction capacitances.

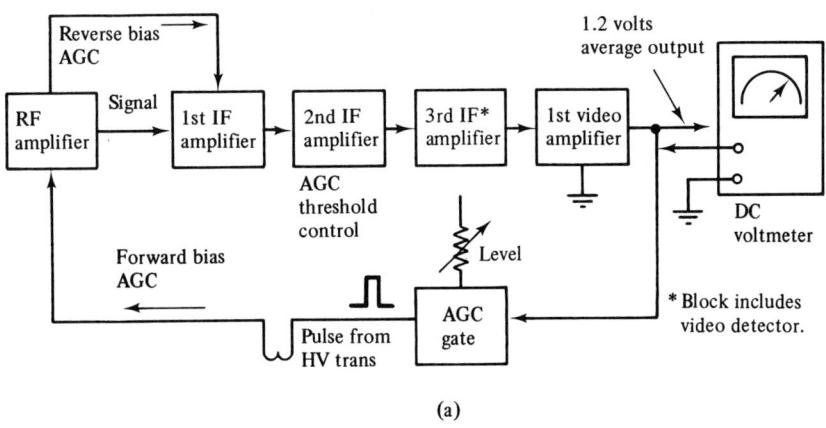

Figure 8-12 An AGC configuration for a TV receiver: **(a)** block diagram; **(b)** skeleton schematic. (*Adapted from an RCA original*)

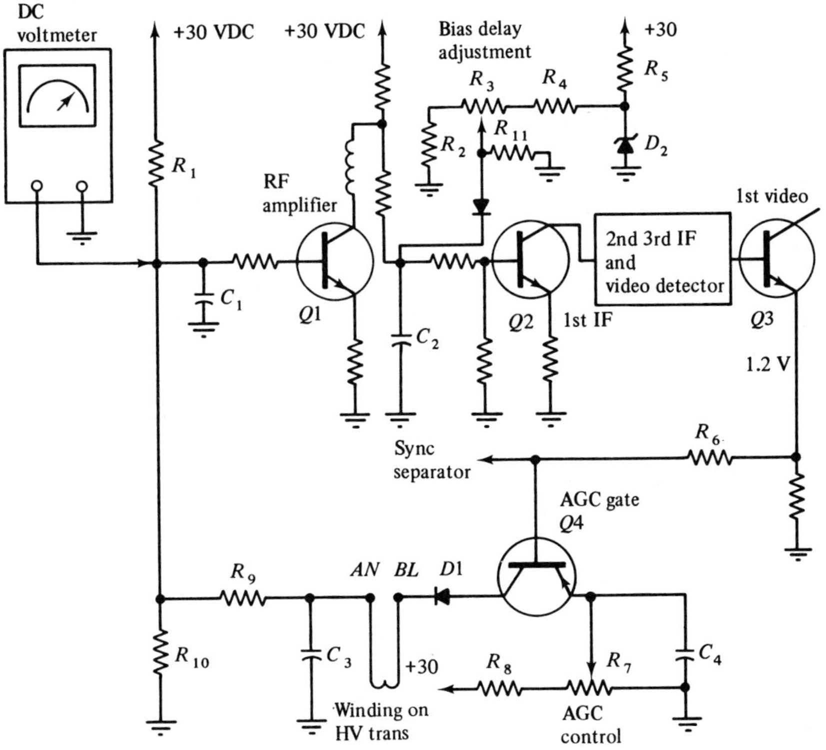

221

To maintain optimum signal/noise ratio in the arrangement of Fig. 8-12, delayed AGC is employed for the RF amplifier, whereby Q1 operates at practically full gain on weak-to-medium signals, with most of the AGC control provided by the IF section. However, with strong incoming signals, the RF amplifier operates at progressively reduced gain. A total AGC range of 70 dB is normally available in this example. Observe that the point at which Q1 starts to operate at reduced gain is determined by the setting of the bias-delay-adjustment control R11. In other words, this control setting establishes the minimum gain of Q2. Filtering of the AGC voltage is provided by C3, C1, R9, and R10. Additional filtering is provided by R2, and C2. From a statistical viewpoint, most AGC malfunctions are caused by leaky filter capacitors. However, if all of the capacitors are found to be in good condition, the gating and delay diodes should be checked for front-to-back ratio. In some cases, the AGC winding on the flyback transformer will be found to be open-circuited, or leakage may develop between the winding and the transformer core. This defect can be checked to best advantage with a high-voltage megohmmeter.

8-5 Multiple Faults

In most trouble situations, only a single failure has occurred. A capacitor may have become leaky, a transistor may have become defective, or a picture tube may have low emission. However, multiple failures can also occur. As an illustration, if a power transistor becomes short-circuited, its load resistor may be burned. Again, if a coupling capacitor to a transistor becomes short-circuited, the transistor may be destroyed. When both sound output and raster are absent, the preliminary assumption is that a single fault is causing both trouble symptoms. In most cases, the failure will be localized in the power supply. On the other hand, if the power-supply voltages measure correctly, it will be concluded that two separate failures have occurred, one of which affects raster development and the other of which affects sound output. In turn, the technician starts by checking for output at the picture detector, and for output from the FM detector. Follow-up voltage measurements are made toward the input or the output ends of the channels, depending upon results of initial tests.

8-6 Video-amplifier Troubleshooting

Video-amplifier troubleshooting often starts by checking for input signal with an ac voltmeter, as depicted in Fig. 8-13. If a TV station

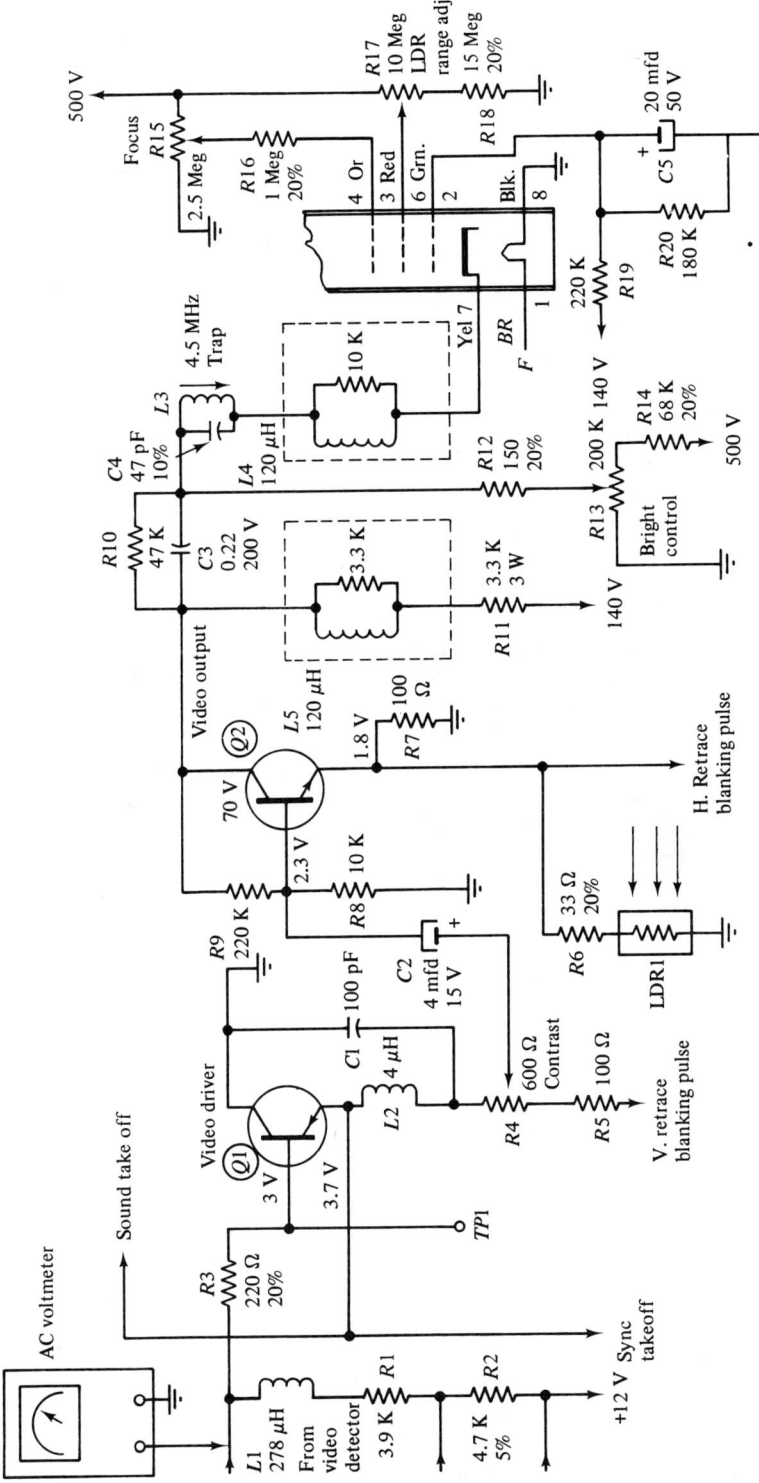

Figure 8-13 A typical video-amplifier configuration.

signal is utilized, a peak-to-peak voltmeter normally indicates approximately 1 volt p-p input signal. There is normally a small amount of signal-voltage decrease from the base to the emitter of Q1. A signal-voltage gain of at least 50 times is normally provided by Q2. After the trouble area has been localized by signal-tracing tests with an ac voltmeter, the technician proceeds to narrow down the possible defective components or devices. As an illustration, a 10-volt p-p signal might be found at the collector of Q2. In turn, if a 1-volt p-p signal is measured at the base of Q2, suspicion falls upon the transistor, or on its load circuit. On the other hand, if a 0.2-volt p-p signal is measured at the base of Q2, the technician dismisses the possibility of trouble in the Q2 stage. Then, if a 1-volt p-p signal is measured at the emitter of Q1, he concludes that C2 is probably defective, and has lost most of its rated capacitance.

Consider the procedure that is followed if the signal voltage of the base of Q2 measures 1 volt p-p, but the collector signal voltage measures 10 volts p-p. The logical procedure is to measure dc voltages at the terminals of Q2. If the collector voltage is comparatively low and the base and emitter voltages are comparatively high, it will be suspected that C2 is leaky and is applying abnormal forward bias to the transistor. However, if subsequent tests clear C2 from suspicion, it will be concluded that Q2 has excessive collector-junction leakage. Although off-value resistors can cause the same trouble symptom, this fault is much less likely than transistor deterioration. To confirm the suspicion of collector-junction leakage, a turn-off test can be made. This requires that the PC conductor between the collector of Q2 and R9 be razor-slit temporarily, so that R9 does not impose a shunt path for collector current during the turn-off test.

Observe that if C3 becomes open-circuited, the picture will become blurry and lack detail, owing to loss of the higher video frequencies. This fault does not change the dc-voltage distribution in the Q2 collector network. However, if the video amplifier is tested with a sine-wave generator, an ac voltmeter at the cathode of the picture tube will show that there is a progressive attenuation of signal voltage as the operating frequency is increased. Although a similar trouble symptom can result from peaking-coil defects, capacitor faults are more common. After a receiver has been in long-continued use, trouble symptoms can be caused by worn and "noisy" controls, such as R4 and R13. A resistance check with an ohmmeter will indicate this defect as erratic scale indications while the potentiometer is being varied through its range.

8-7 DC Voltages Under Signal and No-signal Conditions

Some schematic diagrams specify dc-voltage values under both signal and no-signal conditions. Dual-voltage data are often very helpful when an elusive trouble symptom is being analyzed. In the example of Fig. 8-14, the dc voltages are specified under normal signal conditions. It is helpful to consider what the dc voltages will be under no-signal conditions. If there were no incoming signal, the base and emitter of Q15 would rest at zero volts, and the collector potential would be 9 volts. Let us see why the emitter of Q15 develops 0.5-volt bias, why the base develops 0.2-volt bias, and why the collector operates at 7.6 volts under signal conditions. The circuit action in this stage is summarized as follows:

1. Under no-signal conditions, Q15 is biased for class B operation.
2. Q15 is cut off under no-signal conditions, and the collector voltage will accordingly equal the supply voltage (9.5 V).
3. Under signal conditions, the base of Q15 is driven by an ac waveform, and the transistor conducts on the positive excursion of the driving waveform.
4. During the time that Q15 conducts, an IR drop occurs across R78. In turn, this voltage drop subtracts from the supply voltage, and the average potential of the collector is less than the supply voltage value.
5. When Q15 conducts, electrons flow into the emitter terminal of Q15 from C30. Accordingly, there is an average deficiency of electrons on the left-hand electrode of C30, and the emitter is positively biased at 0.5 volt.
6. Since Q15 is driven into saturation by the base input signal, some of the supply voltage bleeds into the base circuit on positive peaks of the drive signal. This process leaves an average positive bias of 0.2 volt on the base of Q15.
7. Because of the signal-developed base and emitter bias potentials, Q15 is biased beyond cutoff on the average, and the transistor operates in class C under signal conditions.
8. Although the supply voltage is 9.5 volts, the peak-to-peak voltage of the collector waveform for Q15 will normally measure 10 volts. This additional 0.5 volt is provided by the base driving signal during the interval that Q15 goes into saturation.

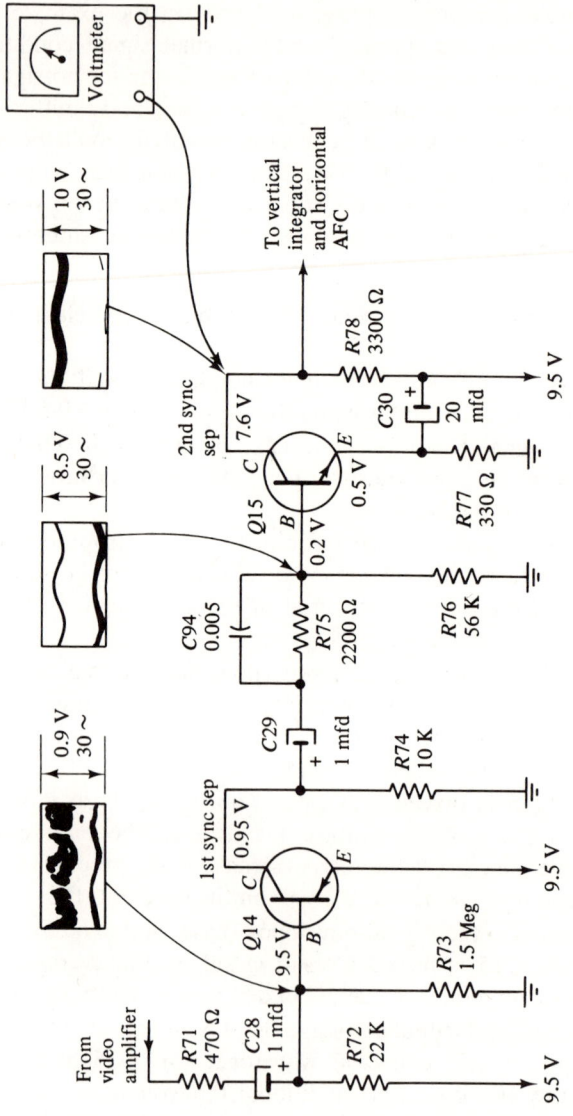

Figure 8-14 Configuration for a sync-separator section. (*Courtesy of Howard W. Sams and Co., Inc.*)

Consider next the circuit action that occurs in the Q14 stage of Fig. 8-14. This transistor is biased for class-B operation, with base and emitter resting at 9.5 volts. Note that the base is driven by a comparatively low-level signal, with an amplitude of 0.9 volt. Accordingly, although Q14 is driven into conduction on the positive peaks of the base-input signal, the transistor is not driven "hard," and does not go into saturation at any time. Therefore, signal-developed bias is negligible in the Q14 stage, and there is practically no difference in dc potentials from no-signal to signal conditions. It is also for this reason that the voltage gain of the Q14 stage is 9.4, although the voltage gain of the Q15 stage is only 1.2.

Under trouble conditions, capacitor defects may not cause a change in dc-voltage distribution. For example, if C28 becomes open-circuited (Fig. 8-14), the base voltage on Q14 does not change. On the other hand, the base voltage on Q15 will be zero, and the emitter voltage on Q15 will be zero. This is in consequence of signal stoppage at C28. Of course, it requires troubleshooting experience to recognize that C28 is open-circuited on the basis of dc-voltage measurements. On the other hand, this defect can be established directly by operating the voltmeter on its peak-to-peak ac-voltage function, and tracing the signal through the network. Again, suppose that C29 becomes open-circuited. In such a case, the base and emitter voltages on Q15 will be zero, as before. In turn, the technician would be in doubt concerning whether C29 is actually open-circuited, or whether C28 is open-circuited, on the basis of dc-voltage measurements only. In other words, the ac signal voltages should be measured with a peak-to-peak indicating ac voltmeter in order definitely to pinpoint the "open" capacitor.

8-8 Signal Tracing the AFC and Oscillator Sections

A horizontal-AFC, oscillator, and driver configuration with operating waveforms is pictured in Fig. 8-15. Pinpointing of defective components and devices is often facilitated by signal-tracing procedures with an ac voltmeter. For example, if an ac voltage with an amplitude of 8.5 V p-p were found on the input terminal of C4, but zero ac volts were measured at the output terminal of the capacitor, it would be concluded that C4 is open-circuited. Again, if a low-amplitude or zero ac voltage were measured at the junction of R3 and R5, it would be suspected that C8 had lost most of its capacitance value. This conclusion would be confirmed if a comparatively high-amplitude ac voltage were found at the collector of Q2. Under this fault condition, all of the ac signal

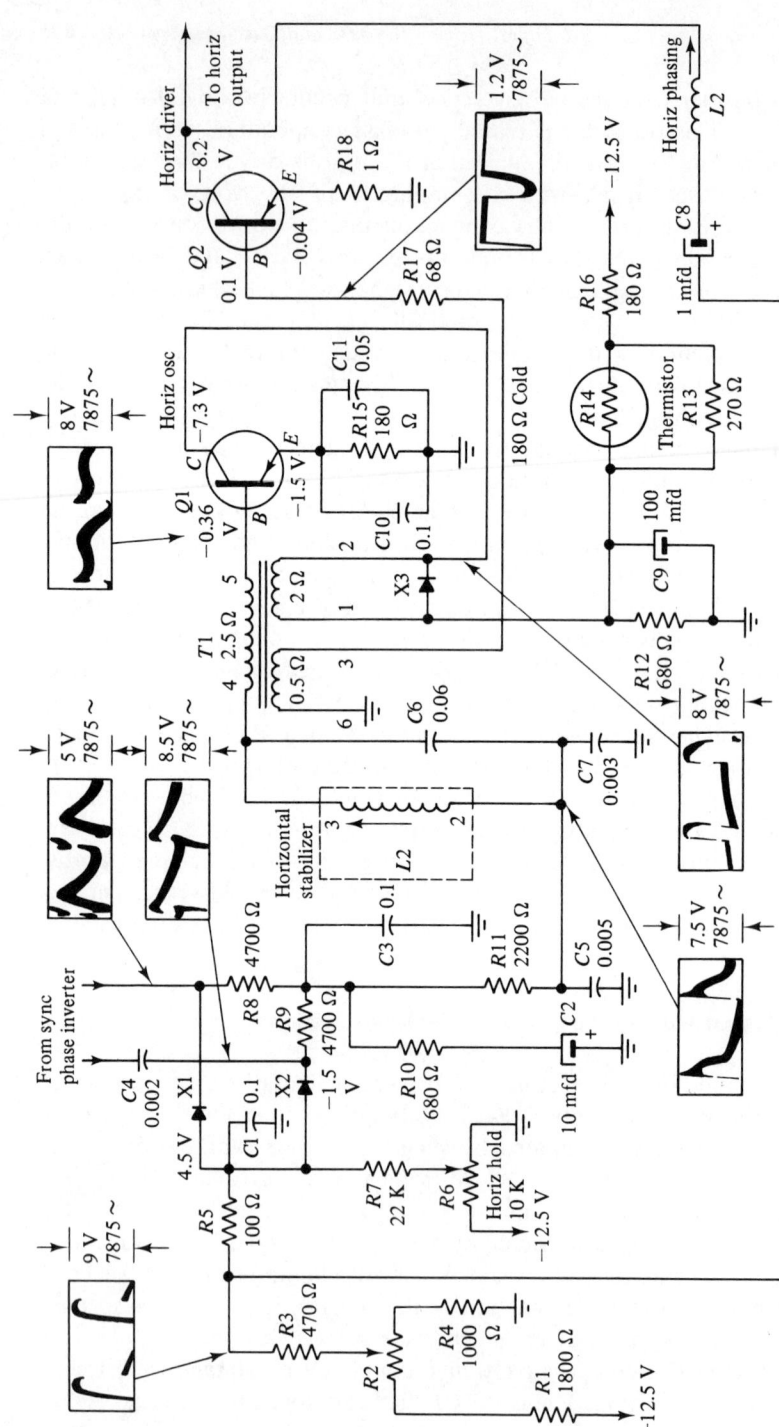

Figure 8-15 A horizontal-AFC, oscillator, and driver configuration. (*Courtesy of Howard W. Sams & Co., Inc.*)

voltages through the network will be affected to some extent, because the comparison waveform at the input to the AFC diodes is a key signal component.

Whereas ac signal voltage is normally measured at both terminals of a coupling capacitor, practically zero ac voltage should be present at the terminals of bypass and decoupling capacitors. As an illustration, virtually no signal voltage is normally found at the junction of C2 and R10 in Fig. 8-15. Note that the reactance of 1 μF is about 1 ohm at 15,750 Hz; this reactance is negligible compared to the associated circuit resistances of 680 ohms, 2200 ohms, and 4700 ohms. Next, observe that C5 and C7 operate as partial-bypass capacitors. In other words, the reactance of 0.005 μF is approximately 2000 ohms at 15,750 Hz. The reactance of this capacitor with the 0.003-μF capacitor in parallel is about 1400 ohms. This reactance is a substantial fraction of the associated 2200-ohm resistor, and in turn an appreciable ac signal voltage is dropped across the partial bypass capacitors.

As in the case of C2, C9 in Fig. 8-15 is a bypass capacitor, and the technician expects to find practically zero signal voltage across its terminals. Next, observe that R15 is shunted by a total capacitance of 0.15 μF. This capacitance has a reactance of approximately 65 ohms at 15,750 Hz. In turn, since R15 has a value of 180 ohms, the reactance is an appreciable fraction of the associated resistance, and only partial bypassing is provided. In turn, the technician expects to find an appreciable fraction of the base ac signal voltage at the emitter of Q1. In the event that these two ac signal voltages were practically equal, it would be concluded that the emitter bypass capacitors were open-circuited. In such a case, current degeneration would cause some attenuation of the collector ac signal voltage. Since determination of reactance values is often troublesome for the bench man, Figs. 8-16 and 8-17 are provided for "looking up" capacitive-reactance values. The charts are self-explanatory, with practical examples of reactance determination.

8-9 Horizontal-output Section

A basic horizontal-output configuration is shown in Fig. 8-18. The dc-voltage distribution is specified for normal drive signal voltage at the base of Q1. Note that the peak-to-peak ac voltages at the bases of Q1 and Q2 can be measured with an ac voltmeter. Since power-type transistors are utilized, they are ready suspects in case of horizontal-sweep malfunction. Collector-junction leakage results in reduced power output from the system, and can often be identified by dc-voltage measurements at the transistor terminals. It is also likely that that output transistor may overheat and burn out if junction leakage becomes

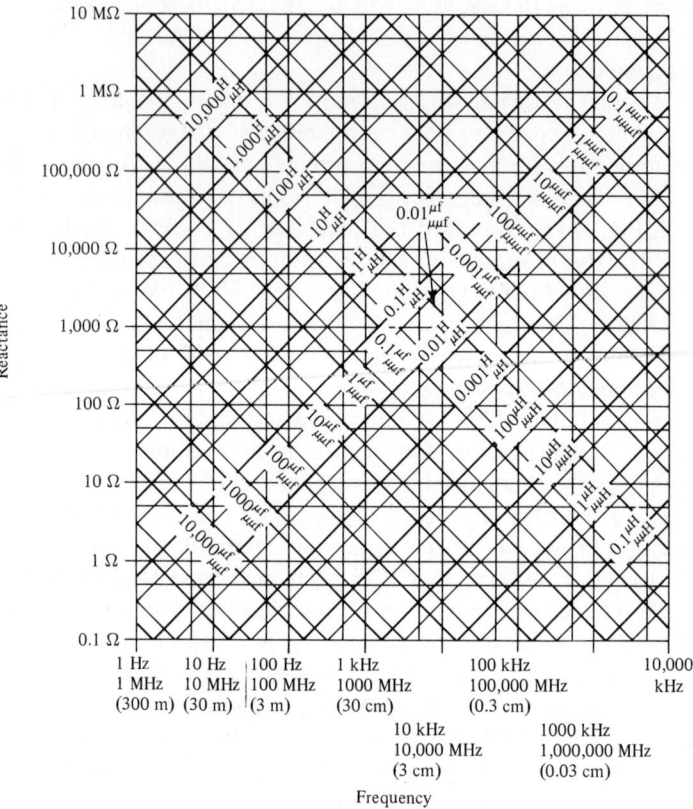

The accompanying chart is used to find:
1. The reactance of a given inductance at a given frequency.
2. The reactance of a given capacitance at a given frequency.
3. The resonant frequency of a given inductance and capacitance.

In order to facilitate the determination of magnitude of the quantities involved to two or three significant figures, the chart is divided into two parts. Figure 8-16 is the complete chart to be used for rough calculations. Figure 8-17, which is a single decade of Fig. 2 enlarged approximately seven times, is to be used where the significant two or three figures are to be determined.

To Find Reactance

Enter the charts vertically from the bottom (frequency) and along the lines slanting upward to the left (inductance) or to the right (capacitance). Corresponding scales (upper or lower) must be used throughout. Project horizontally to the left from the intersection and read reactance value.

To Find Resonant Frequency

Enter the slanting lines for the given inductance and capacitance. Project downward from their intersection and read resonant frequency from the bottom scale. Corresponding scales (upper or lower) must be used throughout.

Figure 8-16 Reactance Chart. *(Courtesy of General Radio Company)*

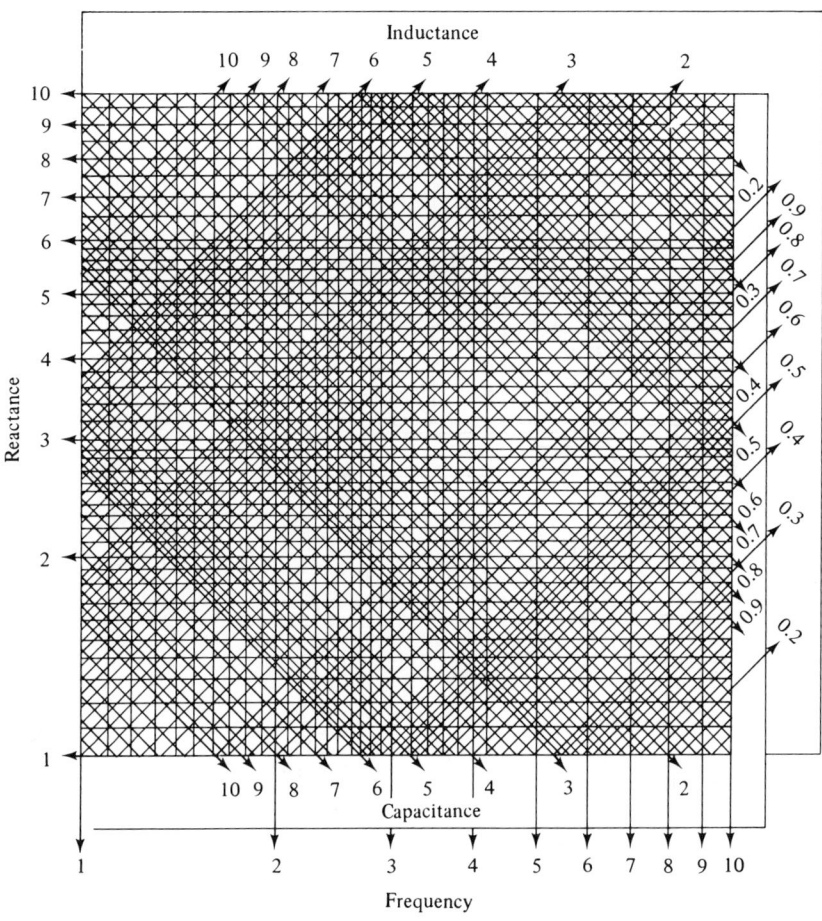

Figure 8.17 is used to obtain greater precision of reading but does not place the decimal point, which must be located from a preliminary entry on Fig. 8-16. Since the chart requires two logarithmic decades for inductance and capacitance for every single decade of frequency and reactance, unless the correct decade for L and C is chosen, erroneous results are obtained.

Typical Results

1. Find the reactance of an inductance of 0.00012 H at 960 kHz.
Answer: 720 Ω.
2. What capacitance will have 265 Ω reactance at 7000 kHz?
Answer: 86 pF.
3. What is the resonant frequency of $L = 21$ μH and $C = 45$ pF?
Answer: 5.18 MHz.

Figure 8-17 Reactance Chart. *(Courtesy of General Radio Company)*

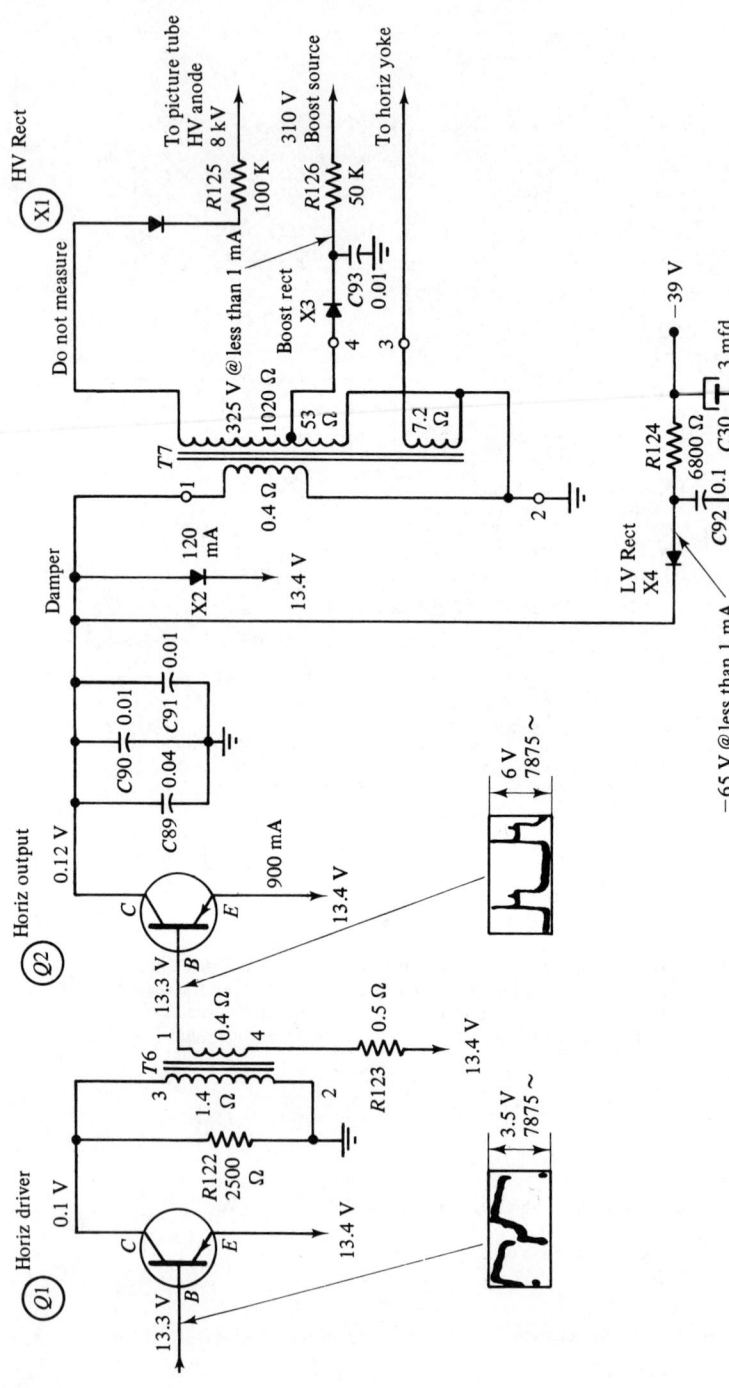

Figure 8-18 A basic horizontal-output configuration.

substantial. To measure the emitter current, cut the associated PC conductor with a razor blade, and connect a milliammeter across the "open." Similarly, the damper and low-voltage rectifier currents can be measured by cutting their PC conductors.

Sometimes an apparent malfunction in the sweep system can be corrected merely by making maintenance adjustments. For example, a typical horizontal-frequency adjustment procedure is shown in Fig. 8-19. Of course, if the maintenance adjustment does not bring the deflection system back on-frequency, attention must be turned to the circuitry that is involved, and routine troubleshooting procedures must be observed in order to localize the malfunction and pinpoint the defective component or device. Figure 8-20 shows an approved method of measuring the high voltage at the picture tube. The TVM must be grounded to the receiver chassis so that there is a complete measuring circuit.

Today's service technician frequently encounters the perplexing problem of locating component failures in the yoke and high-voltage flyback sections of television receivers. These components typically have not been easy to troubleshoot or test. The technician often had to resort to time-consuming substitution to identify a faulty yoke or flyback transformer. Modern receivers are now using a variety of solid-state components, such as transistors, SCR's and triacs. Because of this increasing complexity, a simplified tester that can check all these yokes and flybacks quickly and accurately is a very useful tool on the service bench.

A ringing test allows for testing of any yoke, flyback, larger coils, or vertical-output transformers in minutes. Although ringing tests have been made traditionally with the oscilloscope, a meter such as that illustrated in Fig. 8-21 provides wide capability. The general procedure is to connect the two test leads of the instrument across the windings of the device under test. Then the technician depresses the ringing-test button, followed by each of the six impedance-matching buttons, one at a time. If a "good" readout is obtained for one or more of the buttons, the device is good. However, if none of the buttons offers a "good" readout, the component under test may not necessarily be at fault.

There are a number of conditions that will cause a good component to test "bad." These are detailed subsequently. Note that the yoke and flyback tester is essentially a sweep circuit with analyzer facilities. The horizontal-output circuit of any TV receiver contains the yoke, flyback, and interrelated sections. A ringing test can distinguish whether a malfunction is present in the yoke or in the flyback circuit. However, the operation of the entire horizontal section is also dependent on the drive signals, as well as on the focus voltage and high-voltage levels. In turn, ranges are provided in the instrument for measuring drive signal and high-voltage values. By measuring the drive signal amplitude, the

234 Black-and-White Television Troubleshooting

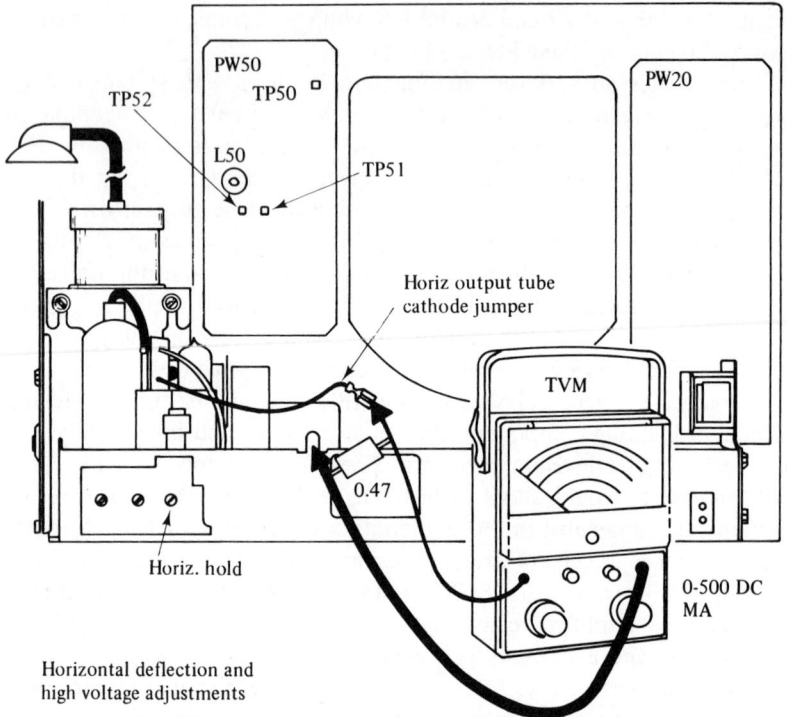

Horizontal deflection and
high voltage adjustments

A. Horizontal Frequency Adjustment

1. Disable sync by shorting TP50 to ground.

2. Short the same wave coil, L50, by placing jumper from TP51 to TP52.

3. Adjust the horizontal hold control so that picture sides are vertical (horizontal oscillator on frequency). Remove short from TP51 to TP52.

4. Adjust sine wave coil, L50, so that oscillator is again on frequency (picture sides vertical).

5. Remove short from TP50.

Figure 8-19 Typical deflection frequency adjustment procedure. (*Courtesy of Radio Corporation of America*)

technician can determine whether or not a malfunction is located in a prior receiver section. Additional trouble localizations can be accomplished by measurement of the second-anode and focus-voltage values.

The yoke and flyback tester provides go/no-go indication of the electrical merit of horizontal and vertical yokes, flyback transformers, vertical-output transformers, horizontal-linearity coils, horizontal-efficiency coils, and numerous general-application mid-frequency coils. This

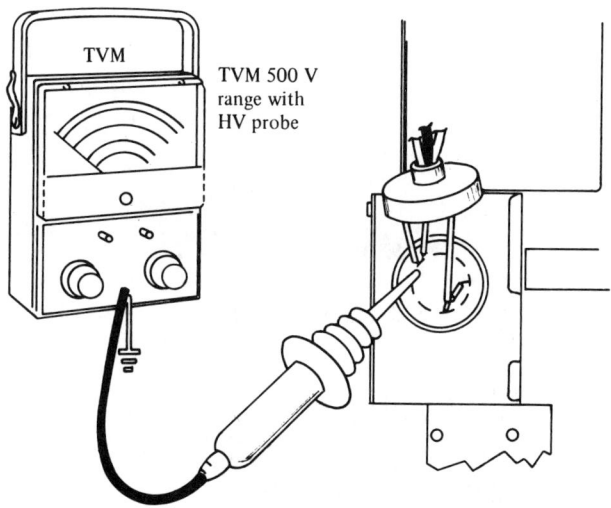

High-voltage check

Connect a high-voltage probe to the picture tube HV anode as shown. Adjust the brightness control to minimum. Read the high voltage. Normally, it should read 21.5 KV, but any voltage between 20 KV and 22.5 KV is acceptable.

Now rotate the brightness control through its range. The high voltage may decrease as much as 3 KV from minimum to maximum brightness level in normal operation.

Figure 8-20 High-voltage measuring procedure. (*Courtesy of Radio Corporation of America*)

indication is developed by a dynamic ringing test wherein an electronic counter determines the number of cycles that a coil rings before reaching a preset damping point after a given driving pulse has been applied. A sensitivity control permits the operator to select the signal level at which the counting circuit terminates its operation. A block diagram of the instrument is shown in Fig. 8-22. Logic circuits (not shown) are employed in the counter section. Instrument functions are evident from the nomenclature included in the block diagram.

8-10 Case-history Approach

In the case-history approach to troubleshooting, the technician observes the picture and/or sound symptoms, but makes no analysis of the con-

236   Black-and-White Television Troubleshooting

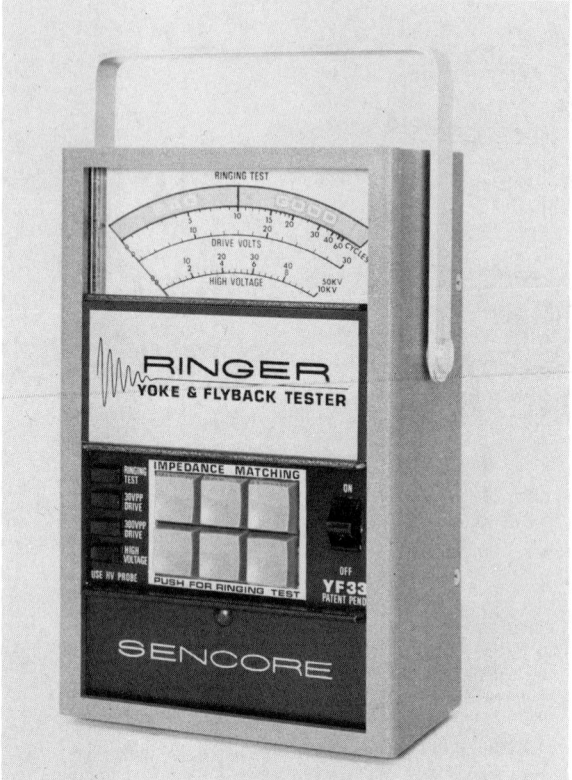

Figure 8-21   A ringing-type yoke and flyback tester. (*Courtesy of Sencore*)

dition. Instead of proceeding to make tests and measurements, the troubleshooter "looks up" the symptoms in a book or file of case histories for the particular brand and model of receiver with which he is concerned. As an illustration, if sync buzz happens to interfere with sound reception on UHF channels for a Zenith 19DC12 receiver, the technician would find the probable "cure" noted in Fig. 8-23. In other words, he would proceed to replace Q8, and if the sync buzz persisted, he would add a 5-pF capacitor from base to ground terminals of this transistor. Of course, if these procedures did not correct the difficulty, the technician would then proceed to analyze the operation of the receiver controls, observe operation on all active channels, and to make tests and measurements in suspected circuits.

The chief disadvantage of the case-history approach is lack of adequate data to apply in the majority of trouble situations. Although

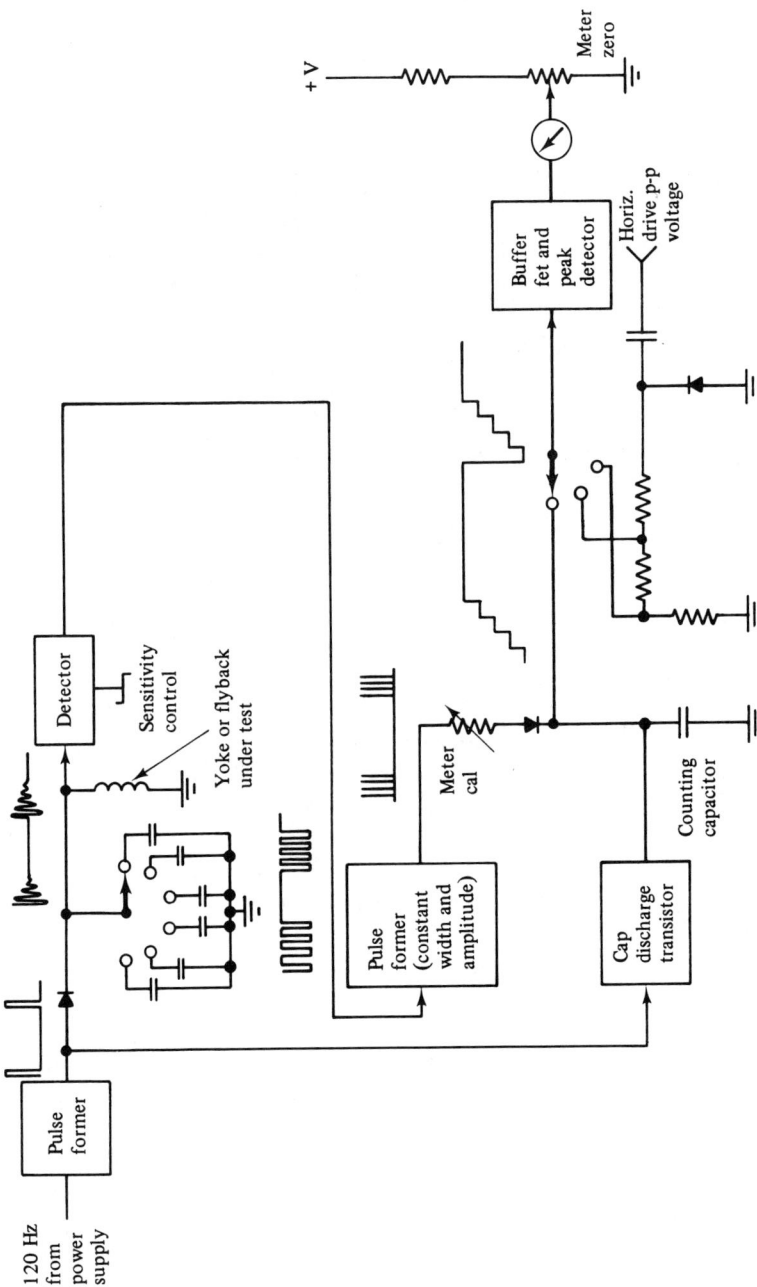

Figure 8-22 Block diagram of yoke and flyback tester. (Courtesy of Sencore, Inc.)

238 Black-and-White Television Troubleshooting

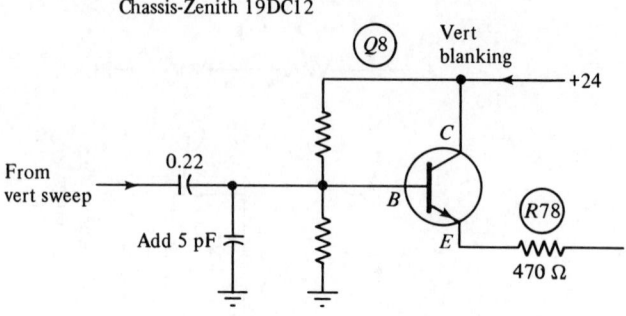

Symptom-Buzz in the sound on UHF
Cure-If Q8 replacement doesn't cure, add 5 pF capacitor as shown

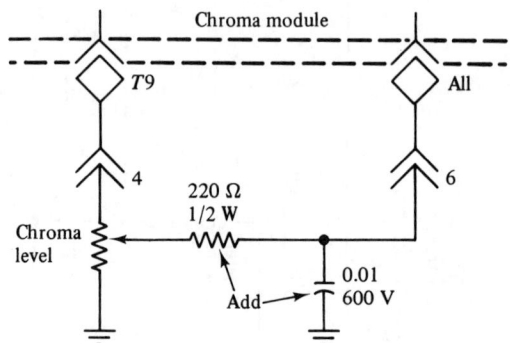

Symptom-Failure of IC2 in early-production sets
Cure-Add a resistor and capacitor, as shown

Figure 8-23 Example of case-history data.

the corrective steps noted in available case-history data are often effective, this is not necessarily true in all trouble situations. A very helpful aspect of case-history data is their inclusion of production changes that were issued after a receiver was marketed. In other words, it sometimes happens that a receiver is found to be inadequately design-engineered after it has been merchandised. In such a case, the factory issues production-change data. (See Fig. 8-24.) One means of publication for these data is to include them in case-history files and books. Some technicians keep a running record of their personal experiences in troubleshooting the more difficult situations that they encounter in their activities. In turn, these records may save time for the crew on future occasions.

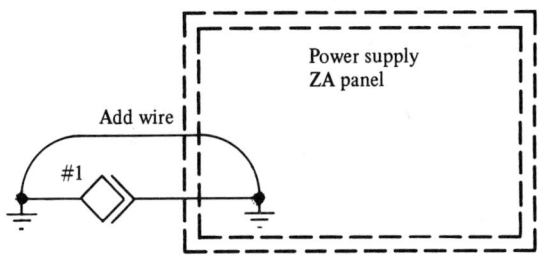

Symptom-Hum bars in picture
Cure-Check #1 pin connection of ZA power panel.
Tighten, or add a jumper wire from chassis to panel ground foil

Figure 8-24 Example of a production-change instruction.

9 Color Television Troubleshooting

9-1 General Considerations

All of the procedures that have been explained for troubleshooting black-and-white television receivers apply equally to color-television receivers. In addition, a color receiver processes a multiplexed chroma signal and has specialized synchronizing, decoding, and matrixing sections. An elaborated picture tube with specialized operating circuits is also included in a color receiver. A general signal-flow chart for a color receiver is shown in Fig. 9-1. Four signals proceed through the IF amplifier: the black-and-white picture signal (also called the Y signal), the intercarrier sound signal, the black-and-white sync signals (controlling signals), and the color signals comprising the color camera signal and the color sync signal (color burst). Internally generated signals are developed for converging the electron beams and for deflecting them on the screen of the color picture tube.

A block diagram for a color-TV receiver is shown in Fig. 9-2. It will be observed that there are approximately twice as many sections in a color receiver as in a black-and white receiver. Malfunctions in color sections cause typical trouble symptoms that are useful in preliminary analysis. For example, a color receiver may lose color sync without losing black-and-white sync. This malfunction results in a normal display of the Y signal image, with the color image broken up into diagonal strips of rainbow colors. Suspicion falls upon the burst amplifier or the chroma sync and phase detector section in this case. Since the burst amplifier is gated from the horizontal sweep system, a relative timing error could be causing the trouble symptom. Thus, readjustment of the horizontal-hold control could possibly restore color sync lock.

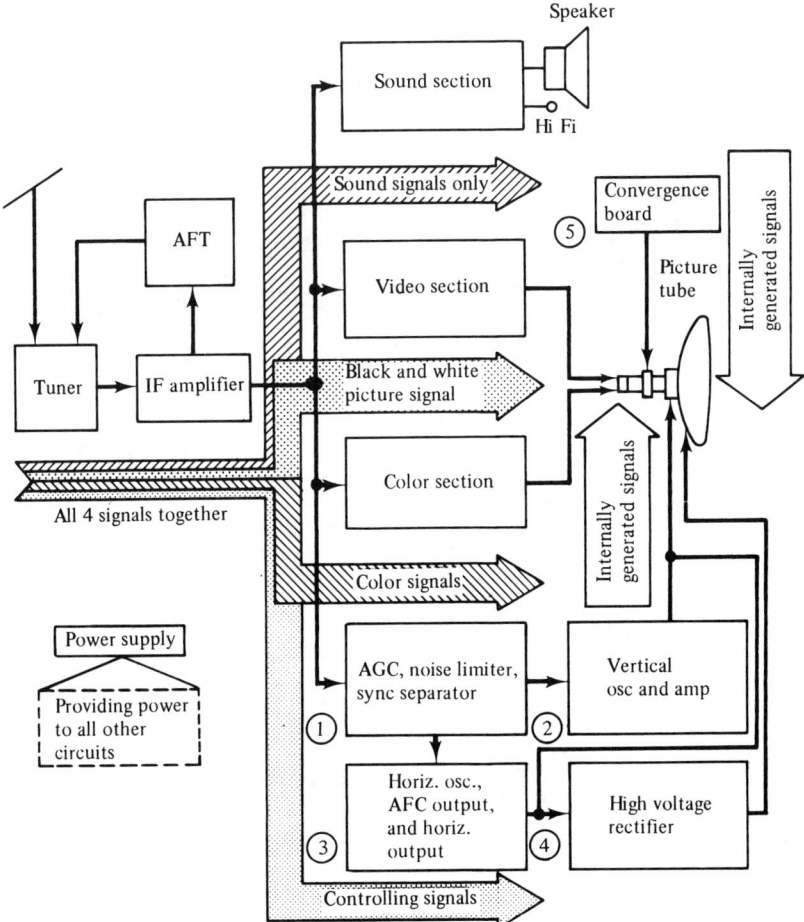

Figure 9-1 General signal-flow chart for a color-TV receiver. (*Adapted from an original by Heath Co.*)

As another example of color picture analysis, normal reception may occur on strong signals, with black-and-white reproduction only on weak signals, or the color component may drift in and out of the image during weak-signal reception. In such a case, suspicion falls on the adjustment of the color killer. It may be found that a slight readjustment of the color-killer threshold will provide normal color reproduction during weak-signal reception. Of course, if adjustment of the threshold control does not correct the trouble symptom, circuit tests must be made to track down a defective component or device. Again, when incorrect

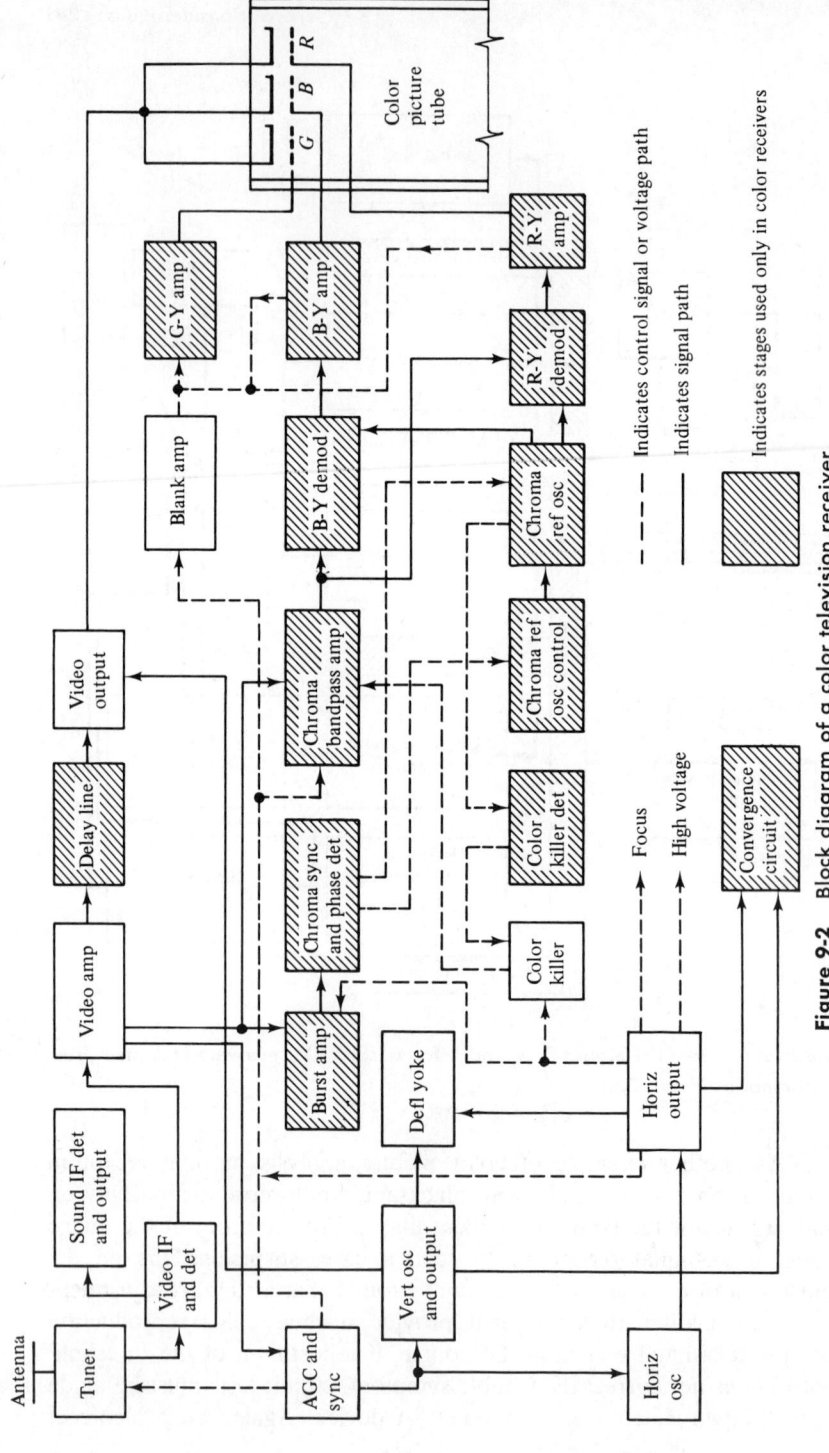

Figure 9-2 Block diagram of a color television receiver.

colors are reproduced in a picture, readjustment of the hue control may provide a normal image. On the other hand, when the hue control is found to be out of range, voltage and resistance measurements must be made in the associated circuitry to localize the fault.

In many situations, trouble symptoms are ambiguous, and preliminary analysis is required to localize the malfunction to either the receiver chassis or to the color picture tube. In such a case, it is helpful to use a picture-tube test jig (rig) to obtain a quick check without removing the color picture tube from the chassis. A jig is connected to the receiver circuitry by means of cables and adaptors. Figure 9-3 illustrates a typical color picture-tube test rig. The unit includes a 40-kV meter that employs a 50-μA movement for low-drain monitoring of the ultor voltage, and a built-in speaker for monitoring of the audio signal. Accessories include a convergence yoke assembly, a blue lateral assembly, a convergence load, a 90-deg picture-tube cable, a yoke cable, and an anode extension cable. In turn, the technician can quickly determine whether a trouble symptom is being caused by a chassis malfunction or by a picture-tube fault.

Figure 9-3 A color picture-tube test rig that includes a kilovolt meter. (Courtesy of TeleMatic Corporation)

9-2 Automatic Tint Control Troubleshooting

A hue (color) control arrangement is shown in Fig. 9-4. This is a comparatively elaborate system that provides automatic tint control (ATC)

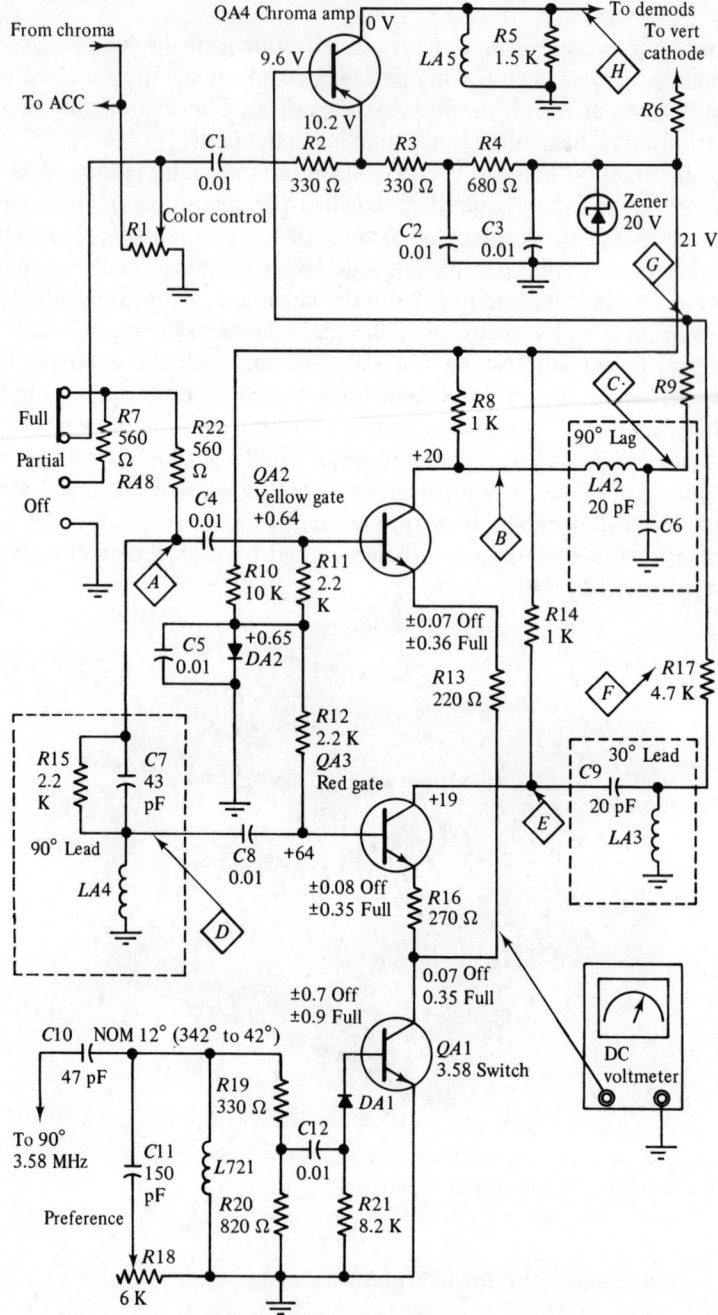

Figure 9-4 An automatic tint control circuit.

action. The purpose of ATC is to introduce judicious hue distortion that provides a wider range for phase error in reproduction of flesh tones than is inherent in a distortionless hue-control circuit. Idealized vectorgrams showing ATC action are pictured in Fig. 9-5. This type of ATC system provides for either partial or full ATC action. When partial ATC is used, the red and yellow-orange demodulation phases are shifted into the vicinity of the orange demodulation phase. Again, when full ATC is utilized, the red and yellow-orange demodulation phases become the same as the orange demodulation phase. Thus, a tradeoff is involved, which is deemed advantageous by most viewers because phase errors in reproduction of flesh tones become less apparent. Chroma phases corresponding to various colors are shown in Fig. 9-6.

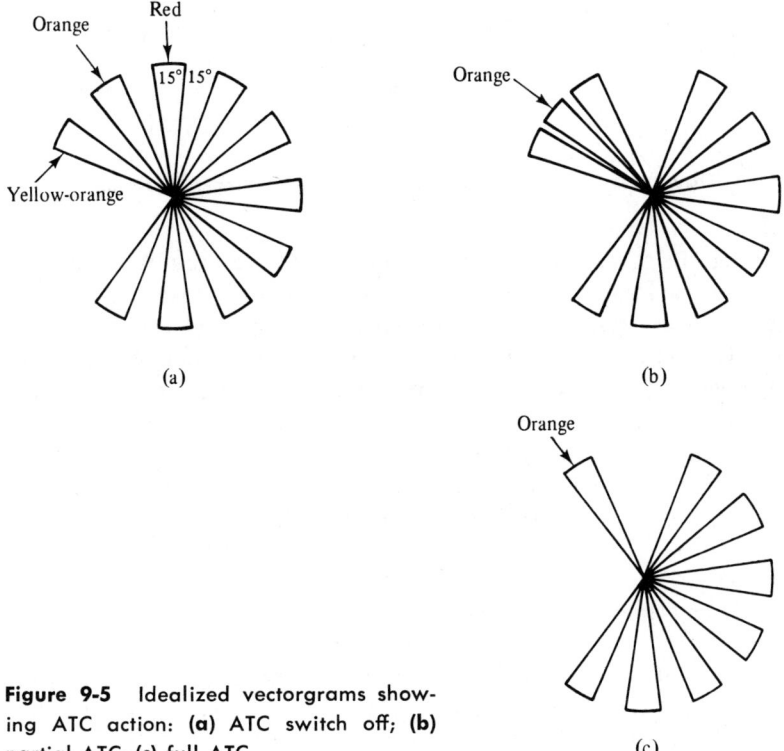

Figure 9-5 Idealized vectorgrams showing ATC action: (a) ATC switch off; (b) partial ATC; (c) full ATC.

Consider the evaluation of dc-voltage measurements in the network of Fig. 9-4. Transistor terminal voltages are specified for the "off" setting of the ATC control, and for its "full" setting. These dual-voltage specifications can be very helpful in pinpointing a defective component

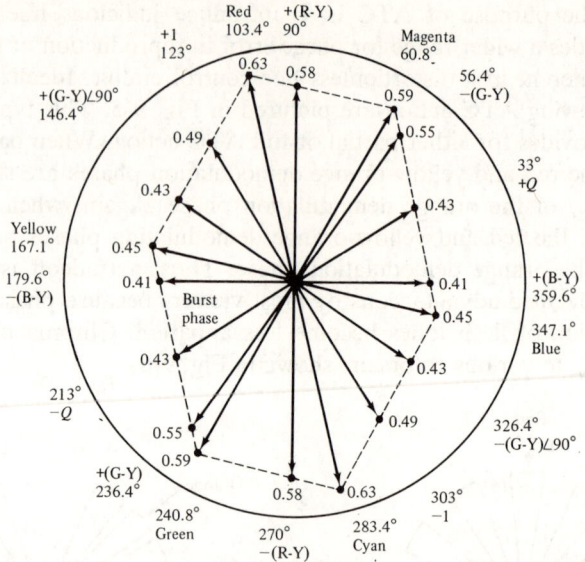

Figure 9-6 Chroma phases corresponding to various colors.

or device. In other words, "off" voltage values are useful in disclosing defective transistors and leaky capacitors. On the other hand, "full" voltage values provide clues concerning open-circuited capacitors and defective diodes. For example, if C10 or C12 is open-circuited, the "off" dc-voltage values for QA1 will remain normal. On the other hand, the "full" dc-voltage values will be little, if any, higher than the "off" values. This same result will occur if C11 is short-circuited. In the event that DA1 is short-circuited, the base voltage on QA1 will be zero for both the "off" and the "full" settings of the ATC control.

9-3 Color Demodulator Tests

Reproduction of correct colors in the image depends also upon normal functioning of the color demodulators. A schematic diagram for a widely used red-green-blue (RGB) demodulation system is shown in Fig. 9-7. Transistor terminal voltages are specified for no-signal conditions and for signal conditions. DC voltages under signal conditions are enclosed in rectangles. Thus, the red video output tube normally has a dc collector potential of 152 volts under no-signal conditions, and a potential of 150 volts under signal conditions. Most technicians use a keyed-rainbow color-bar signal to check color demodulators, as pictured in Fig.

9-8. Failure of the collector voltage to shift when the color-bar signal is utilized indicates that the chroma signal is not passing through the demodulator channel. For example, C145 could be short-circuited in this situation.

If the collector-voltage shift occurs in the wrong amount, or in the incorrect direction, suspicion falls on the associated color demodulator diodes. For example, if the collector voltage of Q1L is found to shift in the wrong direction under signal conditions, it will be suspected that diode E1L is defective. Note that the color signal voltages can be traced throughout the network with an ac voltmeter that has a frequency response through 3.58 MHz. Thus, the signal frequency is 3.56 (or 3.58) MHz from the bandpass amplifier to the color-demodulator diodes. Past the 3.58-MHz traps, the signal frequency is 189 kHz (for a keyed-rainbow test signal). The signal amplitude into the color-demodulator diodes is approximately 1 volt p-p, and the signal amplitude into the color picture tube is approximately 100 volts p-p.

Because direct-coupled circuitry is utilized in the configuration of Fig. 9-7, pinpointing of defective components on the basis of dc-voltage measurements is not as straightforward as in ac-coupled circuitry. For example, if C18L becomes leaky, the emitter voltage on Q1L decreases and its collector voltage also decreases. In turn, the forward bias on Q1M increases, and its collector voltage decreases. Accordingly, if the technician starts to measure electrode voltages on Q1M, he is likely to suspect that the trouble is in the output stage, whereas it is actually located in the driver stage. Similarly, if E1L becomes defective, the same change in dc-voltage distribution for Q1M can occur. In summary, pinpointing of defective components in direct-coupled circuitry requires throughput measuring procedures, owing to the "chain reaction" that is involved.

9-4 Silicon Controlled Rectifier Sweep Circuit

Color receivers often utilize silicon controlled rectifiers (SCR's) in the horizontal-deflection section. An SCR is different from a transistor in that it operates as an electronic switch, and not as an amplifier. Because an SCR is gated into conduction by a pulse, it is sometimes called a gate-controlled switch. With reference to Fig. 9-9, the current-flow sequence in the SCR output circuit is the same as if a power-transistor arrangement were used. Figure 9-10 shows the symbol for an SCR. If a forward bias is applied between anode and cathode of the SCR, the device does not conduct anode current until a gating pulse is applied between the gate and cathode. The SCR then goes into full conduction,

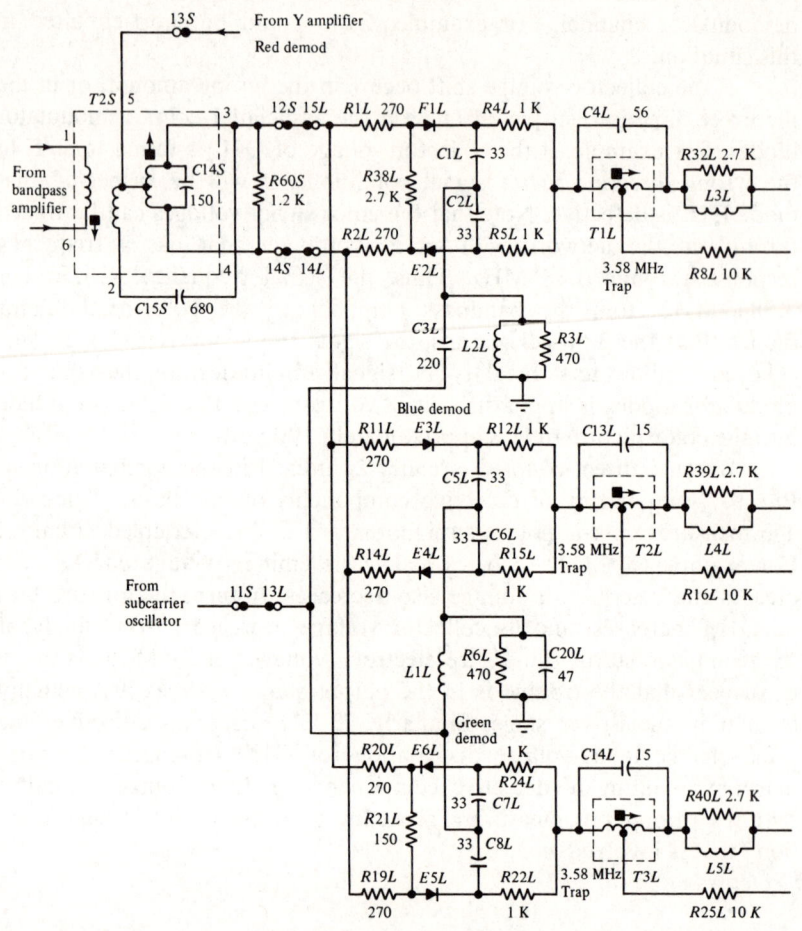

Figure 9-7 A color demodulation configuration. (*Courtesy of Motorola Semiconductor Products Inc.*)

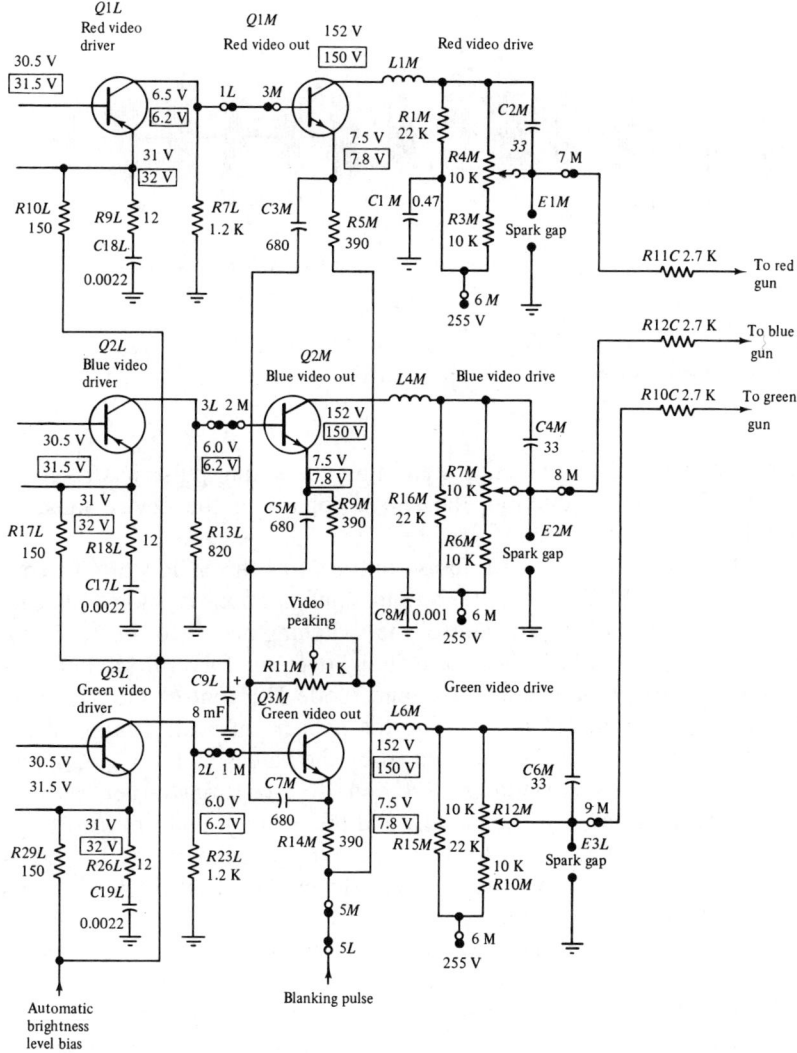

Figure 9-7 Continued

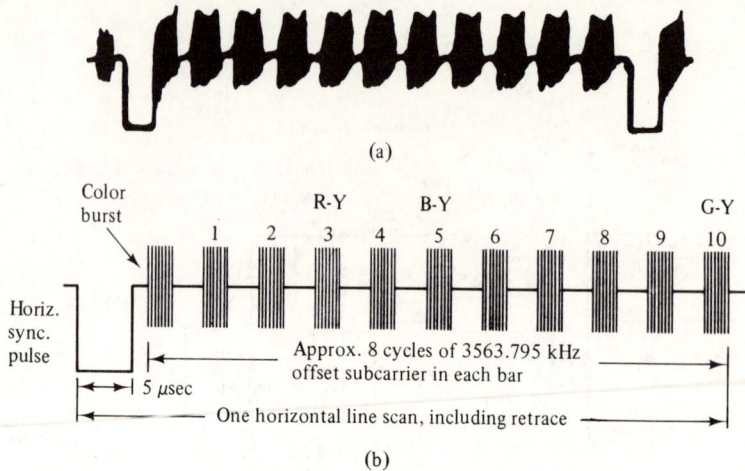

Figure 9-8 Keyed-rainbow color-bar signal waveform: (a) typical generator output waveform; (b) waveform characteristics.

and remains in the conducting state after the gating pulse has been removed. To stop conduction, the anode voltage of the device must be reduced to zero.

It follows that the gate loses control after the SCR starts to conduct. A basic SCR horizontal-output configuration is shown in Fig. 9-11. SCR1 and diode D1 control the switching action for current flow through the horizontal deflection coils L_Y over the forward scanning interval. On the other hand, SCR2 and diode D2 control the switching action for yoke current flow during the flyback interval. Timing of the horizontal-sweep action is determined by the values of L_R, C_R, C_H, and C_Y. The transformer windings LG1 and LG2 on transformer T2 provide a charging path for L_R and C_R, and the gating waveform for SCR1. With reference to Fig. 9-9, forward scanning action starts at the end of the flyback interval; the stored magnetic energy in the yoke has its maximum value at this time.

As the magnetic field in the yoke begins to collapse, current I_1 starts to flow in the deflection circuit as shown in Fig. 9-12. As the forward scanning action proceeds, current I_1 flows through trace diode D1 and charges capacitor C_Y. Flow of current I_1 sets up a reverse bias between anode and cathode of SCR1. Thus, SCR1 does not conduct during this interval. Note that as soon as I_1 decays to zero, C_Y starts to discharge back through the yoke winding. At this time, D1 is reverse-biased, but SCR1 is forward-biased. Also, a gating pulse has been applied to SCR1, so that C_Y finds a closed path for current flow through the yoke winding.

9-4 Silicon Controlled Rectifier Sweep Circuit

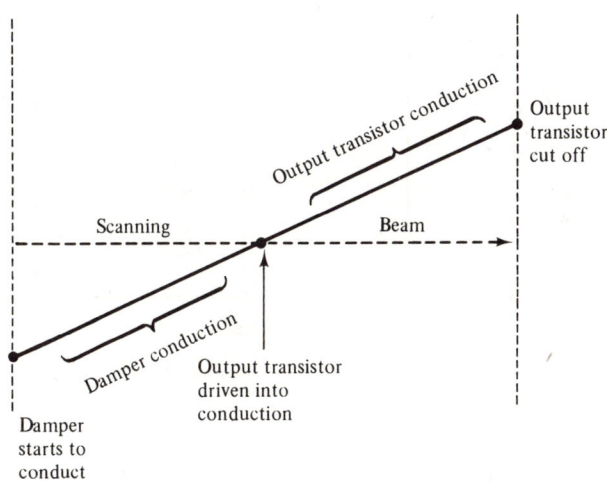

Figure 9-9 Basic scanning action provided by a transistor and damper diode, or by an SCR system.

Thus, the reverse current flow provided by I_2 through the yoke winding provides the second half of the forward scan (Fig. 9-13). At the end of this interval, a pulse from the horizontal oscillator section causes SCR2 to conduct and initiates flyback, as depicted in Fig. 9-14. In turn, a previously stored charge on C_R flows through the path denoted I_3. A large yoke current, I_4, also flows at the onset of flyback.

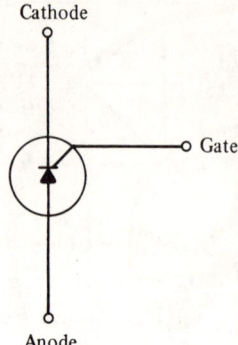

Figure 9-10   SCR schematic symbol.

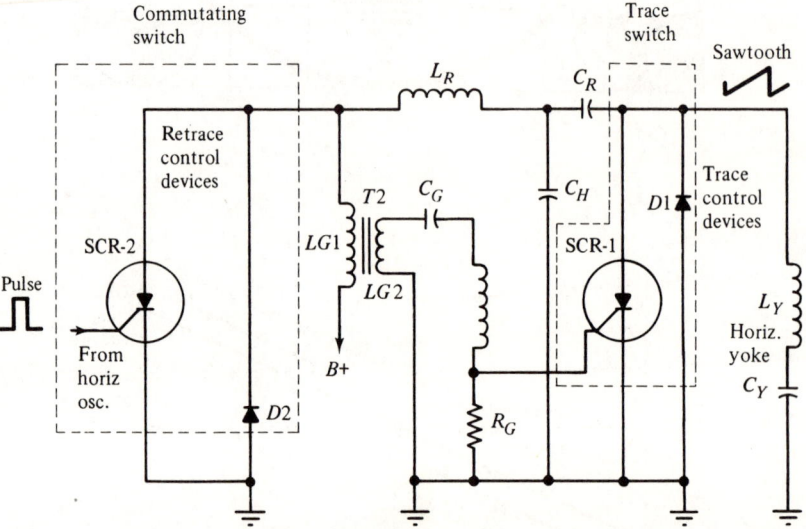

Figure 9-11   A basic SCR horizontal-deflection circuit. (Courtesy of Radio Corporation of America)

I_3 and I_4 have a resultant value that permits SCR1 to remain in conduction. Both SCR1 and SCR2 are conducting simultaneously, but inasmuch as I_3 increases faster than I_4, the resultant current flow through SCR2 after 2 microseconds develops a reverse bias and SCR1 cuts off.

It follows that D1 (Fig. 9-14) becomes forward-biased as SCR1 cuts off. Accordingly, I_4 flows through D1 instead of SCR1 for a brief time. That is, I_3 decreases rapidly in its flow through D1, and D1 soon cuts off. With both SCR1 and D1 cut off, an undamped resonant circuit comprising L_R, C_R, L_Y, and C_Y is established. In turn, this resonant circuit is shock-excited into brief oscillation by sudden removal of its associated damping elements. Thus, the circuit oscillates for one-half

9-4 Silicon Controlled Rectifier Sweep Circuit

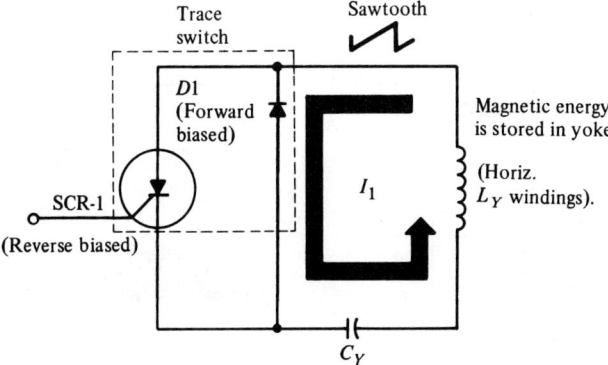

Figure 9-12 First half of forward scan is effected by current I_1. (Courtesy of Radio Corporation of America)

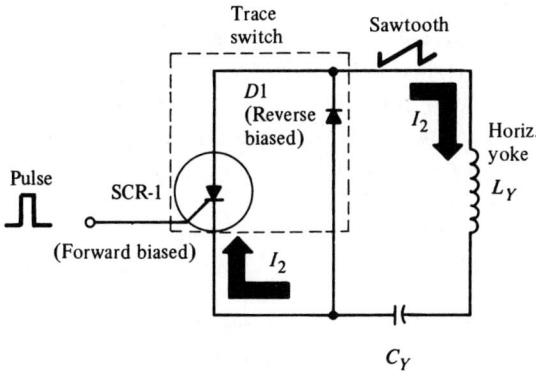

Figure 9-13 Second half of forward scan is produced by current I_2. (Courtesy of Radio Corporation of America)

cycle and produces the retrace current (flyback), quickly returning the scanning beam to the left-hand side of the picture-tube screen. This brief oscillatory action occurs at a frequency of approximately 70 kHz. The flyback current flows through SCR2, inasmuch as the device is forward-biased at this time.

Just as the retrace interval is half completed, the oscillatory current goes through zero and crosses the zero axis. In turn, SCR2 cuts off

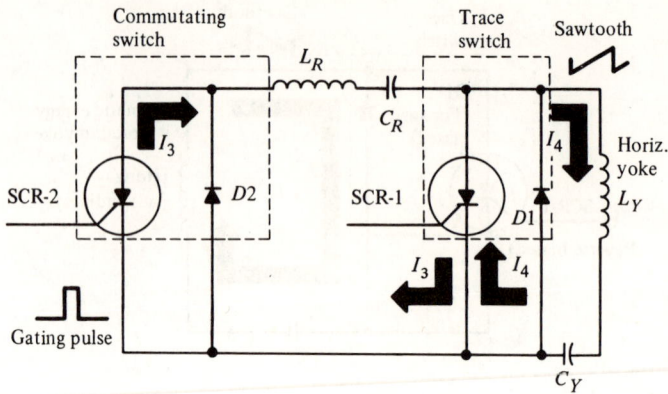

Figure 9-14 Circuit action at the beginning of flyback. (*Courtesy of Radio Corporation of America*)

and D2 is biased into conduction, so that the retrace action is completed through D2. As retrace terminates, the energy previously stored in C_R has been transferred to L_Y as magnetic energy storage. At this time, D1 becomes forward-biased and starts to conduct. In turn, the forward-scanning interval begins, and the stored magnetic energy in L_Y starts to charge C_Y. At the outset, there is some remaining magnetic energy stored in L_R. Next, at the instant that D1 conducts, this stored energy transfers rapidly to C_R through D1 and D2. As soon as the forward scan gets started, D2 cuts off.

Consider now the circuit action associated with LG1 and LG2 in Fig. 9-11. During the retrace interval, LG1 operates between the supply voltage and ground, owing to the conduction of SCR2 and of D2. On the other hand, when D2 cuts off at the start of the forward scan, LG1 becomes effectively disconnected from ground. Accordingly, C_R charges through LG1 from the supply voltage. This charging action continues over the forward-scanning interval. In this manner, the charge on C_R furnishes energy to the yoke circuit when retrace starts.

Observe that the voltage drop across LG1 is coupled to LG2. The voltage from LG2 forward-biases the gate electrode of SCR1, thereby initiating conduction at the appropriate time. This gating waveform is shaped by C_G, R_G, and the circuit inductance in the circuit of Fig. 9-10. This circuitry forms a fast-rise gate pulse for triggering SCR1 into conduction. There is necessarily an I^2R loss in the system. Also, the high-voltage section drains energy from the deflection system. These losses are made up from the supply voltage source. However, this current demand is comparatively small.

Troubleshooting of an SCR deflection system should start with an inspection for evidence of arcing, charring, or overheating. If the technician is present at the time of failure, overheated components or devices will be apparent. DC-voltage measurements serve to show open circuits, such as burned-out coil windings. A systematic checkout involves the following steps:

1. Statistically, capacitors cause more failures in SCR systems than other components or devices. In general, one end of a suspected capacitor must be disconnected for test.
2. Horizontal-output transformers are ready suspects. Unless a high-voltage megohmmeter is available, a substitution test is advisable.
3. In addition to the horizontal-output transformer, other inductive components such as the yoke and the individual inductors should be checked. Resistance measurements are not fully informative, and substitution tests should be made in case of doubt.
4. Device failure is comparatively infrequent. However, if other components are cleared from suspicion, the SCR's and diodes should be checked. A good semiconductor tester will provide reliable out-of-circuit tests.
5. Although off-value resistors are seldom encountered, this possibility should be kept in mind. Ohmmeter checks are adequate.

9-5 Fail-safe Circuit Operation

Arcing is a common trouble symptom in high-voltage systems. In turn, the voltage and current surges can damage trace-switch components such as SCR1 and D1 in Fig. 9-14. With reference to Fig. 9-15, two fail-safe circuits are provided to prevent this eventuality. Thus, diode D4 and the shunt resistor R_s are connected in series with the primary winding of the flyback transformer T_1. Normally, the flyback current will flow through the low forward resistance of the diode D4. On the other hand, if an arc occurs, diode D4 becomes reverse-biased by the resulting surge; in turn, the surge energy is dissipated chiefly as an I^2R loss in R_s. However, R_s cannot absorb the entire surge, and a supplementary fail-safe circuit is provided.

In Fig. 9-15, the supplementary fail-safe circuit comprises D3, C_T, and R_T. Normally, D3 conducts on the peak of the flyback pulse with the result that C_T acquires a charge. Note that resistor R_T provides a time constant such that the voltage on C_T will remain sufficient to reverse-bias D3 during the forward scanning interval. On the other hand,

256 Color Television Troubleshooting

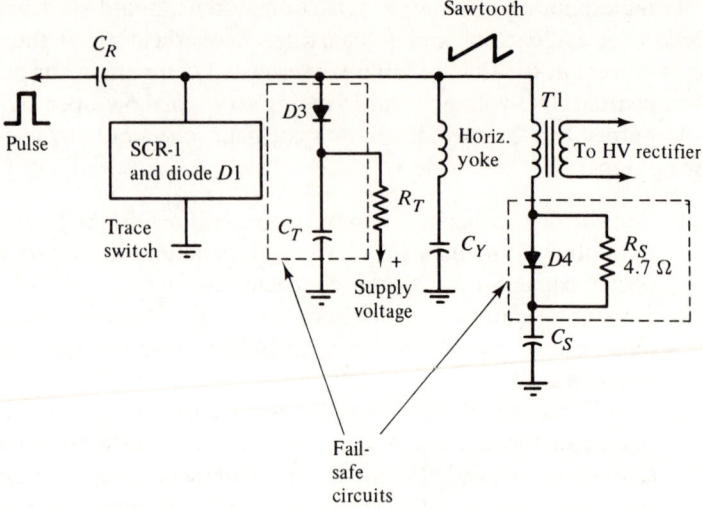

Figure 9-15 A fail-safe arc protective configuration. (*Courtesy of Radio Corporation of America*)

if an arc occurs, D3 becomes forward-biased; the energy of the surge is largely absorbed by C_T. Thus, SCR1 and D1 are protected from the surge voltage. When a trouble symptom such as a narrow picture or dark screen occurs, there is likely to be leakage present in one or more of the capacitors C_R, C_T, C_Y, or C_S. A poor front-to-back ratio in D3 will also cause a narrow picture or dark screen.

Next, if D4 becomes short-circuited, it will provide no arc protection, although flyback operation remains unaffected. On the other hand, if D4 develops high forward resistance, or an open circuit, the picture will be narrow. If D4 is short-circuited, the trace-switch components are very likely to break down in the event of an arc. Since D4 is called upon to withstand considerable voltage in case of an arc, it is good practice to check the diode for front-to-back ratio after replacement of arc-damaged components such as the high-voltage rectifier.

9-6 Deflection Linearity Correction

An SCR deflection system requires a horizontal-linearity control to obtain a linear sawtooth current waveform through the yoke windings. In other words, the generated current waveform has an exponential shape that requires correction. Exponential curvature in the current sawtooth is linearized by means of circuitry such as that shown in Fig. 9-16. Basically, it is a shock-excited sine-wave generating arrangement that has a comparatively low Q value. The ringing circuit comprises L_A, C_A, and R_A. It is shock-excited by the flyback voltage reversal and rings for one-half cycle over the forward-scanning interval. At this time, another flyback voltage reversal occurs, and the circuit is shock-excited over again. The half-sine output from the ringing circuit is applied to C_Y, which is simultaneously charged by the SCR network. Combination of these waveforms occurs in the capacitor, and provides an essentially linear rise in the sawtooth current waveform.

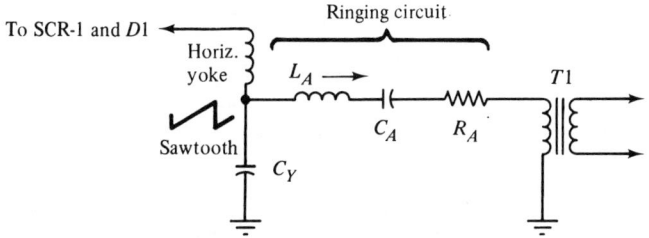

Figure 9-16 Horizontal linearity waveshaping circuit. (Courtesy of Radio Corporation of America)

After a horizontal-deflection system has been in operation for an extended period of time, component and device aging can cause nonlinear scanning. In turn, the slug in the horizontal-linearity coil L_A will require adjustment. If satisfactory scanning linearity cannot be obtained by slug adjustment, a component defect or a device defect either in the horizontal-sweep network or in the linearity circuit is indicated. Ringing capacitor C_A is a ready suspect. If the slug is out of range, a marginal defect is likely to be found in one of the trace-switch devices. Deterioration of SCR1 or D1 will produce a narrow-picture trouble symptom.

9-7 Summary of Trouble Localization

A summary of trouble localization approaches is given in Fig. 9-17. Picture and sound reproduction are checked first; if a raster is present without an image, the Y and chroma sections are checked as required, with dc-voltage measurements to narrow down the trouble area. Waveforms are checked for peak-to-peak voltage with an ac voltmeter. If

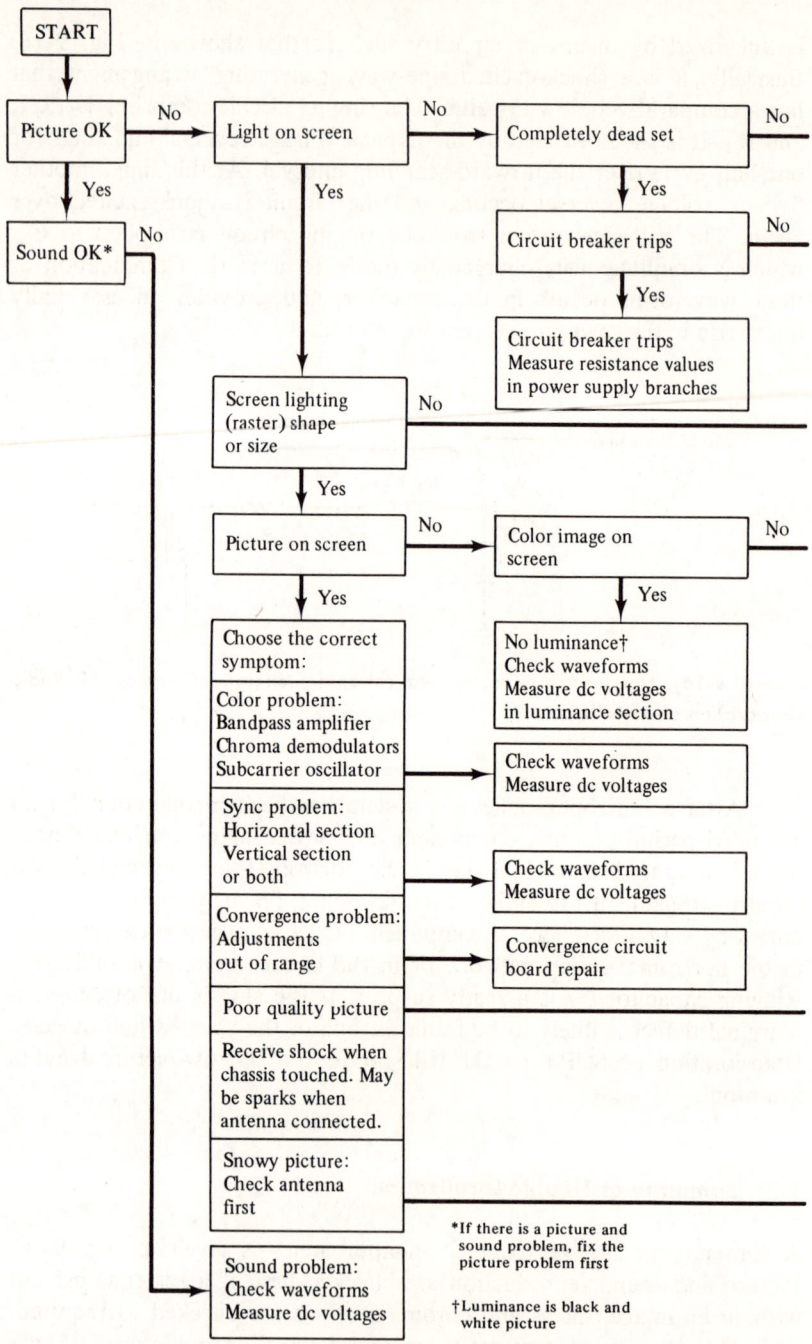

Figure 9-17 A trouble area chart. (*Courtesy of Heath Co.*)

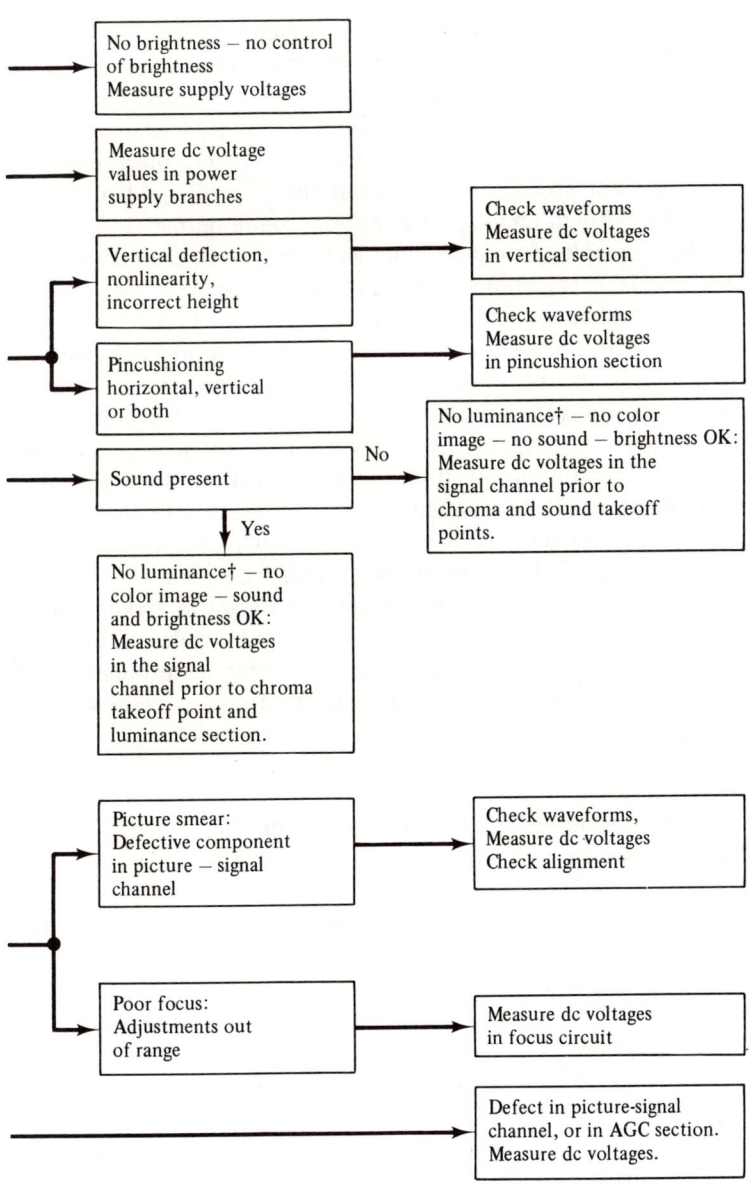

Figure 9-17 Continued

sound reproduction is present, it is known that the signal channel is operating as far as the sound-takeoff point. If a black-and-white image is present, but color is not reproduced, the technician turns his attention to the chroma section. On the other hand, if a color image is present without black-and-white reproduction, the fault is evidently in the Y amplifier. If the receiver is completely dead, without a raster and without any sound output, the power supply falls under immediate suspicion. The general procedure is to repair the picture problem first, if there is a raster without an image and without sound output.

Another example of troubleshooting procedure is seen in Figs. 9–18 and 9-19. This is a typical modularized color-TV receiver; each module is enclosed by dashed lines in the diagram of Fig. 9-18. A total of nine modules is utilized. Solid-state design is employed throughout, with integrated circuitry in the chroma module. Modularized construction facilitates troubleshooting procedures, because a suspected module can be quick-checked by plugging in a new module. In turn, a defective module can be repaired at any future time, or, if a catastrophic failure has occurred, the defective module may be discarded. Step-by-step troubleshooting procedure for a symptom of no raster and no sound is presented in Fig. 9-19. As a practical note, when one is measuring voltages in a modularized receiver, it is very important to avoid accidental short circuits between adjacent terminals on modules or on integrated circuits. If a test probe short-circuits a pair of terminals, even for a fraction of a second, the semiconductors on one or more modules may be instantly destroyed.

9-8 Notes on Automatic Frequency-Phase Control Troubleshooting

A modern automatic frequency-phase control section is diagrammed in Fig. 9-20. Most AFPC malfunctions are localized to good advantage by signal tracing procedures with a wide-band oscilloscope supplemented by a color-bar generator. Many malfunctions can also be localized effectively with a TVM supplemented by a color-bar generator. AFPC alignment checks are used to locate an inoperative circuit. With reference to Fig. 9-20, connect a keyed-rainbow generator to the antenna-input terminals of the receiver. With the hue control set to the center of its range, turn the color control to maximum and turn the color-killer threshold control to its minimum position. The fine-tuning control is adjusted for minimum sound-beat interference in the color-bar pattern (the sound-carrier function of the generator is utilized in this test). There are four adjustments to be checked in this example:

1. The 3.58-MHz oscillator transformer T2 (Fig. 9-20).
2. Burst-phase transformer T1.
3. The 3.58-MHz oscillator coil L1.
4. The 3.58-MHz oscillator adjust control R1.

First, a TVM is connected to the cathode end of diode D3. This connection is made with the positive dc lead through a 470-kilohm resistor. The burst signal is temporarily eliminated by short-circuiting the base terminal of Q1 to ground. Then T2, the 3.58-MHz oscillator output transformer, is adjusted for maximum output. If the voltmeter indicates a small voltage, or zero volts, it is probable that the oscillator or phase detector is defective. On the other hand, if T2 can be peaked for normal output, proceed to the next test.

Second, remove the short circuit from the burst-amplifier transistor. Then adjust T1 for maximum output voltage. Note that an increase in the meter reading when the short circuit is removed shows that the burst amplifier is operative. On the other hand, failure to obtain an increase in the voltage reading when the short circuit is removed, or the absence of voltage variation while T1 is being adjusted indicates that there is a fault in the burst amplifier. In this situation, the technician should check for the presence of horizontal pulses at the base of Q1. If T1 peaks normally, proceed to the next step.

Third, ground the test point TP2 and adjust R1 for optimum free-running color sync (zero beat). In this procedure, the color bars should display solid hues that change slowly, but with no barber-pole effect. When T1 is grounded, the correction voltage from the phase detector is removed and a condition of zero correction voltage is simulated. An ability to zero-beat the color bars shows that the 3.58-MHz oscillator can operate at correct frequency when there is no correction voltage. If color sync is lost when the short circuit is removed from TP2, it is indicated that a fault in the phase-detector section is producing a steady and false correction voltage. In turn, this false correction voltage shifts the oscillator frequency away from the near zero-beat condition that was established previously.

To check for phase-detector imbalance, measure the correction voltage at TP2 with the base terminal of the burst-amplifier transistor short-circuited to ground. A zero reading is normally obtained. An indication other than zero points to some component or device fault that has caused the detector to become unbalanced. For example, the technician may find an open secondary winding on T1, a defective diode, or a change in value of a detector load resistor. Transformer windings can usually be checked effectively with an ohmmeter. Diodes can be checked for front-to-back ratios. Load-resistor pairs should be closely matched.

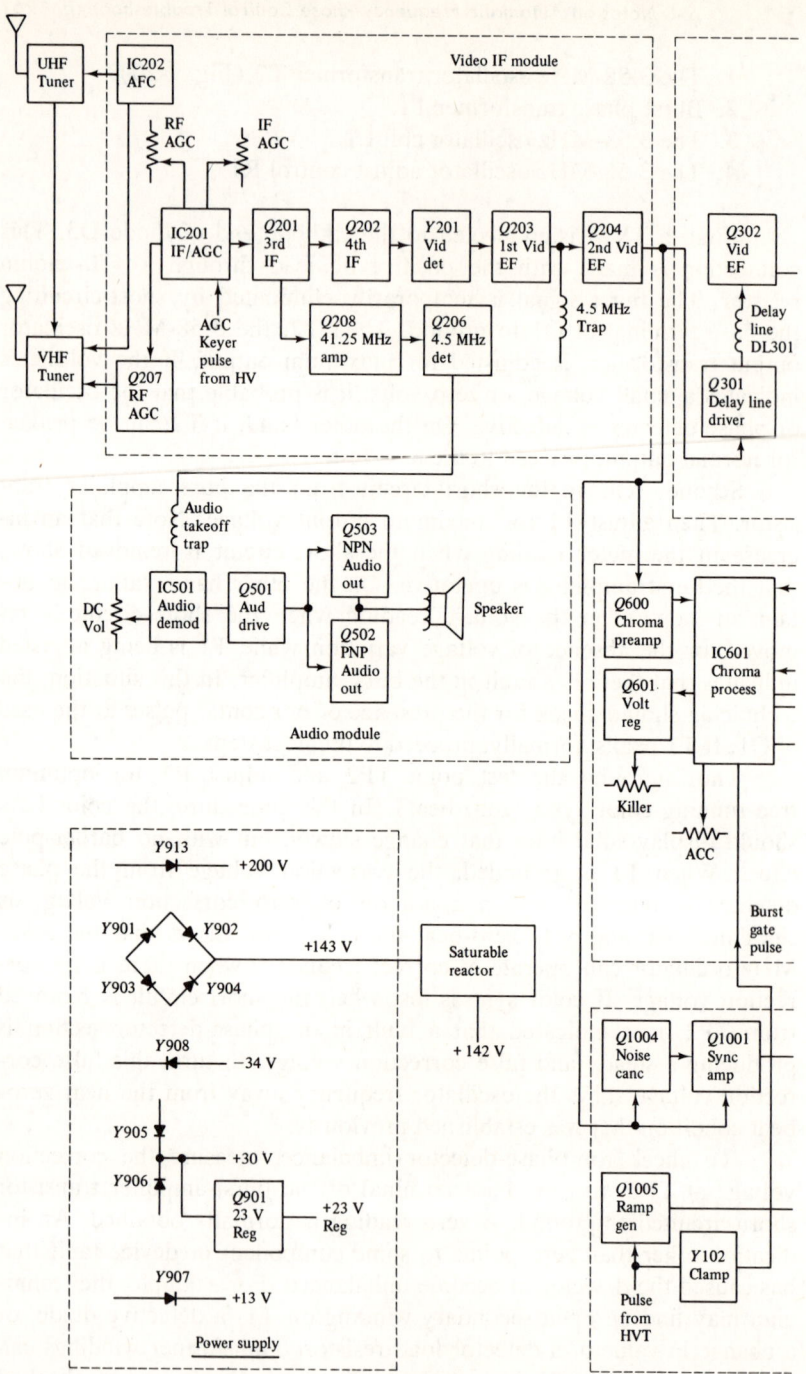

Figure 9-18 Block diagram for a modularized solid-state color-TV receiver.

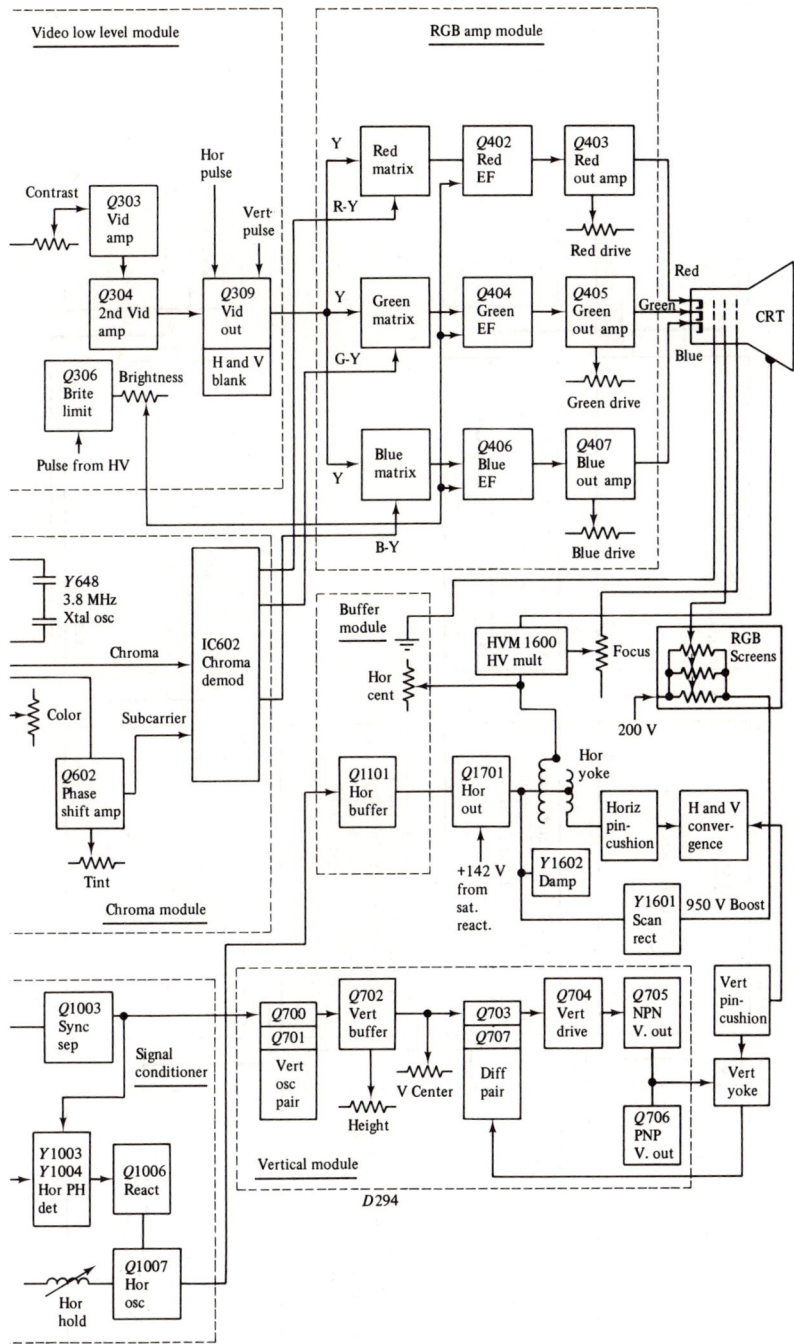

Figure 9-18 Continued. (Courtesy of General Electric)

No raster, no sound
Caution: Do not remove or insert modules with power on

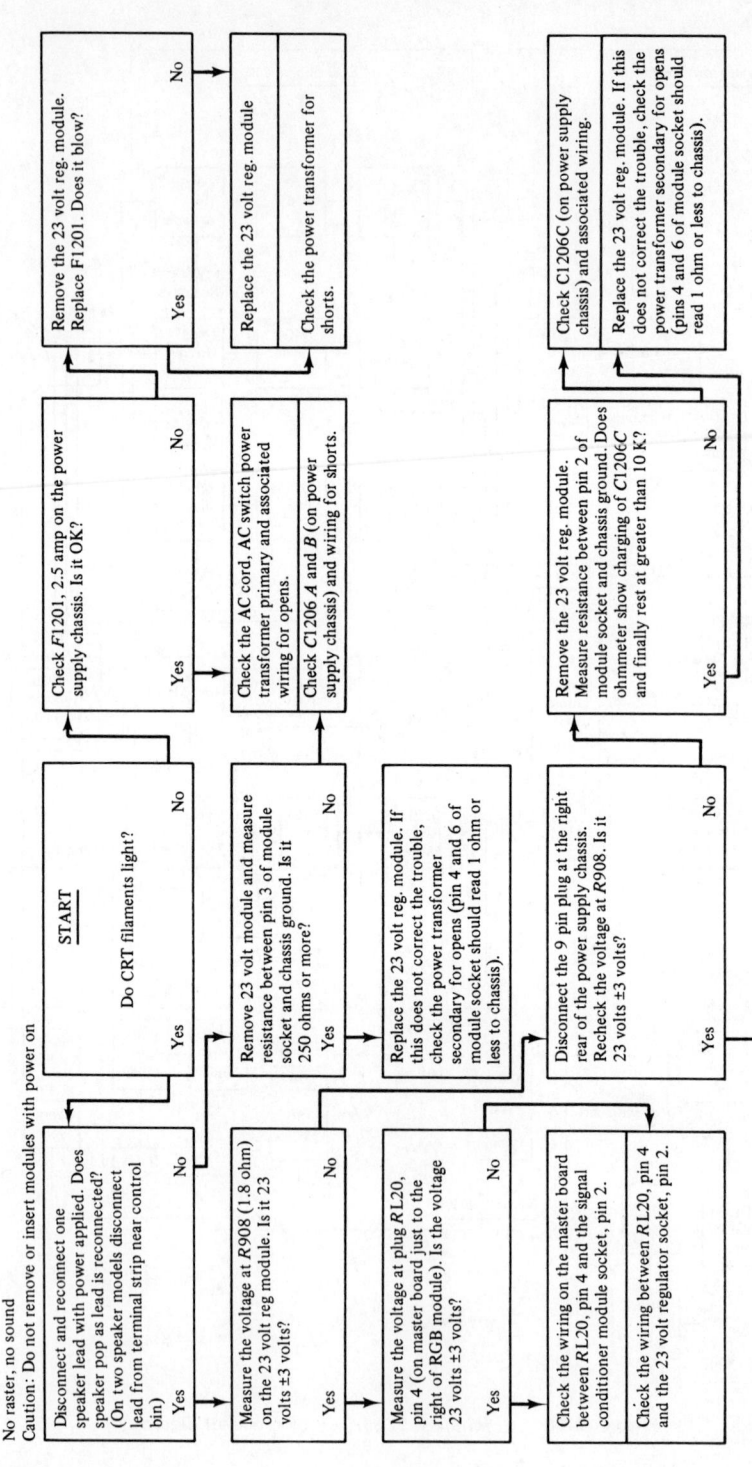

264

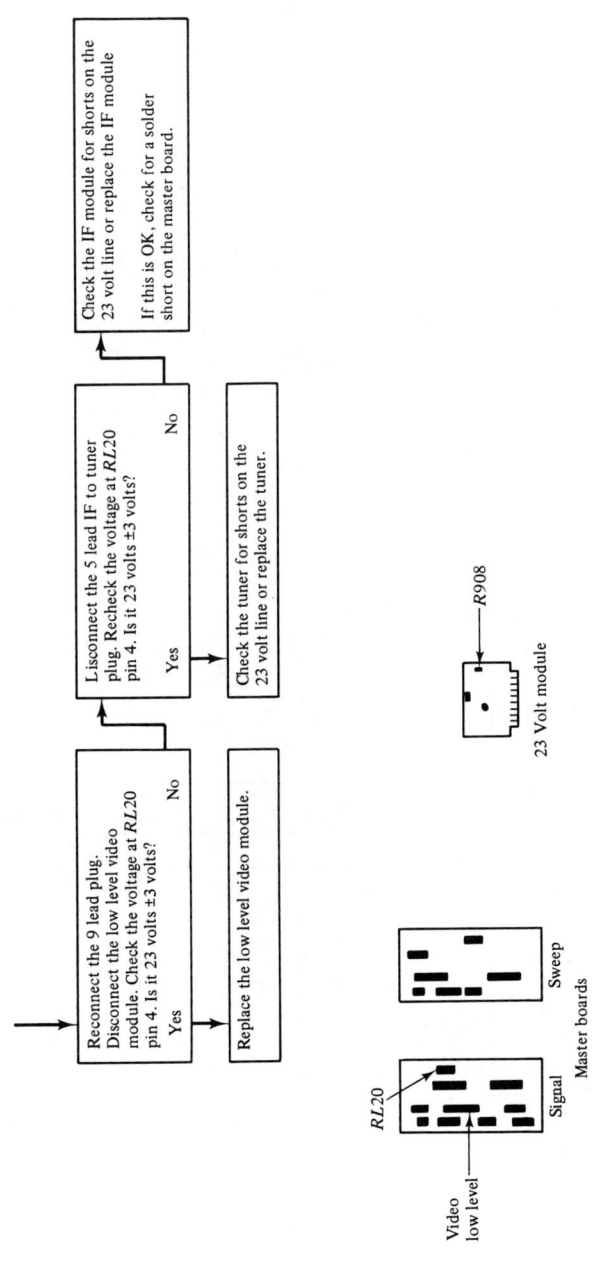

Figure 9-19 Troubleshooting chart. *(Courtesy of General Electric)*

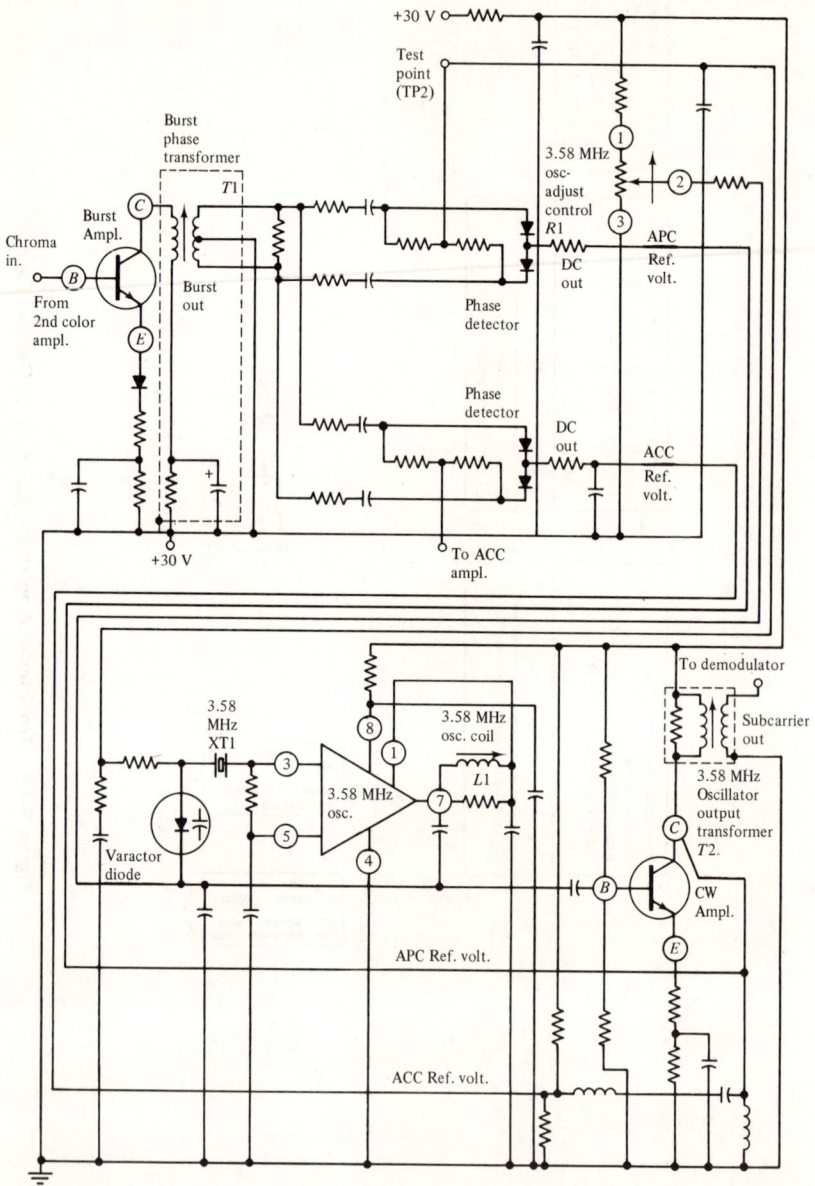

Figure 9-20 Color-sync circuits with AFPC control in an RCA color receiver.

9-9 Automatic Fine Tuning

Automatic fine tuning (AFT) action in a television receiver is accomplished in the same manner as automatic frequency control (AFC) action in FM radio receivers, as explained in Chapter 7. In other words, a varactor diode is reverse-biased by a control voltage from a discriminator, and its resulting capacitance variation is used to correct the frequency of the local oscillator in the event that it starts to drift to a higher or to a lower frequency. Note that varactor tuning is also employed in various front-end designs. A schematic diagram for a varactor-tuned VHF front end is shown in Fig. 9-21. It is a four-circuit tuner comprising a single-tuned input, double-tuned interstage, and a local oscillator. Tuning over channels 7-13 and 2-6 is accomplished by the varactor diodes D1, 2, 3, and 4. The varactor used in this configuration has a capacitance of 2.15 pF at 25 volts. Band-switching circuitry is utilized which requires a positive voltage only with respect to ground. A different value of bias is provided for the oscillator on the high band and the low band.

A simplified diagram showing the dc circuitry for band switching is shown in Fig. 9-22. Diodes D7, 8, and 9 are connected in parallel. If +18 volts are applied to the high-band switching terminal, a 50-mA current divides equally among the three diodes, thereby turning them on and supplying power and bleeder currents for the RF amplifier and the local oscillator. A current of 15 mA also flows through D6 and D5 in series, thereby turning them on and supplying a bias of 8 volts to the oscillator base through R22 and R23. Next, if the +18 volts are removed from the high-band switching terminal, and are applied to the low-band terminal, power is applied directly to the RF amplifier and to the local oscillator. Also, diodes D7, 8, and 9 have +18 volts applied to their cathodes, and are reverse-biased. Diodes D5 and 6 are reverse-biased by means of the positive voltage on the oscillator base via R22. Leakage resistors R4 and R3 across the diodes complete the reverse-bias circuit.

One power supply is employed. With reference to Fig. 9-21, R8 and R7 comprise a divider that brings the source of Q1 +3.6 volts above ground. In turn, a single polarity AGC supply is used to control this depletion-mode MOSFET. The 75-ohm incoming signal following the high-pass filter is tapped down on the tuned circuit. Note that L4 and L5 are low-band coils, and L6 and L7 are high-band coils. For high-band operation, D5 and D6 are turned on, placing the high-band coils in the circuit; the high-band coils operate in parallel with the low-band coils.

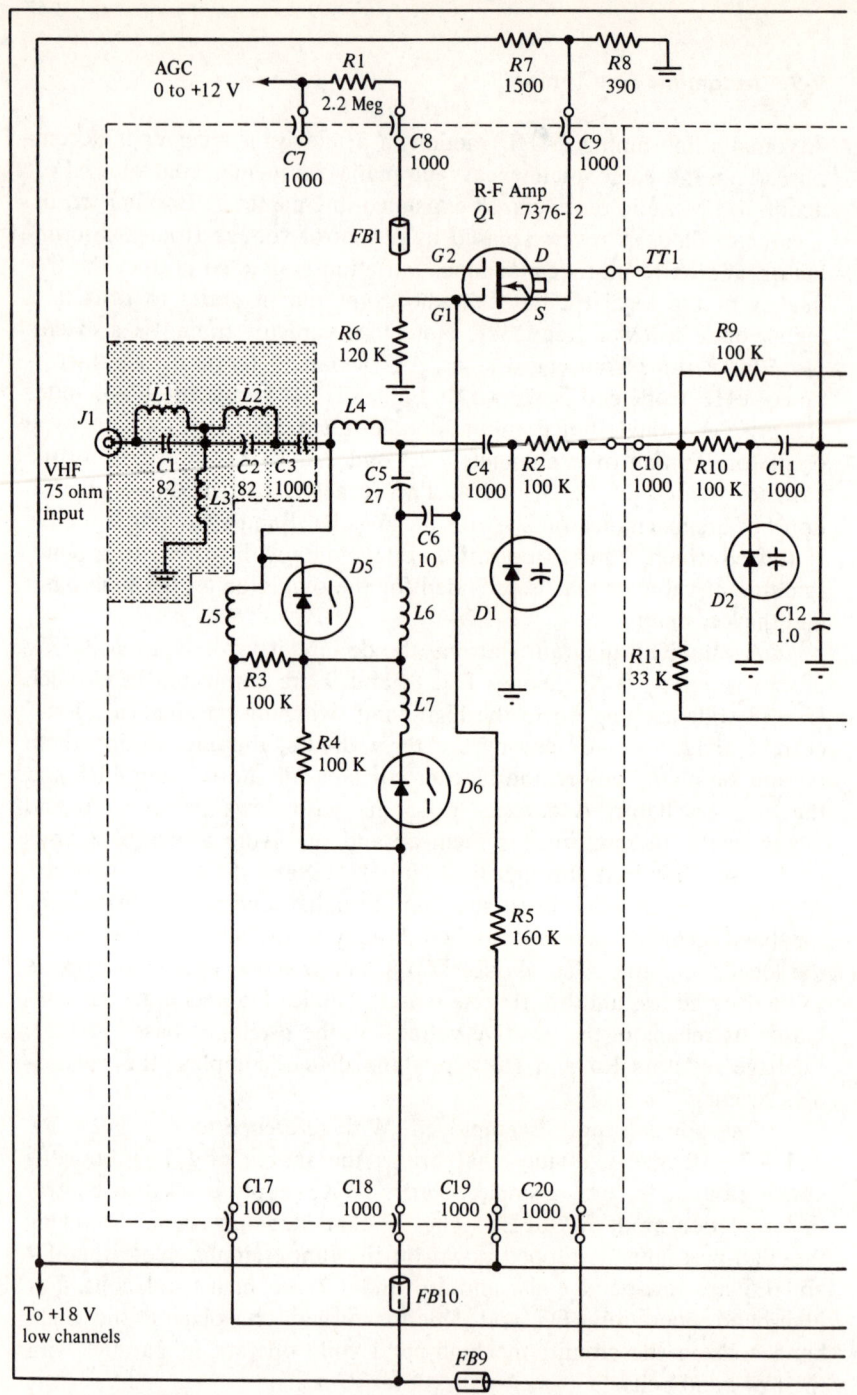

Figure 9-21 Schematic for a varactor-tuned VHF front end.

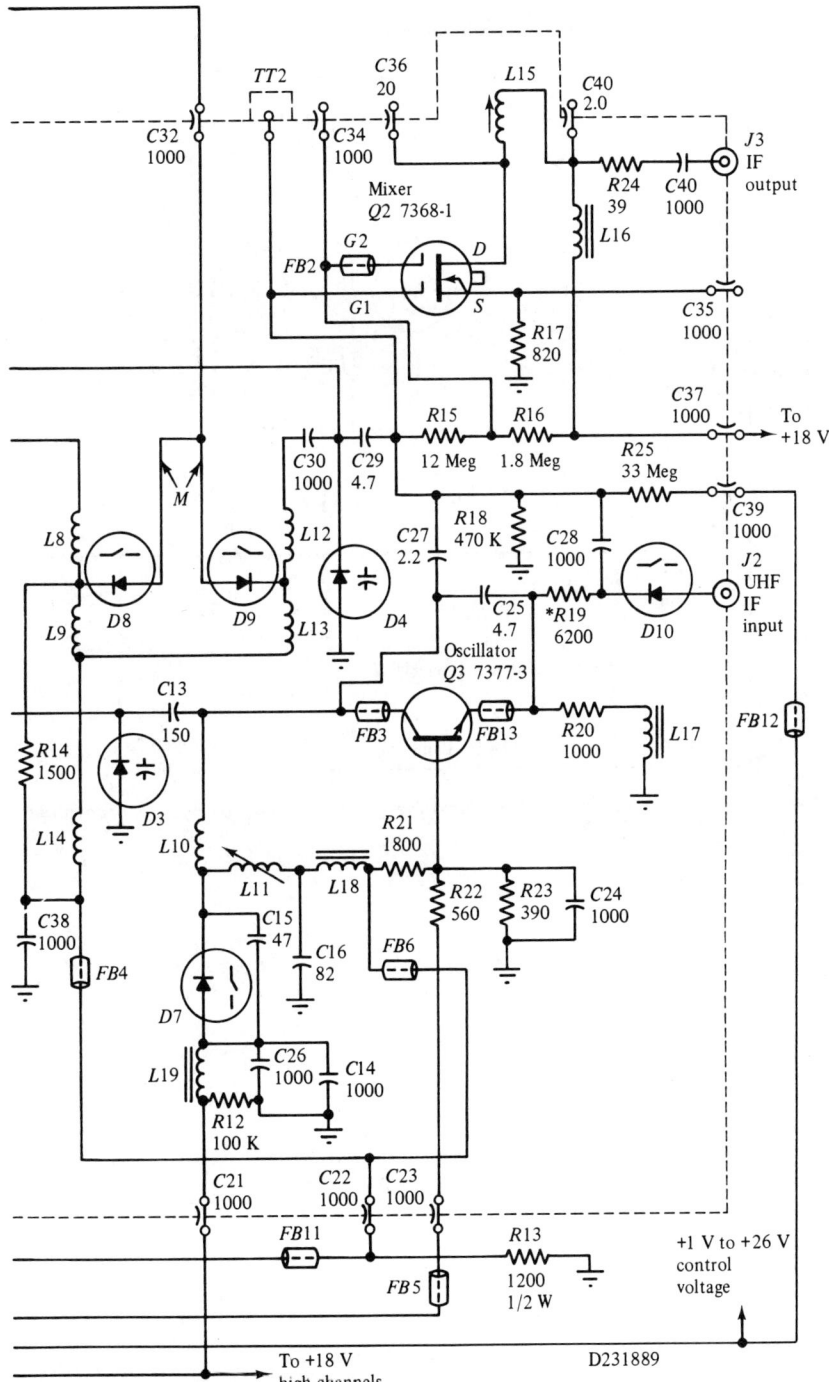

Figure 9-21 Continued. (Courtesy of Radio Corporation of America)

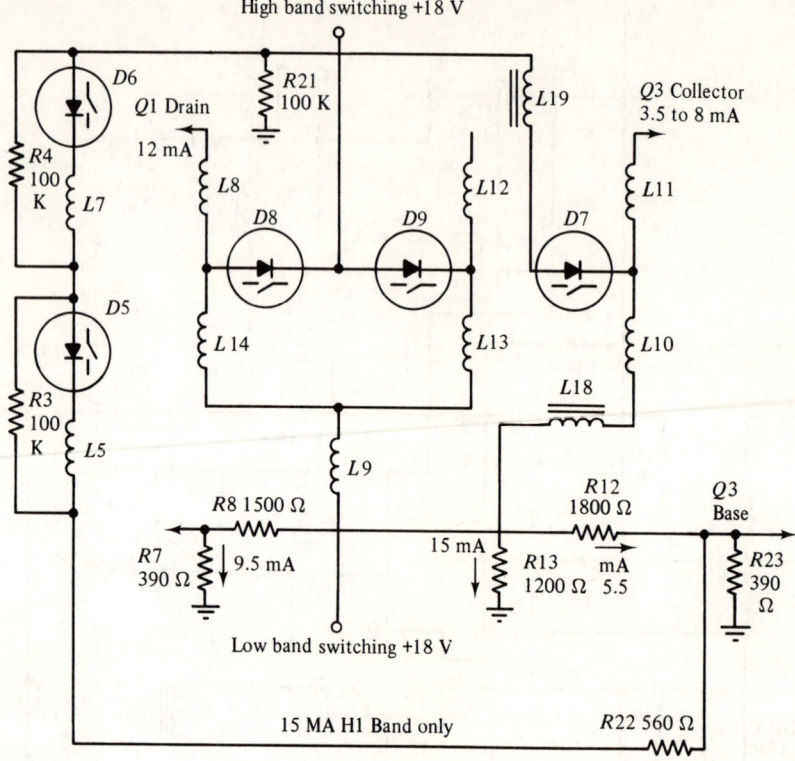

Figure 9-22 VHF band-switching arrangement. (Courtesy of Radio Corporation of America)

10
Miscellaneous Applications

10-1 Oscilloscope Troubleshooting

Various trouble symptoms are encountered in oscilloscope operation, including a "dead" instrument symptom, dark CRT screen, vertical-centering control out of range, horizontal-centering control out of range, no vertical deflection, no horizontal deflection, poor focus, erratic pattern, loss of sync, no retrace blanking, poor frequency response, low sensitivity, off-frequency time-base operation, vertical nonlinearity, horizontal nonlinearity, dim trace, astigmatism, and distorted transient response. It is helpful to evaluate trouble symptoms in relation to the block diagram and to the schematic diagram for the oscilloscope. Block diagrams for typical service-type oscilloscopes are shown in Figs. 10-1 and 10-2. If the malfunction cannot be localized on the basis of the trouble symptom(s), it is usual procedure to check the dc-voltage distribution against the values specified on the schematic diagram. Signal waveforms can be traced through the vertical amplifier with the ac-voltage function of a TVM. Horizontal-deflection waveforms can be similarly traced through the horizontal channel.

A typical vertical-amplifier configuration for a 5-MHz oscilloscope is shown in Fig. 10-3. Specified dc voltages may vary as much as ± 20 percent in normal operation. Because the circuitry is direct-coupled, it is virtually impossible to evaluate malfunctions in a "cause and effect" form. As an example, a saturated transistor on one side of a differential amplifier may appear to be a defect on the other side of the amplifier. Observe in Fig. 10-4 that the vertical-input signal is coupled through R1 and C1 to the gate of Q1. R1 serves as a protective resistor to avoid overload damage to Q1. D1 and D2 have the appearance of transistors,

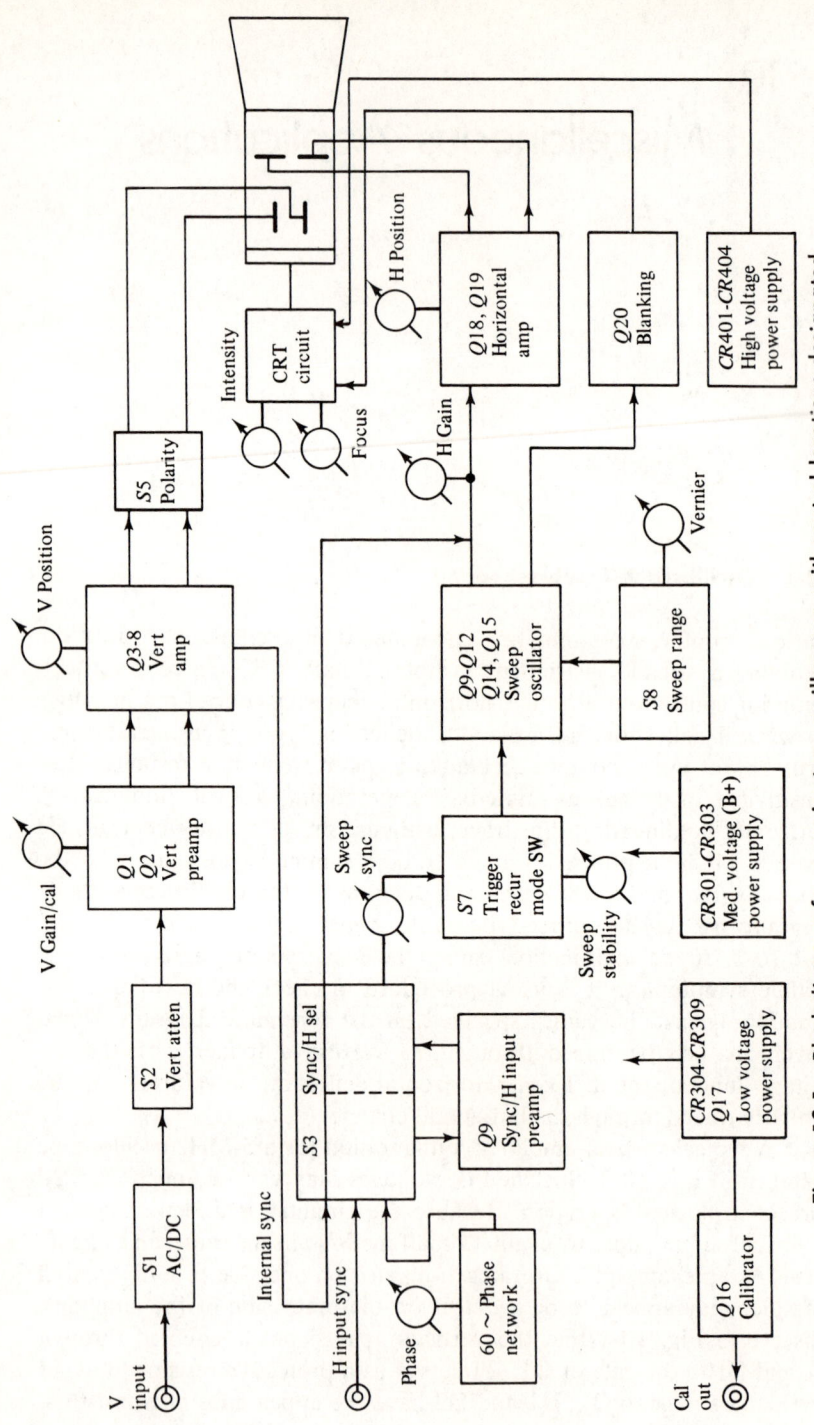

Figure 10-1 Block diagram for a service-type oscilloscope, with control locations designated. (Courtesy of Radio Corporation of America)

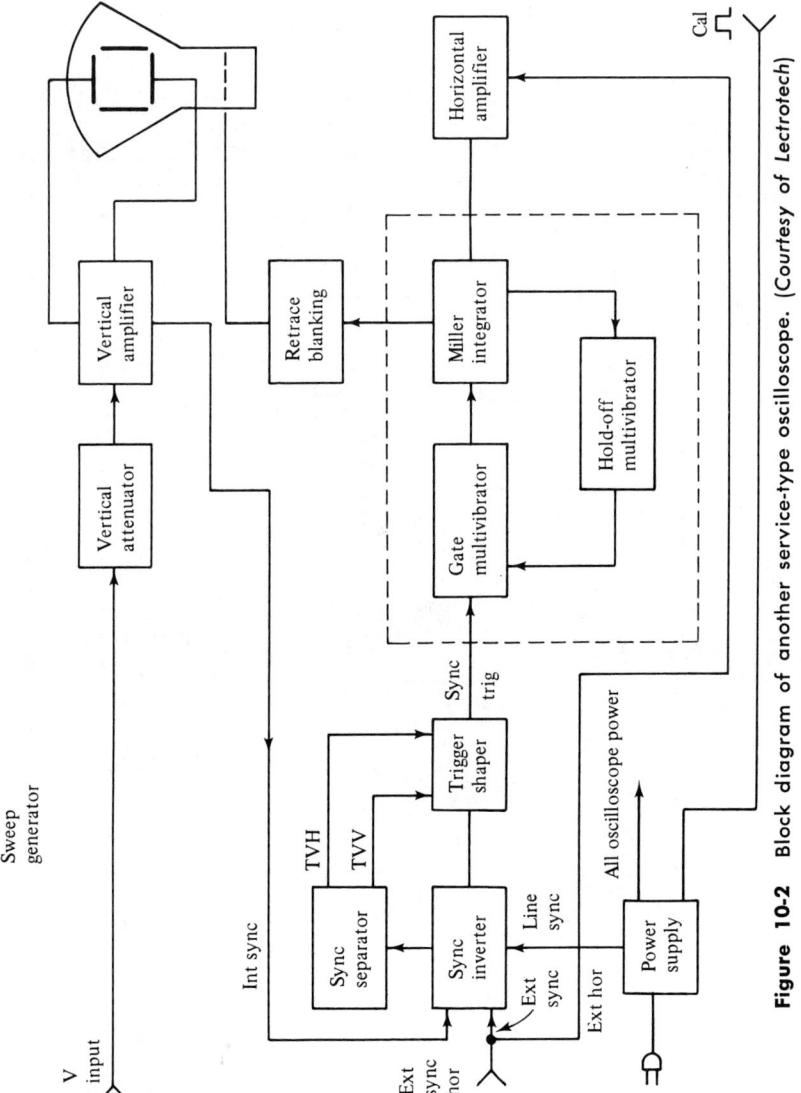

Figure 10-2 Block diagram of another service-type oscilloscope. (*Courtesy of Lectrotech*)

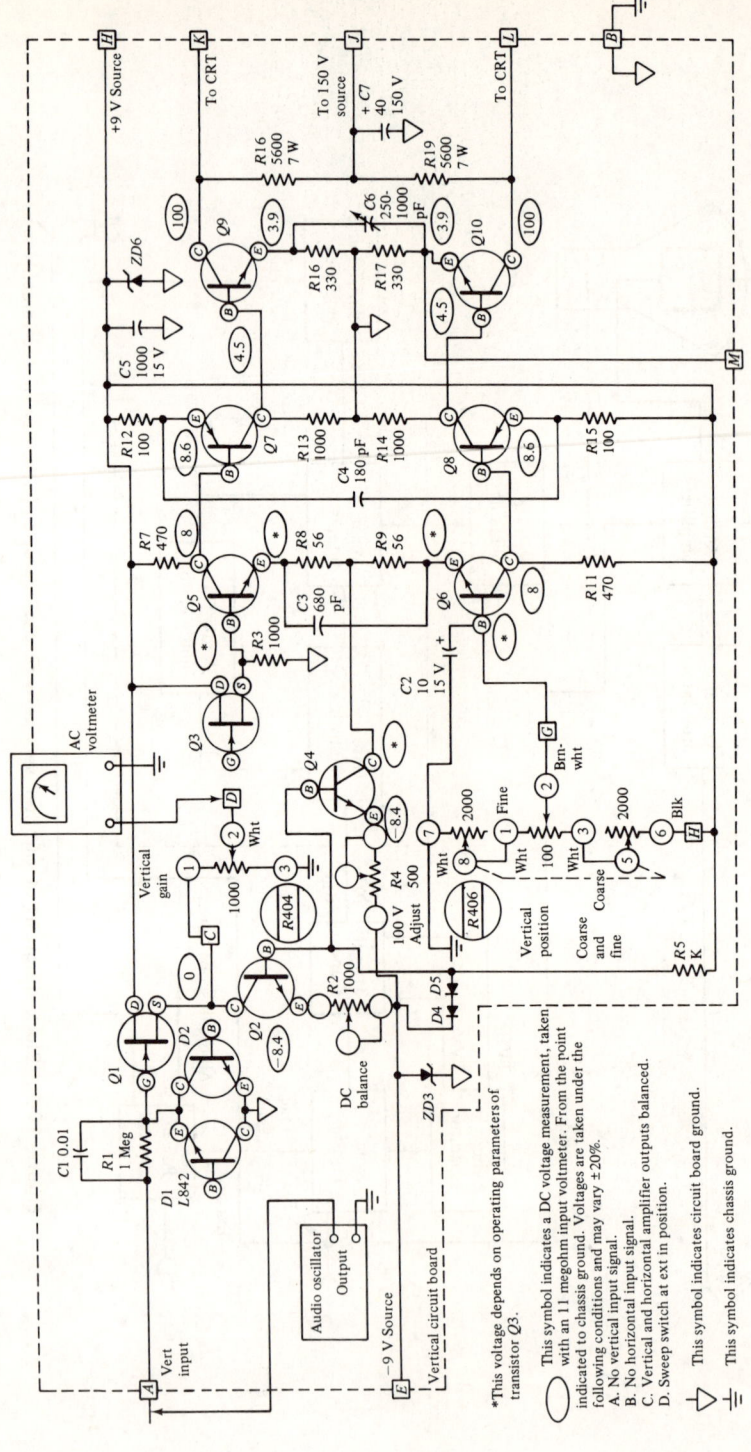

Figure 10-3 A typical vertical-amplifier configuration. (*Courtesy of Heath Co.*)

but they are connected to provide zener action. Thus, they limit the gate voltage on Q1 to ±9 volts. C1 provides improved high-frequency response around R1. Q1 functions as a source follower and provides a very high input impedance. Q2 operates as a constant-current source for Q1 to stabilize its operation. D4 and D5 each provide a 0.6-volt drop (a total of 1.2 volts) to hold the base of Q2 at a constant voltage. Inasmuch as the circuit for Q2 is basically an emitter-follower configuration, and the emitter voltage is dependent on the base voltage, the emitter voltage also remains constant in normal operation. In turn, the current through R2 is constant. R2 is adjusted so that the source voltage of Q1 is zero when no input signal is applied.

A signal applied to the gate of Q1 will cause only voltage changes at the source, because the current through Q1 is constant. These voltage variations are applied across the gain control R404, and a portion of this signal is applied to the gate of the source follower Q3. Q4 provides a constant-current source for Q5 and Q6. Since the emitter of each transistor is connected to the constant-current source, the current source serves as an emitter-follower resistance and fixes the operating point for the following stages. The output from source-follower Q3 is amplified by Q5. A portion of the signal applied to the base of Q5 appears at its emitter. Because transistors Q5 and Q6 have a common-emitter resistance, the signal at the emitter of Q5 is effectively coupled to the emitter of Q6. In turn, Q5 and Q6 function as a differential amplifier.

Transistor Q6 alone functions as a common-base amplifier, with its base potential determined by the setting of the vertical position control R406. This control positions the trace by applying a dc voltage to the base of Q6, thereby establishing a certain dc unbalance in the vertical-amplifier system. When the collector output voltage of Q5 decreases, its emitter voltage increases. In turn, the forward bias on Q6 decreases and its collector voltage increases. The signal at the collector of Q6 is 180 deg out of phase with the signal at the collector of Q5, thereby providing push-pull amplification. C3 is an emitter partial-bypass capacitor that functions to increase emitter degeneration at low frequencies, thereby providing improved high-frequency amplification. R8 and R9 establish the dc gain of the amplifier. Driver transistors Q7 and Q8 operate as common-emitter amplifiers. They provide gain and also isolate Q5 and Q6 from the changing current demand of the output transistors. C4 is an emitter partial-bypass capacitor that improves high-frequency response. Output transistors Q9 and Q10 provide final amplification and drive the vertical-deflecting plates of the CRT.

Since the vertical-position control is at the front of the differential amplifier, and affects each succeeding stage, it serves as a troubleshooting aid. When troubleshooting the vertical amplifier, for example, first

276 Miscellaneous Applications

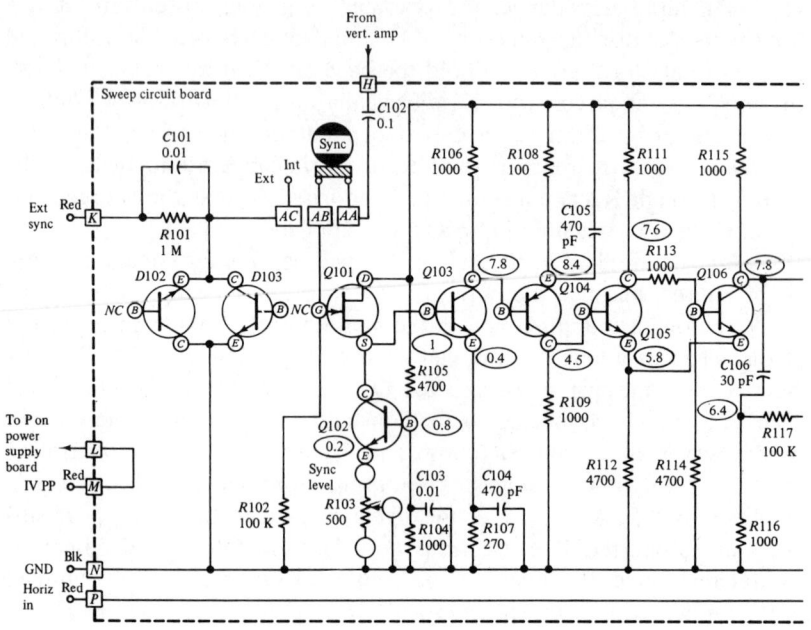

Figure 10-4 A typical time-base configuration. (*Courtesy of Heath Co.*)

check its power-supply voltage. Then check the collector voltages on transistors Q9 and Q10. These voltages should vary as the vertical-position control is turned. If these voltages change accordingly, the trouble may be located in the CRT circuit. On the other hand, if these voltages do not change, the fault will be found in either Q9 or Q10, or the preceding stages. In such a case, move the voltmeter to the preceding stage (Q7 and Q8) and repeat the procedure until the defective component or device is pinpointed.

Troubleshooting the sweep generator of an oscilloscope involves many of the procedures noted above. The sync switch selects either a portion of the amplified vertical-input signal or of a signal applied to the external-sync connector. The selected signal is then coupled to the gate of source follower Q101. (D101 and D103 are transistors that function

10-1 Oscilloscope Troubleshooting

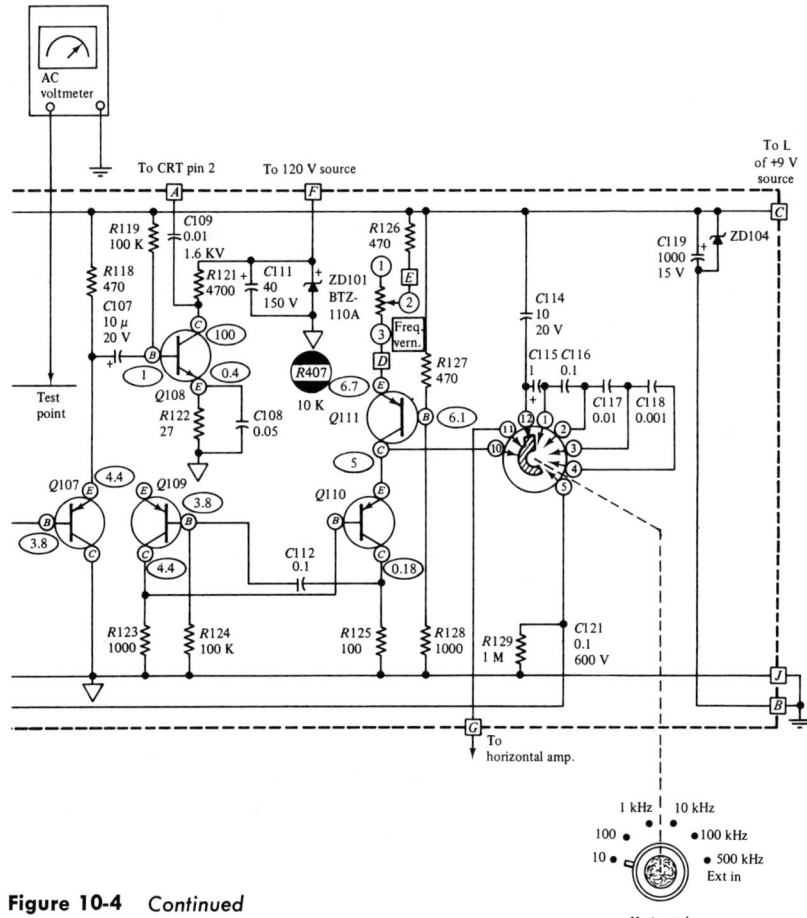

Figure 10-4 *Continued*

as diodes to protect Q101 from high voltage by means of zener action at ±9 volts.) Constant-current source Q102 is adjusted by the sync-level control R103 to provide proper biasing of the sync circuits. This ensures that even a small signal can sync the sweep generator.

Transistors Q103 and Q104 amplify the signal and apply it to the Schmitt trigger circuit consisting of Q105 and Q106. This Schmitt trigger circuit is a regenerative bistable circuit which produces a rectangular pulse each time that it is triggered and reset. Transistors Q109 and Q110 form an astable multivibrator. When Q110 is conducting and Q109 is cut off, one or more of the timing capacitors (C114 through C118) are charged through transistor Q110. As the voltage at the emitter of Q110 approaches the voltage at its base, in consequence of the charge on the capacitor, Q110 will cut off and drive Q109 into conduction. The charged timing capacitor will discharge through the constant-current source circuit of Q111. The setting of the frequency-vernier control R407 determines the current flowing through Q111, which in turn determines the discharge current (and discharge time) of the timing capacitor. As the timing capacitor discharges, a positive-going ramp voltage (sawtooth) is generated and is coupled to the horizontal amplifier. The frequency of the horizontal sweep is determined by the particular timing capacitor selected by the frequency-range switch and the discharge current.

Since transistors Q107 and Q109 have a common emitter resistor, a signal applied to the base of Q107 is emitter-coupled to transistor Q109. The pulse output (sync signal) of the Schmitt trigger Q106 is coupled to Q109. This causes Q109 to turn on and Q110 to cut off, thereby starting the sweep just prior to the time that it would normally begin the unsynchronized mode. When the signal at the emitter of Q109 goes positive, a positive pulse is coupled through C107 to the base of blanking amplifier Q108. A negative-going output pulse is coupled through capacitor C109 to the grid of the CRT. This pulse turns off the electron beam during retrace and thereby prevents the retrace from being displayed on the screen of the CRT.

Troubleshooting the horizontal amplifier in an oscilloscope involves the same general procedures employed in vertical-amplifier troubleshooting. In the example of Fig. 10-5, the major difference between the horizontal configuration and the vertical configuration shown in Fig. 10-3 is that the former does not include a PNP amplifier stage (Q7 and Q8 in the vertical arrangement). The positive-going ramp voltage (sawtooth) from the time base is amplified and applied to the horizontal plates of the CRT. This increasing voltage causes the electron beam to sweep across the face of the CRT and to produce a visible trace. The sweep rate of the electron beam is determined by the sawtooth frequency. Nonlinearity may be caused by malfunction of the time base, or by a defect in the horizontal amplifier. To distinguish between these sources, a Lissajous figure may be displayed on the CRT screen. This test eliminates the time base from consideration. In turn, if an undistorted Lissajous figure is displayed, it is indicated that the source of nonlinear deflection is in the time base.

Consider the power-supply circuit shown in Fig. 10-6. Low voltage is connected through the slow-blow fuse and the on-off switch to the primary windings of the power transformer. The dual-primary transformer windings may be connected in parallel for 120-volt operation, or in series for 240-volt operation. A high-voltage secondary winding of the power transformer is connected to the voltage doubler circuit comprising D301, D302, C302, and C303. Capacitor C301

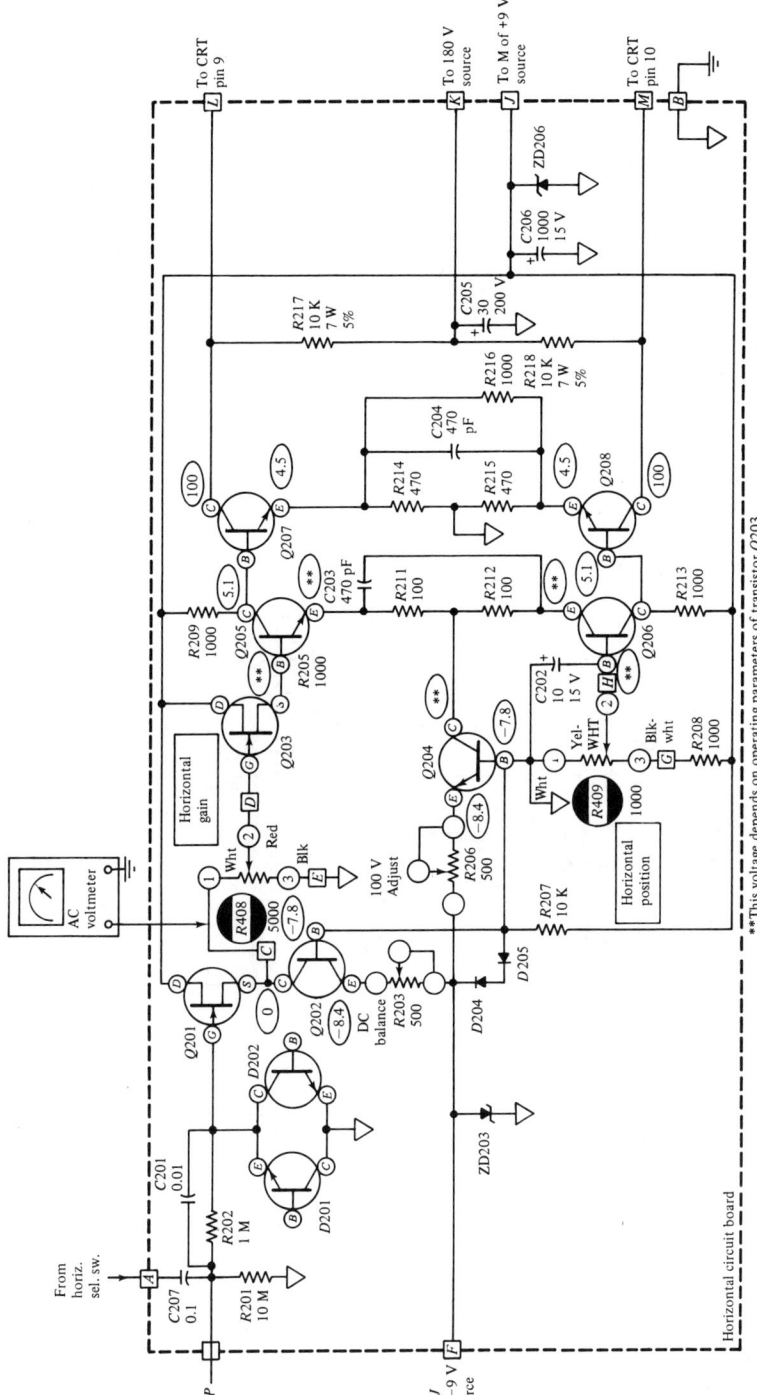

Figure 10-5 A horizontal-amplifier configuration. (*Courtesy of Heath Co.*)

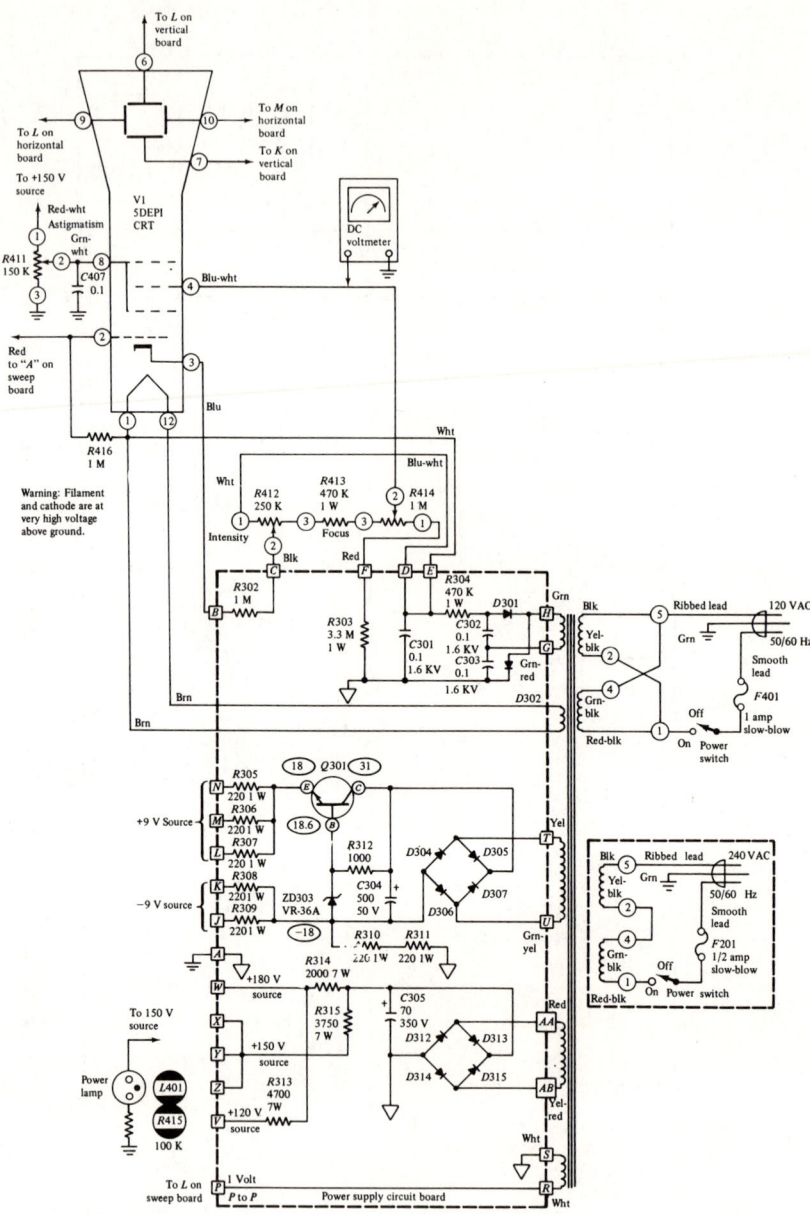

Figure 10-6 A power-supply configuration for an oscilloscope. (*Courtesy of Heath Co.*)

filters this negative high voltage, which is fed through R416 to the grid of the CRT. The intensity and focusing voltages are also applied to the CRT from the voltage-divider network comprising R412, R413, R414, and R303. A separate 6.3-volt winding supplies the CRT filament voltage.

Optimum focus is obtained when the CRT deflection plates and the astigmatism grid are at the same potential. Since the vertical-deflection plate voltages operate at 100 volts, the astigmatism voltage is also adjusted to 100 volts. A low-voltage secondary winding is connected to the full-wave bridge-rectifier circuit comprising D304, D305, D306, D307, and capacitor C304. Zener diode ZD303 and resistor R312 maintain a constant bias voltage to the base of pass transistor Q301. Fig. 10-7 shows a skeleton schematic of this power supply. The output voltage is regulated at 36 volts by the series pass transistor Q301 and zener diode ZD 303. By the connection of equal loads from each side of the power supply to ground, two separate supplies are provided with outputs of $+9$ volts and -9 volts dc.

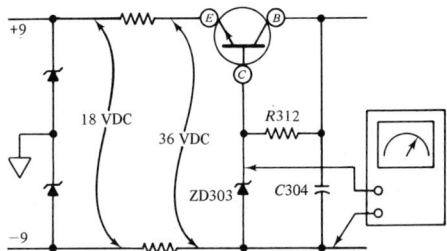

Figure 10-7 Skeleton schematic of a regulated power supply. (*Courtesy of Heath Co.*)

Another secondary winding is connected to the full-wave rectifier circuit comprising diodes D312, D313, D314, and D315. Capacitor C305 filters the rectified voltage. Dropping resistors R315 and R314 provide output voltages of $+150$ V dc and $+180$ V dc, respectively. Resistor R313 and zener diode ZD101 (on the time-base circuit board) reduce the 180 volts to regulated value of $+120$ V dc. A separate secondary winding provides a 1-volt p-p reference voltage to a jack on the front panel of the instrument. A summary of basic troubleshooting procedures is presented in Fig. 10-8.

10-2 CCTV Troubleshooting

Closed-circuit television equipment appears to be superficially different from television receivers and instruments such as oscilloscopes. How-

282 Miscellaneous Applications

Difficulty	Possible Area of Trouble
Neither pilot lamp nor CRT filaments light.	1. Fuse blown. 2. On-off switch. 3. No AC power from outlet.
Pilot lamp lights, CRT filament does not light.	1. Power transformer. 2. CRT.
No spot or trace on CRT.	1. Positioning or intensity controls improperly adjusted. 2. High voltage power supply. 3. CRT.
Dot cannot be centered vertically.	1. Vertical position control and associated circuit.
Dot cannot be centered horizontally.	1. Horizontal position control and associated circuit.
No vertical deflection.	1. Vertical amplifier.
No horizontal deflection.	1. Horizontal amplifier.
Poor focus.	1. CRT. 2. Focus control. 3. Astigmatism control. 4. Resistors $R412$, $R413$, $R414$, and $R303$.
Trace acts erratic when the window is touched.	1. Clean the window with detergent to eliminate static charge.
Cannot synchronize input signal with sweep generator frequency.	1. Sync switch in the EXT position. 2. Control $R103$ misadjusted.
No retrace blanking or poor retrace blanking.	1. Transistor $Q108$. 2. Diode $AZ101$.
Pilot lamp changes intensity from bright to dim.	1. This is normal operation.

Figure 10-8 Troubleshooting chart. (*Courtesy of Heath Co.*)

ever, the differences are basically more apparent than real. Thus, a CCTV camera contains a video amplifier, a sync generator, horizontal and vertical deflection sections, an RF modulator, and a power supply. The video amplifier is similar to its TV receiver counterpart, except that more stages are employed and higher gain is provided. High gain is required, because the vidicon camera tube has a low-level output. A sync generator can be compared in a general way with a digital pulse generator. Horizontal and vertical deflection circuitry is much the same in CCTV equipment and in TV receivers. An RF modulator is essentially an oscillator with a supply voltage that varies at a video-frequency rate. Thus, a technician who is familiar with television and electronic-instrument troubleshooting can apply his knowledge directly to CCTV cameras.

A video section configuration for a CCTV camera is shown in Fig. 10-9. It consists of a video preamplifier, a high-peaker amplifier, an emitter follower, a driver (feedback pair), a clamp (dc restorer), a focus regulator, a sync adder, and output stages. In case of a no-output trouble symptom, an audio voltmeter can be used to trace the input signal stage by stage through the video amplifier. An input signal is required, of course. This input signal can be obtained from the vidicon tube in the camera, if the camera is focused on a moving scene. Otherwise, a low-level signal from an audio oscillator can be used to energize the amplifier.

After the point of signal stoppage has been located, the technician proceeds to make dc-voltage measurements to close in on the defective component or device. In case of doubt, resistance measurements can be made with a low-power ohmmeter, or by temporarily slitting printed-circuit conductors and using a conventional ohmmeter. Open capacitors can be pinpointed by checking the ac voltage at both terminals. Leaky capacitors upset the dc-voltage distribution in the associated circuitry. Transistor turn-off and turn-on tests are not straightforward in direct-coupled circuitry. For example, a turn-off test of Q1 in Fig. 10-7 cannot be made unless the conductor from the collector of Q1 to the base of Q2 is temporarily slit. However, a turnoff test of Q2 can be made without expedients. However, the possibility of collector-junction leakage can be confused with leakage in C14. Accordingly, additional tests may be required to clearly identify the fault.

Technicians often make quick checks in CCTV troubleshooting procedures, just as in radio and TV troubleshooting. For example, if a quick check of the video amplifier in Fig. 10-7 is desired, the filament voltage (6.3 volts) may be jumpered through a 10-megohm resistor to the target terminal in the camera. In turn, an output of 1 or 2 volts will normally be measured at the output of the video amplifier with an ac voltmeter. Another quick check that takes the vidicon tube into account is to remove the camera lens and to touch a finger to the face of the vidicon tube. An ac meter normally indicates a substantial noise-voltage output from the video amplifier. Of course, a 60-Hz voltage or a noise voltage can be traced through the video amplifier stage by stage with an audio voltmeter.

10-3 Fuel-vapor Detector Troubleshooting

A fuel-vapor detector is a comparatively simple example of marine electronic equipment. It indicates the presence of gasoline vapor in the bilge, which could cause an explosion. With reference to Figs. 10-10 and 10-11, the arrangement consists of a sensing element, control unit, and interconnecting cable. The element is formed from a small platinum-wire filament housed in a glass tube. This tube has openings at the top and bottom to let air pass through. It is placed in a stainless-steel mesh shield. Air passes through the shield and glass tube, past the filament.

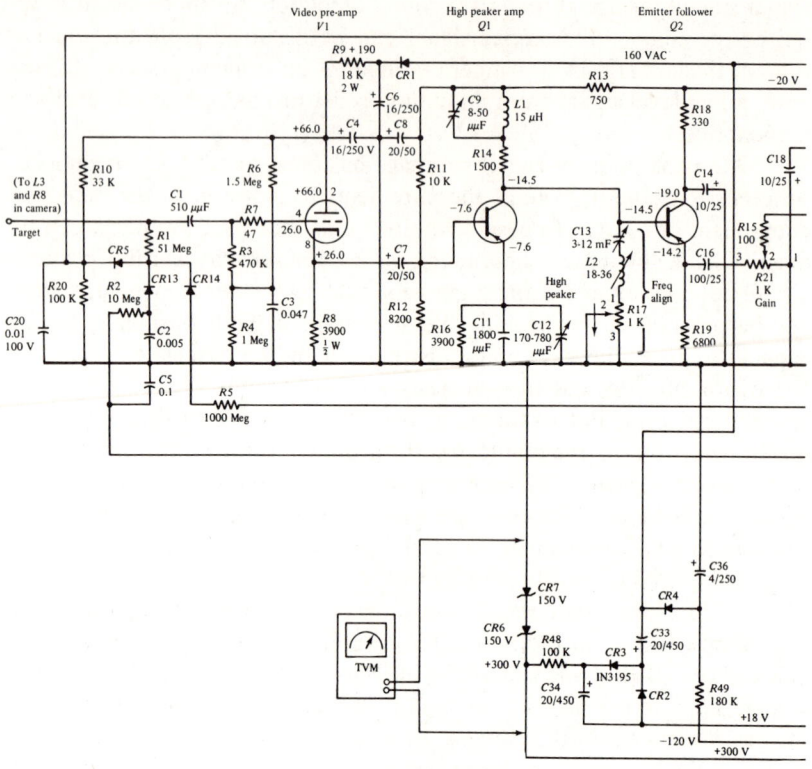

Figure 10-9 Video section configuration for a CCTV camera.

In operation, the wire filament is heated by current flowing through it from the control unit. If an explosive vapor comes into contact with the sensing element, a chemical reaction occurs that causes the temperature of the filament to increase. In turn, the filament resistance increases, and the voltage drop across the filament changes. This change is applied to the control unit through the interconnecting cable. The control unit consists of a relay amplifier Q1, sensing amplifier Q2, and series voltage regulator Q3. When the switch is turned on, a positive voltage is applied from the battery through D2 and D3 to the voltage regulator circuit, consisting of Q3, R7, and D1. This circuit regulates the voltage to Q1 and Q2 and to the sensing element. Diodes D2 and D3 prevent the circuit components from being damaged in case the battery voltage is accidentally reversed.

10-3 Fuel-vapor Detector Troubleshooting

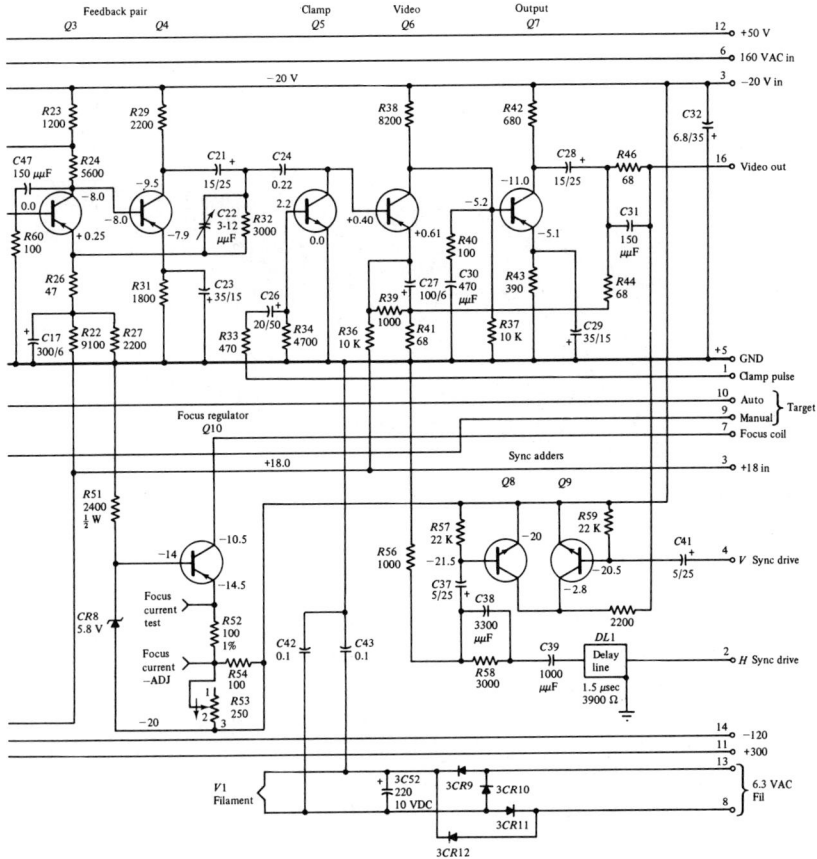

Figure 10-9 Continued. *(Courtesy of General Electric)*

Transistors Q1 and Q2 operate as a two-stage trigger circuit. In normal operation (when an explosive vapor is not present at the sensing element), Q1 conducts more than Q2. When Q1 conducts, current flows through the relay winding and opens the relay contacts, disconnecting the battery voltage to the alarm and any external warning device. On the other hand, when an explosive vapor is present at the sensing element, the voltage drop across the element increases. The base of Q2 becomes less negative than the emitter. This causes Q2 to conduct, and the meter deflects. At the same time, the emitter of Q1 becomes more positive than its base. In turn, no current flows through the emitter-collector circuit and the relay contacts close, connecting the alarm circuit to the battery.

When the switch is turned to its Test position, R6 is shorted out,

286 Miscellaneous Applications

Figure 10-10 Fuel vapor detector cabinet and external sensor. (*Courtesy* of *Heath* Co.)

and Q2 then conducts more than Q1. This simulates an explosive condition, in order that the control-unit circuits may be checked and the oxide may be cleaned from the filament of the sensing element. If the result of the test is unsatisfactory, the battery supply voltage should be checked first. If the supply voltage is normal, the circuit voltage and resistance values should be measured. Figure 10-12 shows voltage and resistance charts for the configuration of Fig. 10-11. If the pilot lamp fails to light, it may be burned out. Other possibilities are a defective function switch, or R6 open-circuited. If the meter deflects full-scale but the pilot lamp fails to light, the sensing element may be open, or there may be an open circuit in the cable between the control head and the sensing element. If the pilot lamp lights but the meter does not deflect and the sensitivity control does not work, the meter may be burned out, transistor Q2 may be defective, or control R3 may be open.

If the meter indication changes when the battery voltage changes, transistor Q3 may be open-circuited, R7 could be open, or the zener diode may be short-circuited. Note that if the meter indicates properly but the external alarm does not function, Q1 is possibly defective, or the relay contacts might be bent, dirty, or burnt. If the meter starts to indicate toward "dangerous" after the unit has been turned on for some time, Q3 may be defective, or the zener diode may have become unstable. DC-voltage measurements, supplemented by resistance measurements specified in Fig. 10-11, are usually sufficient to pinpoint the defective component or device.

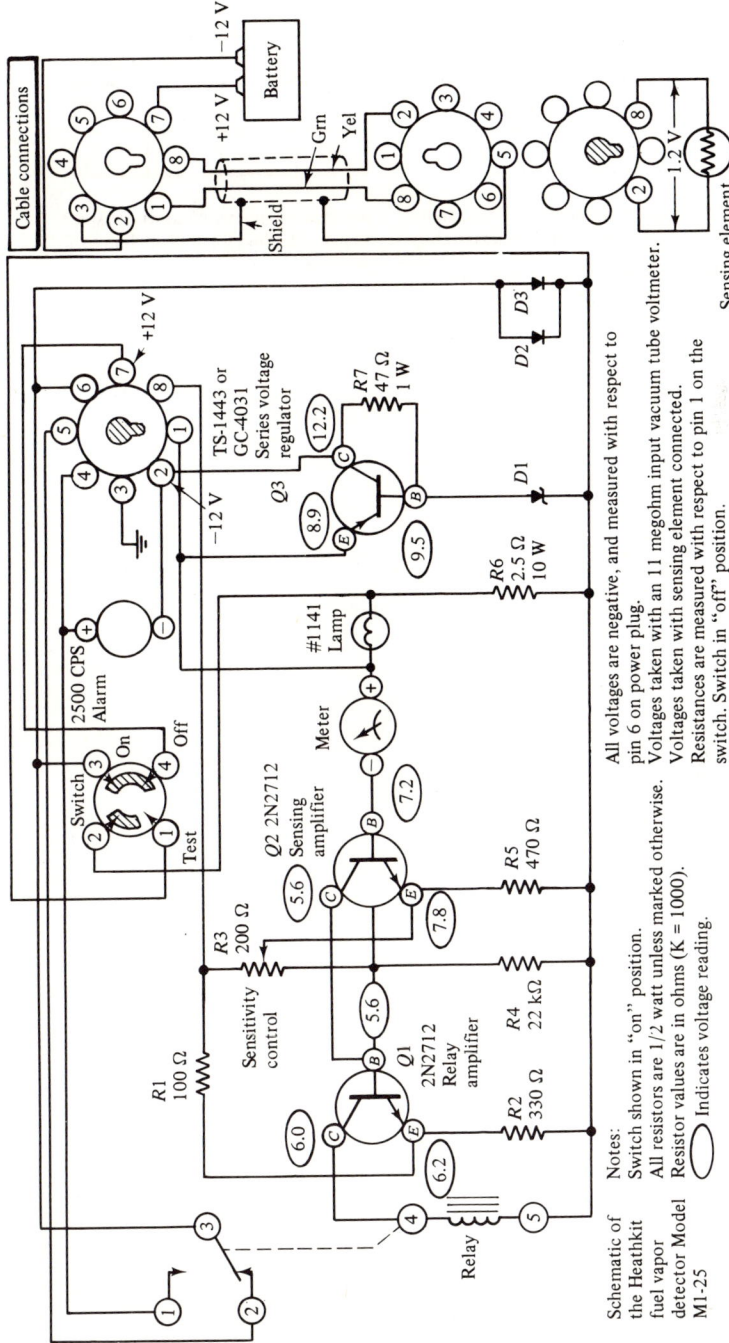

Figure 10-11 Circuit of fuel-vapor detector. (*Courtesy of Heath Co.*)

Voltage chart

	B	E	C
Q1	5.6	6.2	6.0
Q2	7.2	7.8	5.6
Q3	9.5	8.9	12.2

Resistance chart

	B	E	C
Q1	800 Ω	80 Ω	620 Ω
Q2	530 Ω	62 Ω	800 Ω
Q3	5 Ω	660 Ω	620 Ω

All voltages are negative, and measured with respect to pin 6 on the power plug.

Voltages taken with an 11 megohm input vacuum tube voltmeter.

Voltages taken with sensing element connected.

Resistances are measured with respect to pin 1 on the switch. Switch in 'off' position.

Figure 10-12   Voltage and resistance charts.

Appendix I
Resistor Color Codes

RESISTOR COLOR CODES

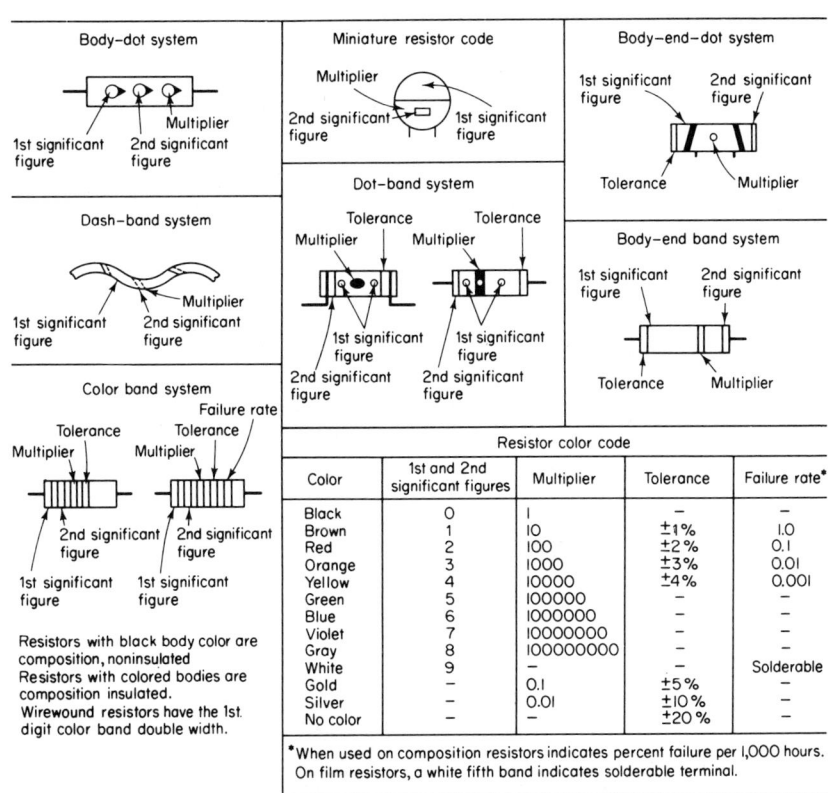

Appendix II
Capacitor Color Codes

CAPACITOR COLOR CODES

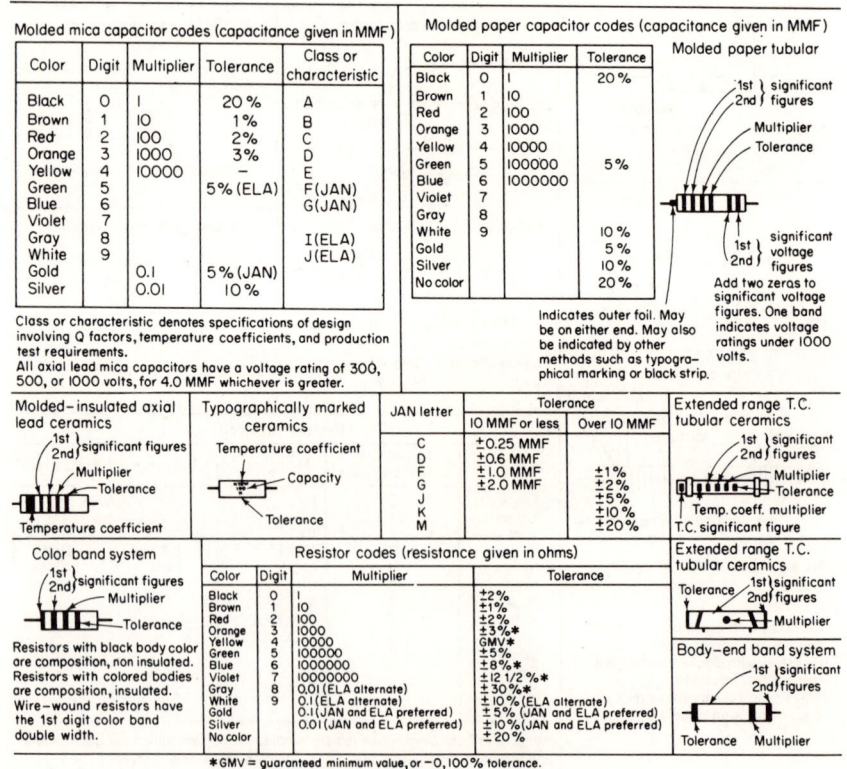

Appendix II Capacitor Color Codes

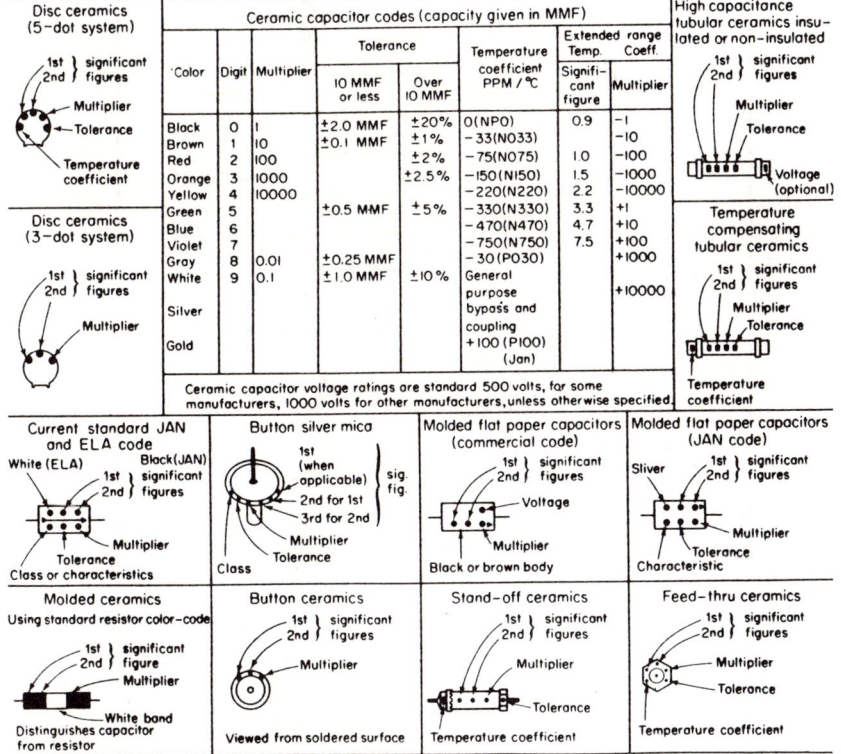

Appendix III
Diode Polarity Identification

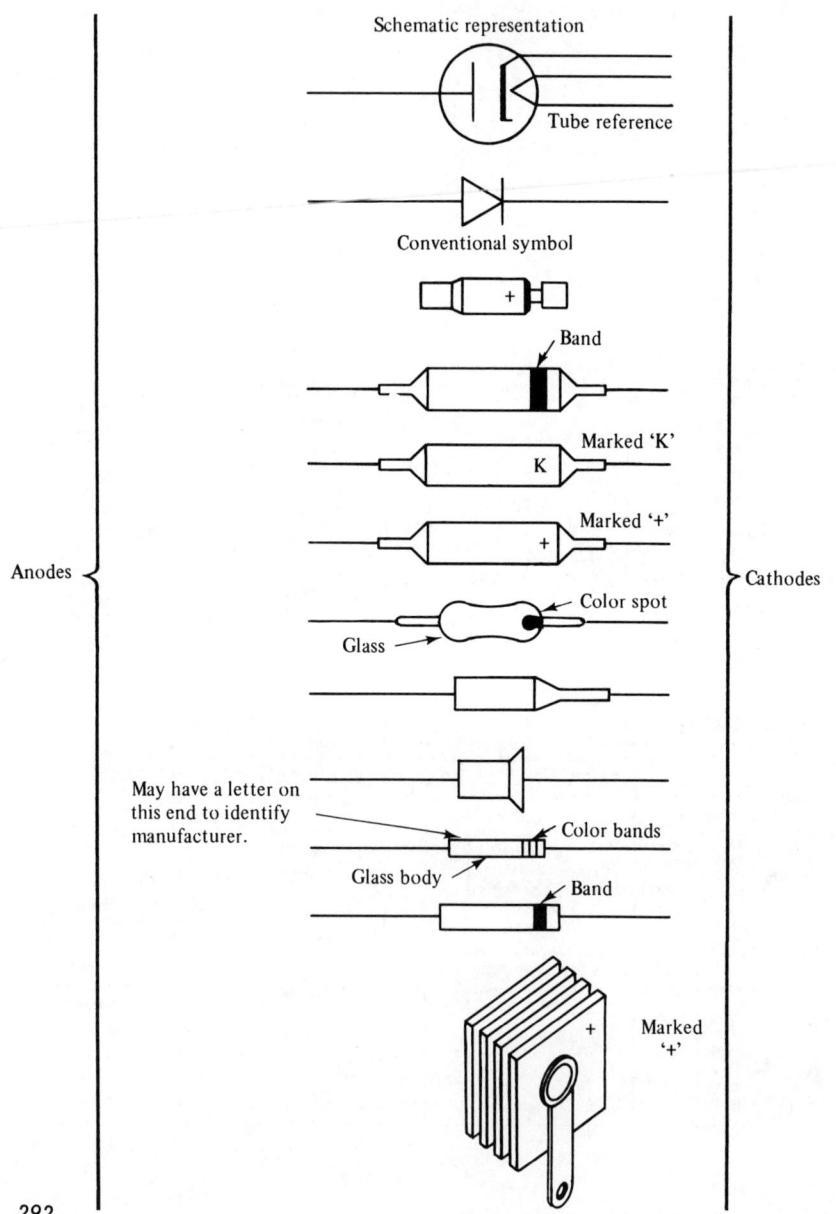

Index

A

Absolute
    dB, 124
    error, 65
AC converters, 137
Accuracy, 11
Action
    demodulator, 200
    limiting, 201
A-D, 34
AFC, 201, 267
    action, 204
    filter, 204
    horizontal, 215
    signal tracing, 227
AFPC, 260
AFT, 267
AGC, 178
    delayed, 222
    forward, 181
    keyed, 220
    reverse, 181
    source, 198
Alignment, 184
AM detector, 198
Ammeter, 1
Amplifier
    burst, 240
    final, 192
    Y, 240
Analog
    instruments, 9
    meter, 35
Apparent power, 134
Arc of error, 65
Arcing, 255
ATC, 243
Attenuator, 9
    calibrated, 182
    step, 184
Average
    response, 11
    value, 30, 55
Audio voltmeter, 126

B

Balance adjustment, 32
Ballast resistor, 35
Base, time, 271
Beat, zero, 261
Beta, 86
BH curve, 31
Bias
    adjustment, 32
    stabilization, 153

Bipolar transistor, 78, 83
Blanking amplifier, 278
Blocking capacitor, 182
Board, circuit, 191
Bridges, 2
Bridging tests, 191
Broad tuning, 184
Burden, voltage, 14
Burst amplifier, 240
Buzz, sync, 236
B/W signal, 240
B-Y, 246

C

Carrier level meter, 201
CB, 192
CCTV, 281
Center scale
    indication, 73
    value, 24
Chroma signal, 240
Circuit
    boards, 91, 174
    loading, 48
    tester, 92
Class
    A, 148, 178
    AB, 148
    B, 148, 178
    C, 148, 210
    D, 148
Click test, 174
Clipping distortion, 142
Compensating strips, 5
Complex waveforms, 32, 133
Compression, 143
Constant
    current, 102
    voltage, 103
Continuity, 1
Continuous balance, 33
Control-action tests, 90
Conversion gain, 198
Converter, AC, 137

Corona field, 48
Counter emf, 43
Cross checks, 21
Current
    balance, 103
    calibration, 28
    distribution, 177
    drain, 177
    probe, 95

D

d'Arsonval meters, 3, 12
Data, service, 186
dB, 9
    absolute, 124
    error, 124
dBm, 126
DC component, 49
"Dead" receiver, 174
Decibels, 9
    measurement, 119
    table, 122
Decoder, 159
Deemphasis, 140
Degrees of arc, 65
Delayed AGC, 222
Demodulator action, 200
Depletion, 81
Difference voltages, 45
Digital
    multimeter, 19, 116
    readout, 9, 34
    voltmeter, 6, 33
Diode probe, 11
Distorted output, 174
Distortion, 11, 186
Dropout, oscillator, 174
Dual
    gate, 32
    ramp, 33
    voltage specifications, 245
Duty cycle, 98
DVM, 6, 34
Dynamometer, 3

E

Electrodynamometer, 4, 99
Electronic
    counter, 34
    multimeter, 61
    voltmeter, 61
Emitter follower, 275
Energy, magnetic, 254
Enhancement, 81
Equalization, 142
Equivalent circuit, 90
Error
    estimation, 34
    frequency, 233
    indication, 39
    parallax, 34
    phase, 245
    waveform, 23, 30
Extension cable, 243
External sensor, 286

F

Factors, offset, 11
Fail-safe, 255
Faults, multiple, 222
Feedback, negative, 142
FET, 80, 267
Field-strength meter, 132, 182
Figures, Lissajous, 278
Filament, wire, 284
Final amplifier, 192
Fine tuning, 213
Flyback, 233
    interval, 250
    tester, 237
FM
    capture effect, 205
    oscillator, 190
    troubleshooting, 174
Forward
    AGC, 181
    bias, 23

Frequency
    counter, 192
    error, 233
    modulation, 205
Front-to-back ratio, 89
Fuel vapor detector, 282
Full scale
    accuracy, 41
    percent of, 9
Full-wave bridge, 23
Functions, 37
Fundamental, 25

G

Gain
    conversion, 198
    control, 220
    mixer, 197
    stage, 180
Galvanometer, 2, 96
Gate
    controlled switch, 247
    insulated, 80
Gating pulse, 250
Generator
    keyed rainbow, 247
    signal, 181
    stereo, 159
    sweep, 276

H

Half bridge, 23
Harmonic
    distortion, 24
    meter, 30, 129
    percentage, 30
Harmonics, 30
Hay bridge, 170
Heater strip, 60
HF circuits, 58
High fidelity, 139
High-low ohmmeter, 190, 214

High-ohms adapter, 73
High-resistance circuits, 39
High voltage
    megohmmeter, 222, 255
    ohmmeter, 71
    probe, 43
    terminal, 243
Horizontal
    AFC, 215
    amplifier, 279
Hour
    kilowatt, 2
    watt, 2
Hue control, 243

I

IC, 86, 201
I_{CEO}, 106, 109
I_{CES}, 106
I_{DGO}, 110
I_{DSO}, 109
I_{EBO}, 106
I_{ECS}, 107
IGFET, 80
I_{GSO}, 109
IM analyzer, 129
Impedance bridge, 168
Indication
    accuracy, 22, 27, 61
    error, 39
Indirect measurement, 94
Inductive load, 43, 47
Input resistance, 29
Instrument
    accuracy, 4
    circuitry, 37
    rectifiers, 22
Integrated circuit, 86, 201
Intermittent, 20, 174
Intermodulation, 129
Internal resistance, 28, 45, 97
Iron third harmonic, 31
Iron vane meter, 4
I_{SDO}, 110

I_{SGO}, 109
Isolation transformer, 46

J

JFET, 80
Jig, test, 243
Joule's law, 29
Jumper capacitor, 216
Junction
    field-effect transistor, 80
    resistance, 80, 90

K

Keyed
    AGC, 220
    rainbow signal, 247
Killer, color, 241
Kilovoltmeter, 44
Kilowatt-hour meter, 2

L

LDR, 74
Level, signal voltage, 175
Light dependent resistor, 74
Limiting action, 201
Linear
    circuits, 50
    log scale, 9
Linearity, 256
Lissajous figures, 278
Logarithmic scale, 10
Lo-pwr ohmmeter, 70
Loopstick, 174
Losses, deflection, 254
Low-ohms adapter, 73, 74
Low-pass probe, 47

M

Magnetic circuit, 5
Maxwell bridge, 171

Index

Megger, 72
Megohmmeter, 1
Mercury batteries, 14
Meter
    carrier level, 201
    dip, 195
    field strength, 182
    movement, 5
    overload, 42
Microammeter, 1, 28, 45, 95
Microphone, 195
Microvolts, 131
Milliammeter, 1
Misalignment, 184, 220
Mixer gain, 197
Modulator, 195
Module, 260
MOSFET, 80, 267
Multimeters, 14
Multiple faults, 222
Multiplier resistors, 44

N

Narrow picture, 257
NBS, 1
Negative feedback, 142
Neutralizing capacitor, 217
Noisy controls, 224
Noise voltages, 127
Nonlinear
    circuits, 50
    resistance, 80, 90
Nonlinearity, 278
No-load condition, 54
Null point, 97

O

Observational error, 34
Offset factors, 11
Ohmmeter, 1, 63
    high-low, 190, 214
    polarity, 69

Op amp, 104
Open-circuit testing, 54
Operating point, 175, 220
Operational amplifier, 104
Oscillator
    crystal, 192
    dropout, 174
    injection voltage, 197
Output transformer, 177
Overdrive, 186
Overload, 27, 178
    diode, 191

P

Parallax error, 34
Partial digit, 35
PC board, 20
Peak
    response, 11, 185
    voltage, 57
Percent
    of full scale, 9
    of reading, 9
Phase error, 245
Phases, chroma, 246
Photovoltaic cell, 57
Platinum filament, 284
Polarity reversing switch, 41, 65
Poor selectivity, 174
Poor sensitivity, 184
Potentiometric calibration, 18
Precision resistors, 14, 24
Preemphasis, 140
Printed circuit, 20
Production change, 238
Pulsating DC, 49, 199
Pulse voltage, 49

Q

Quartz crystal, 192
Quasi-complementary amplifier, 155
Quick check, 178, 217

R

Radio receivers, 174
Ramp, 33
Ranges, 37
Raster, 213
Reactance, 118, 229
Reactive power, 100, 135
Real power, 100, 134
Realignment, 185, 220
RF wattmeter, 192
Regeneration, 186
Reluctance, 5
Residual
    error, 5
    magnetism, 5
Resistance
    chart, 76
    comparison, 92
    variation, 53
Resistor decade, 24
Resolution, 35
Reverse
    AGC, 181
    bias, 53
Ring shunt, 29
Ringing test, 233
RMS, 30

S

Scale
    factor, 121
    plate, 37
S curve, 56
Sectionalization, 153
Semiconductor rectifiers, 22
Sensitivity, 35
Separation, 161
Service data, 1
Shunts, 94
Signal
    chroma, 240
    injection, 181, 190
    level meter, 12
    tracing, 224
    Y, 240
    voltage, 175
Stacked diodes, 67
Standard resistors, 24
Standards, 1
Staircase ramp, 33
Stereo
    analyzer, 163
    generator, 159
    indicator, 162
    multiplex, 159
    separation, 161
Subsonic, 11
Switching bridge, 161
Sync, 213

T

Test
    jig, 243
    leads, 29
Thermistor, 75
Thermocouple, 5, 99
Third harmonic, iron, 31
Time
    base, 271
    constant, 200
Tolerance, 35
Torque, 6
Trace switch, 256
Transfer characteristic, 199
Transformer, isolation, 46
Transistor
    detector, 200
    field effect, 80, 267
    unipolar, 83
Troubleshooting, FM, 174
True rms, 11, 98
Tuned AC VTVM, 129
Turn-off test, 214
TV receiver, 206

U

UHF, 131
Ultor, 243
Unipolar transistor, 83

V

Vane, iron, 4
Varactor, 267
Variational analysis, 174
Varistor, 27, 171
VARS, 101, 134
Vectorgrams, 245
VHF, 130, 267
Video
    amplifier, 206, 222
    detector, 206
Vidicon, 282
Visual inspection, 174
Voice coil, 213
Voltage burden, 14
Voltmeter, 1
Volume control, 178
VTVM, 119
VU, 128

W

Walkie-talkie, 192
Watthour meter, 2
Wattmeter, 2
    RF, 192
Waveform errors, 23, 30
Weak reception, 174
Weston cell, 18
Wheatstone bridge, 96
Wire filament, 284

Y

Yoke
    circuit, 254
    convergence, 243
    current, 251
Y signal, 240

Z

Zero
    axis, 253
    beat, 261
    center, 55
    ohms, 37
    set, 20
Zener
    action, 277
    conduction, 83
    diode, 58, 59, 281